THE ALCHEMY OF LOVE AND LUST

*D*ISCOVERING OUR SEX HORMONES

AND HOW THEY DETERMINE WHO

WE LOVE, WHEN WE LOVE, AND

HOW OFTEN WE LOVE

G. P. PUTNAM'S SONS
NEW YORK

T H E
ALCHEMY
O F
Love
A N D
Lust

THERESA L. CRENSHAW, M.D.

"Dear Abby" excerpts taken from the DEAR ABBY column by Abigail Van Buren.
Dist. by UNIVERSAL PRESS SYNDICATE. Reprinted with permission.
All rights reserved.

G. P. PUTNAM'S SONS
Publishers Since 1838
200 Madison Avenue
New York, NY 10016

Book design by Gretchen Achilles

Library of Congress Cataloging-in-Publication Data
Crenshaw, Theresa Larsen.
The alchemy of love and lust : discovering our sex hormones and how they
determine who we love, when we love, and how often we love /
Theresa L. Crenshaw.
p. cm.
ISBN 0-399-14041-7
1. Sex (Psychology)—Endocrine aspects. 2. Sexual excitement—
Physiological aspects. 3. Hormones, Sex. I. Title.
QP251.C925 1996   95-36864 CIP
612.6—dc20

Printed in the United States of America
1 3 5 7 9 10 8 6 4 2

This book is printed on acid-free paper. ∞

ACKNOWLEDGMENTS

Many minds and moods influenced the ebb and flow of this manuscript. Through the Linda Chester Agency I found the perfect partner in Putnam, my publisher, and my editor, Laura Yorke, who was not only a pleasure to work with, but made me stretch myself beyond anyplace I had intended to go. The destination we reached together turned out to be much more intrigı ing than the original. By pressing me to speculate, and asking questions that I thought could not be answered, information reorganized itself and answers formed.

Foremost, I must express my appreciation to Linda Chester, my literary agent, whose enthusiasm for this book, encouragement, and unwavering support made even the most difficult challenge feasible. Her able accomplices in this effort, Laurie Fox and Billie Fitzpatrick, were always available with their time, their thoughts, and their help.

What can I say about Caron Golden? Only that her orderly mind, consistency, and systematic approach helped me to present this material in the clearest and most organized fashion. Without her encouragement and humor, I doubt I would have dared to try some of the fiction or racy vignettes that somehow found their way into these chapters in the wee hours. She worked with me intensely to meet deadlines and struggled valiantly to decipher my writing as it circled the page.

Then there is Beth Deahl, a skilled librarian, who contributed on numerous levels, undaunted by computer glitches when the bibliography had to be painstakingly compiled.

Tim Swezey kept our computers performing, here at the drop of a hat to help us out whenever we needed him, adding to this project his knowledge of graphics, charts, and software programs to help us work smarter.

Helen Kaplan, friend, mentor, and guardian angel, was exceptionally valuable to me in every respect.

ACKNOWLEDGMENTS

My thanks also to Joe and Terry Graedon for their enthusiasm, support, advice, and good example; also to Jim and Brandy Price, Joe and Gloria Shurman, Annharriet Buck, David Gellar, and Bonnie Gorman, who persisted through multiple drafts, and many others who helped me manicure and improve the material. A special thanks to Dr. James Goldberg, my coauthor of *Sexual Pharmacology*, who reviewed this manuscript with a critical eye to make sure I didn't take too many liberties, or mispronounce anything.

Allison Booth kept the rest of my life running smoothly, providing peace of mind so I could write, and orchestrated again the delivery of another manuscript.

To Ulrika Ehrenborg, my mother, and Brant Crenshaw, my son, who as always were by my side, there are no words.

I would also like to express appreciation to John and Terry Gregoricus, Alan Zelon, Brian Sebeckis, John Mainelli, Bob Morey, Lenny Thompson, Roger Crenshaw, Lorraine Day, Ingrid Rimland, Doak Davison, and Tony and Suzanne Story, for reasons they know so well; to Ahimee Cazares and Peggy Benning, who established order and consistency in key areas; to Chuck Hammond, who brought me music; and to Jeffrey Dean Mumma and Harry Butter, who enhanced my health.

To William Masters and Virginia Johnson, who personify courage and fine scholarship, I remain grateful for their early training, insights and guidance.

I also thank Sally McCartin for special PR projects, and Diane Glynn and Marilyn Duckworth in advance for the big PR yet to come. Cathy Fox, Jill Sansone, and Dan Harvey at Putnam have done a superlative job on subsidiary rights and sales and marketing, respectively, as well as Joanna Pulcini at the Linda Chester Agency, who joined this project toward the end, and David Groff, my editor's able assistant.

Most of all, to the men and women I worry about, who have to figure it all out on their own, beginning with whom to believe.

My patients, whose hormones taught me most of what I know. And to the one man who taught my hormones a thing or two.

In honor of Ulrika Ehrenborg

and

Brant Crenshaw

CONTENTS

Introduction   xiii

CHAPTER 1: *Sexual Cycles and Peaks*   1

CHAPTER 2: *Passages: Sexual Stages*   18

CHAPTER 3: *Romance: Love, Limerance, and Lust*   53

CHAPTER 4: *Touching: Attachment, Bonding, and Commitment*   90

CHAPTER 5: *The Aggressive Sex Drive*   118

CHAPTER 6: *Your Receptive Sex Drive*   164

CHAPTER 7: *Menopause and Viropause*   203

CHAPTER 8: *Preventing Menopause and Viropause*   231

CHAPTER 9: *Lasting Longer*   276

Selected Bibliography and References   310

Index   333

INTRODUCTION

It's appalling, but a seemingly trivial lack of information about hormones can destroy a marriage:

For the first two years of Janet and Richard Miller's marriage, the young couple enjoyed passionate sex. In fact sex got even better with time—the opposite of what Richard had been led to believe based on the woeful tales of his not-so-happily married buddies. Then Janet got pregnant. They were both overjoyed, but Janet had a hard time of it physically, usually feeling too green to think about sex, much less shake the bed with any enthusiasm whatsoever. Throughout her pregnancy, Richard was sympathetic and patient, knowing that her condition was only temporary. But after their child was born, things did not improve. They actually got worse. Janet had absolutely no interest in sex. Instead, she went to considerable lengths to avoid it.

At first Richard attributed the problem to the fatigue and anxiety of new motherhood. He assumed it would pass. It didn't. The situation upset Janet, too. She loved her husband and, until she got pregnant, adored being close to him. Richard was just going to have to be patient until she got back to her old self. She wished he would be more understanding, rather than trying to force the issue. At the same time, Janet threw herself completely into motherhood, not realizing how neglected Richard was feeling.

By the time three months had gone by, Richard was so jealous of the baby that he couldn't even bear to watch his wife nurse. He wanted to be at her breast instead, and blamed the destruction of their sex life on the child. He became increasingly alienated, critical, and withdrawn. Janet resented his attitude, accusing him of being insensitive to her feelings. She was also anxious about his attitude toward the baby. She thought it was foolish and immature for a grown man to be jealous of his own child—infantile, actually. In what seemed like no time, the tension grew into open hostility punctuated by loud arguments. By this point sex was no longer the issue; they were talking divorce.

As a last resort, they went to a marriage counselor. The therapist advised them that childbirth had brought to the surface previously repressed relationship conflicts. With all good intentions, she explored Richard's jealousy toward the baby and Janet's reservations about sex. The counseling helped to restore better communication, but did nothing for Janet's libido. At this point, the counselor referred them to me.

It took less than five minutes to diagnose the problem. Nothing deep and pernicious was at play. No involved psychotherapy was in order. The cause of all this turmoil was strictly biological. Nursing!

A nursing mother produces unusually high levels of prolactin, the chief chemical in charge of milk production. Secreted by the pituitary gland, prolactin stimulates the growth of mammary tissue and triggers the production of milk. When a baby suckles, the hormone surges to approximately ten times its normal value. After breast-feeding, prolactin gradually decreases, returning to its original level over the next two to three hours. Women who suckle their infants on a regular basis have altered prolactin levels—and severely reduced sex drive.

Presumably, nature thought it wise for new mothers to wait until they finish nursing one child before conceiving another. What better way to guard against pregnancy than to reduce a new mother's desire for sex, and in case that wasn't enough, suppress ovulation as well? In fact, some women trust nursing as a form of contraception. It isn't foolproof, probably not much better than the rhythm method, but it does reduce the odds against getting pregnant. But these women who appreciate prolactin's contraceptive protection don't realize that it also dampens their desire. Libido and prolactin are dose related: the more frequently a mother nurses, the more prolactin she produces, and the less interest she has in sex.

The same prolactin that reduces sex drive in nursing mothers reduces desire in men with prolactin-producing pituitary tumors. These men may lose their libido completely, along with their erections, until their prolactin levels are brought back to normal.

Janet's obstetrician had not explained this relationship between nursing and sex drive, and her psychotherapist was not even aware such a connection existed. Imagine all the heartache and misery that could have been avoided had the Millers only been informed of what was really going on.

Fortunately, they got treatment in time. After reviewing their history, I ordered blood studies to demonstrate exactly how Janet's prolactin levels were affected by nursing. Once they realized that the adverse sexual effect was inevitable, predictable, and *temporary*, their relationship began to mend. Their feelings of guilt, anger, and inadequacy, their mutual accusations—all these toxins dissolved in light of this new understanding. Instead of seeing themselves as a dysfunctional couple besieged by dark subconscious forces, they were once again on the same side, a team dealing with a manageable biological/medical challenge.

My next piece of good news for the Millers was that halting nursing wasn't necessary to solve the problem. While prolactin does decrease desire, it does not decrease enjoyment of sex. I told Janet that although her libido had been chemically neutralized, she could still enjoy sex with her husband if she was willing to play by new rules until her chemistry was back in balance. If she just got started, she would have fun. I was not advising her merely to accommodate Richard's needs; she had already tried that out of guilt and obligation, which just caused more resentment. On the contrary, she needed to adapt her attitude to the situation. Knowing that she could not count on her lust to inspire her, Janet needed to have sex for different reasons for a while: she loved to be held and touched still. She also loved her husband and the physical intimacy. Instead of waiting for the mood to strike her, she could choose to have sex periodically just to be close. If she approached him with the right attitude, she would discover that she could enjoy making love just as much as ever, even though her lust was muted. Fortunately, prolactin doesn't seem to interfere with a nursing mother's orgasms.

Janet reached out to Richard with love and affection. He welcomed her back like a thirsty man. Their intimacy revived, the fighting stopped, and sex sizzled. Had they been deprived of this critical information much longer, their problems would no doubt have persisted even after Janet stopped nursing. So much emotional damage would have occurred that restoring normal hormone balance would not have been enough to end the cold war between them.

WHO'S THE BOSS?

The Millers' case is only one example of the formidable number of ways in which fluctuating hormones influence our relationships. Hormones are dictators; their job is to tell other substances, including each other, what to do and how to behave. Some of them can be bullies, wreaking havoc with our moods and behavior. But these molecular rascals can operate with abandon only to the extent that they are not recognized, respected, and understood.

A little strategic information, given at the right time, has the power to protect individuals and relationships from experiencing devastating problems. Sometimes, as in the case of prolactin and nursing, just knowing the facts about these hormones aborts an enormous problem, replete with misunderstandings. A little common sense and careful planning does the rest. In regard to other hormones, some with very complex profiles, we still need to learn to recognize their effects and develop techniques to modulate them. In some cases this can be as simple as changing our environment. In other cases, it might be necessary to take supplements or medication. Regardless of the situation, there are dozens of hormones and related substances that we would do well to become better acquainted with for our own protection—and, of course, enjoyment.

Like the Millers, scores of patients have come to me confused and doubting themselves, only to learn that they were unknowingly in the grip of some subversive hormonal tyranny. Empowered by this knowledge, they have learned how to neutralize the damage and turn biochemistry to their advantage, instead.

DEBUNKING THE GENDER MYTHS

My interest in hormones and sexuality came about naturally in the course of my work. I had decided to become a physician when I was eight years old. However, I had not settled on a specialty until I was already halfway through medical school. Between my sophomore and junior years, I had the opportunity to attend a postgraduate California Medical Association course in Pebble Beach, California. One of the keynote presentations was given by the renowned sex therapists Masters and Johnson. Human sexuality did not appear on my medical school curriculum, nor did the majority of medical schools across the country offer any such courses. Fascinated by their research, I was struck by the neglect of this important arena of the human condition, when it seemed to me that it applied in some way to every aspect of modern medicine. At that point, I decided to get some training in it, no matter what I ultimately decided to choose as my specialty. In time, the idea grew on me, and I decided to give it my full attention.

In the course of my practice, which is devoted to sex therapy, sexual medicine, and human relationships, I became intrigued by the sexual side effects of commonly prescribed drugs, and had the opportunity to study certain drugs to evaluate their positive sexual side effects. Since many of these substances influence hormones, they became central to my research. The more I learned, the more persuaded I became of the profound effect of our respective hormones on our brains, our behavior, and each other. Yet I was shocked at how little research had been conducted in these areas.

Indeed, back when I was a medical student in the 1960s, the idea that biochemical changes in the body could influence someone's behavior and emotions was hardly taken seriously. We were taught, for example, that the mood swings associated with menopause and premenstrual tension (as it was then called) were neurotic tendencies rooted in childhood programming.

Like many other women, I felt this way of thinking was just plain wrong. Once a month, along with terrible cramps, I would get depressed and moody. Often I didn't realize what was happening to me until I checked my calendar. I would go to bed one night feeling perfectly normal and wake up the next morning with a strong family resemblance to a blowfish. Clearly, my emotional low was due to a molecular onslaught, not a traumatic upbringing. As an aspiring physician, I considered the prevailing attitude not only absurd but harmful. Millions of women, in fine mental health, were being dismissed as neurotic instead of being treated for a genuine, and uncomfortable, physical condition. I determined to help women become wiser medical consumers, capable of standing their ground when a diagnosis superimposed on them contradicted their own instinct and experience.

My first task was to debunk the myth that what we now call premenstrual syndrome was merely a psychological affliction. I set out to find an appropriate way to study PMS. My idea was to learn whether nonhuman primates also experienced behavioral changes prior to estrus (their version of menstruation). As luck would have it, one of the few men who shared my interest in the subject was on the faculty of my medical school. He arranged a research grant for me. I spent a summer canvassing zoos and primate centers, asking whether female primates exhibited any significant physical or behavioral changes prior to estrus.

To my surprise, I learned very little from the scientists I interviewed. On the whole, they seemed to be amused by my questions (what else could one expect from letting women into medicine?). However, the zookeepers who handled the animals on a daily basis, and knew them as well as a parent knows a child, had a lot to say about premenstrual monkey business. One handler's story tells it all: "I know when she's coming into estrus without even checking the charts or her daily weights," he said proudly of one rhesus. "She's usually tame, cooperative and affectionate—easy to handle. However, just before estrus she starts chattering at her mate, scolding her young, and causing a general ruckus. Her infant cringes in terror, her mate keeps his distance, and so do I. Once she got loose from me, tore up the lab, and did this." He held up his right hand. The middle finger had been bitten off. "The rest of the month she's a pussycat," he added.

The conclusion was obvious: either female rhesus monkeys have penis envy and castration anxiety (note the severed digit; somebody might try to make a case for that) or their hormones change their behavior dramatically. If such behavior was apparent in our closest primate relatives, it seemed reasonable to surmise that our menstrual moods were influenced by the whims of our chemicals as well.

But what about men? This is the gender that has long gloried in the reputation of being logical, unemotional, and stable, with the possible exception of their sexual judgment. But there are suggestions to the contrary. For one thing, men have problems with aggression; they seem to be pretty good at starting wars and statistics show that our prisons are overcrowded with male perpetrators of violent crimes. This makes sense when looked at from a biochemical perspective: aggression is often a by-product of testosterone—a sex hormone that men have much more of, generally, than women.

These along with other signs lead us to suspect that hormones and related molecules are wreaking havoc with men as well as women. We just haven't looked deeply enough yet to understand it all. Only in recent years, for example, have researchers been taking a serious look at viropause—commonly referred to as male menopause—to determine whether it is just a psychological syndrome or whether there are biochemical factors involved. The information that results from these studies may just reverse the way we have traditionally viewed midlife male behavior, revealing a depth of chemical complicity about which we can now only hypothesize.

As my interest in the influence of hormones on mood and behavior grew stronger over time, I inevitably became more interested in male-female differences. This was brought home to me personally during my years in medical school. As a woman in a male-dominated profession in the early days of feminism, I knew that in order to excel I would have to work hard to fit in. It was my conviction at the time that both sexes were fundamentally alike, except for our upbringing and social conditioning. So, I did my best to walk, talk, and think like a man while downplaying my femininity. I used no makeup and wore horn-rimmed glasses, practical shoes, and bras that flattened my chest. But, no matter how hard I tried and how much I believed otherwise, I was constantly struck by how different I was from my male counterparts.

Eventually I relaxed into my natural biology and psychology. I concluded that although I made a lousy man, I had real talent as a woman. I began to observe men and women more objectively. I noticed with admiration their distinct and special qualities. Clearly, these differences could not be attributed entirely to cultural factors. The hormones that cause one to have breasts and the other a hairy chest, that give one soft curves and the other chiseled muscles—these chemicals must also affect the different ways in which men and women think, behave, and experience the world, sexually, emotionally, and intellectually.

However, it wasn't until after I had a child of my own that I was hit by the full force of the distinctions between us. Until then I still thought nurture overwhelmed nature. I was determined to raise a New Age man—Barbie dolls, unisex toys, no guns. My son played with dolls all right—but he used them as weapons. On the day he was born I began my formal education in the opposite sex with my son as my teacher.

To my dismay, my passionate interest in hormones and gender differences still was not shared by the rest of the scientific community. I was constantly amazed by how little was known about these issues and perplexed by the lack of curiosity in exploring them.

Then came psychoactive drugs in the early 1960s. Antidepressants, antipsychotics, lithium, and other medications proved so effective in treating certain psychiatric problems that it became increasingly difficult to argue that biochemistry had no influence on emotions and behavior. Highlighted by landmark discoveries a little later, such as the link between genetics and certain mental disorders, this trend accelerated through the seventies and eighties. Now, for instance, schizophrenia, obsessive-compulsive disorder, and certain types of depression are widely regarded as medical conditions and treated accordingly.

While the psychiatric community was uncovering the chemical underpinnings of various emotional disorders, laboratory scientists were analyzing the multitude of substances that course through our bodies. Since the time I was a student, when we knew about relatively few hormones, we have come to identify numerous others, along with dozens of closely related substances, such as peptides and neurotransmitters. We know that the relationship between all these chemicals is quite involved. Indeed, so complex is the interaction between hormonelike substances that it is often impossible to isolate the effects of any single one.

The very definition of a hormone has changed since most of us learned about them in high school biology. We once defined hormones as more or less single-purpose substances produced by specific glands, then delivered by the bloodstream to specific target tissues. We then discovered that hormones are manufactured not just in the endocrine glands but in a variety of places throughout our bodies. They have a multitude of responsibilities and a variety of functions. One biologist now defines a hormone in these broad terms: "anything produced by one cell that can get to another cell by any means and change what it does."

When it was discovered in the 1980s that nerve cells have specific receptor sites for chemicals such as endorphins, a flood of research followed. We soon learned that the vast tributaries of the brain are awash in hormones, some sent there from other locations and some actually produced in the brain itself. This means, for example, that the actions of testosterone, estrogen, and other sex hormones are not limited to the reproductive system as was once believed, but can act directly on the control center—our brain.

Indeed, hormones are so intricately involved in neural events that some researchers no longer picture the brain as a machine or computer but consider it instead to be *a gland*. This is a radical departure from past perceptions, tantamount to accepting that the earth revolves around the sun instead of vice versa. It is becoming increasingly clear that our control center is not only a place, but a process involving the interrelationship between the brain, our hormones, our biochemistry, and the environment. Given a model like that, the question becomes, how can these substances *not* affect our minds, moods, and behavior?

The overwhelming weight of evidence in the last two decades has confirmed my early convictions about the influence of biochemistry in our lives and loves. Clearly, our mating dance, the desires that drive us and frustrate us, the bonds we make, the love we give and take, the hearts we break, the differences that delight and infuriate us, the mystery of attraction, "sexual chemistry," and the agonies and ecstasies of intimacy—all this and more is influenced by the ever-changing bouillabaisse of chemicals in our bodies that I think of as "sex soup." *The Alchemy of Love and Lust* is the culmination of three decades of research and analysis in these areas of hormones, sexuality, and gender differences.

GENDER CODES

Back in the days when I studied with Masters and Johnson, I particularly appreciated the emphasis they placed on relationships, and their awareness of the subjective warp between one sex and the other. As I continued my work in my own practice, however, it became increasingly clear that men and women spoke in different codes and communicated on different frequencies, making rapport difficult to say the least, and disastrous at the extremes. In a book that I authored ten years ago, called *Bedside Manners*, I described the different sex-linked communication patterns, attributing these differences to various hormones and newly appreciated anatomical and histological differences between the male and female brain. Since then, numerous best-selling books have expanded upon these sex-related styles, suggesting that we are drastically different, but not offering much in the way of explanation or solution, other than attributing these sex-linked differences to social and psychological factors.

The Alchemy of Love and Lust goes well beyond just describing the problem or highlighting these sex-linked differences; it shakes up our long-embedded belief that men and women are biologically the same, giving us instead a physiological, in fact logical, basis for better understanding our sometimes illogical behavior. But more than that, *The Alchemy of Love and*

Lust gives us a proactive way of confronting our differences, so we can address the challenges sparked by errant hormones as they arise.

You are about to meet a dynamic cast of characters. Some, like testosterone and estrogen, you are already familiar with to some degree. However, you will discover fascinating new things about them you never knew before, and by the end of this book you will see them in an entirely different light. Others you have probably never heard of, even though they have been living inside your body and exerting a profound influence on your life since before you were born. I recommend that you greet them with an open mind, learn all you can about them, accept them for what they are and embrace them as allies in the pursuit of love, health, and happiness.

The Alchemy of Love and Lust first introduces the key hormones directing our love lives, followed by a tour of our sexual peaks and cycles in the context of the sexual stages and passages we encounter throughout our lives. Individual chapters are then devoted to the roles of each hormone.

Throughout this book you will find two prevailing themes: the biological, sexual, and emotional differences between men and women throughout their life cycles, and the identification of the hormones that influence these disparate characteristics. Perhaps the most important aspect of this book is the recognition that the very same differences that attract men and women to one another are also the cause of the majority of the conflict between them—that how you handle these differences makes all the difference.

The information and recommendations you will find here will help you get past the age-old sex-linked conflicts and behavior patterns that have the power to destroy relationships. You will learn how to appreciate one another's complementary modes of being. This book offers specific recommendations on how you can modulate these biochemical forces to your greatest advantage along with practical suggestions that will turn the tables on your hormones, putting you in charge for the first time.

In a sense, this is a guidebook to intriguing places within yourself that you didn't know existed. It will acquaint you with their zip codes, show you how to maneuver around in them, help you understand and appreciate this territory and point you to treasures and resources that can enrich your life— especially your relationships with each other.

One point should be made clear at the outset: This book does not argue that everything can be reduced to or explained in terms of biology. In introducing the biochemical element, I do not intend to downplay the importance of mental and emotional forces. Romantics need not worry that science is going to eliminate the wonder and mystery of love in favor of a set of equations or charts with long, unpronounceable words.

Indeed, far from destroying romance, the information in this book can help you discover it, rekindle it, and keep it alive. It gives you power. By developing insight into the basic biological differences between men and women, you will learn how to make these differences complement one another as profoundly and intensely as they might otherwise clash.

While modern research into hormones did not develop momentum until as recently as the 1940s, witch doctors and budding sex therapists have been grinding up animal testicles and prescribing them for sexual purposes for centuries. Attempts to influence love and lust centered on witchcraft, spells, voodoo, and folklore. But as with herbs and other medicines, some of the primitive remedies of yesterday have become the foundation of treatment today. Then again, today's science often becomes the alchemy of tomorrow.

Medieval alchemists were dedicated to discovering a substance that would turn base metals into gold, and that would also cure any ailment and prolong human life. According to *The World Book Dictionary*, "Alchemy dealt not only with the mysteries of matter, but . . . it sought to harmonize the human individual with the universe surrounding him." Today we have the rudiments of knowledge to do just that—not to change cheap metal to gold, but to recognize and even influence the dynamic relationship between our hormones and our environment, the alchemy, that is, of human love and lust.

My years of work in this field led to the publication of an extensively referenced medical text titled *Sexual Pharmacology*, which I wrote with Dr. James Goldberg. *The Alchemy of Love and Lust* grew out of that endeavor and gave me the freedom to translate its most fascinating concepts for the general public. By taking a few liberties and speculating on what we may discover in the future, based on what we know today, a remarkable picture emerges, casting a very different light on our dating and mating patterns than that we have accepted to be true.

The Alchemy of Love and Lust represents the most up-to-date information available on the biology and psychology of relationships, along with a healthy dose of speculation about what we might discover down the road as data continue to accumulate. When only animal research is available, it is not possible to definitely conclude that men and women will respond in the same manner as their primate counterparts. However, often these studies are all we have to guide us. In some cases, I have taken the clues found in animal research and extended them to human behavior just to provoke thought.

As it happens, most of the research on hormones in humans has been done on women. Endocrinology is one of the few fields of scientific research where women are not only included, but have received the lion's share of attention, particularly when it comes to reproduction and its prevention, and more recently, the influences of estrogen and other hormones on breast cancer. For these reasons—opportunity (an abundance of data) and importance—I have devoted a good deal of space in the book to a discussion of estrogen.

As a personal aside: During the course of writing this book, my mother developed breast cancer and underwent a bilateral mastectomy, followed by chemotherapy and reconstruction. A year later, I discovered a lesion in my own breast that turned out to be malignant. I also chose to have bilateral mastectomies and fortunately did not require chemotherapy. Using impersonal statistics with over 90 percent probability, the doctors consider me cured.

I had to make a personal decision whether to continue hormone replacement therapy or to abandon it. In doing so, I reread my own advice from a very invested point of view, first to see if I could profit from it, and second, to determine if I still agreed with my earlier point of view given my subjective involvement. It is with great care, therefore, that I address estrogen issues for other women. You can look forward to a thoughtful, provocative, and indeed, controversial discussion, designed to help you make your own decisions about estrogen replacement therapy in regard to your overall health, and cancer considerations in particular.

Don't mistakenly conclude that *The Alchemy of Love and Lust* is for women only. Almost all of the information is essential for any man who likes, loves, or works with women. The wealth of material in this book will give men tremendous insight into the minds, hearts, and souls of women, along with how best to deal with the common issues that put men and women so often at odds.

There is also startling new material on men and their moods, linking their emotions and behavior to hormone fluctuations—both their own and others. Men don't have many resources available where they can learn about themselves. They certainly don't share very much about themselves with each other. Defining the interplay between hormones and their emotions will clarify actions and reactions that men otherwise tend to find vague and confusing.

A NEW FRONTIER

We stand at the edge of a new frontier in human research. While we have barely scratched the surface, what we have discovered in recent years promises to transform the way we view ourselves and live our lives. It is now known that the mix of hormones circulating in your body at any given time will determine, at least in part, how you respond to a new attraction or an old love, how sexy you feel, how you react emotionally, if you are inclined to make an emotional commitment, and much much more. In turn, when you fall in love, touch, or even think about your object of desire, certain hormones surge while others decrease, and this new concoction alters how you feel, what you think, and what you do.

Some people worry that such discoveries will lead to a Brave New World in which we modulate our most fundamental and ethereal emotions with pills. They fear we might reduce love, joy, romance, sex, and ecstasy to a course in Chemistry 101, and lose touch with what it really means to be human. I take the opposite view. Unchecked, our hormones are subversive dictators with tremendous influence to sabotage our lives. Once their activity is recognized and understood, these forces can be enjoyed, and/or influenced to our benefit.

What you will find as you read *The Alchemy of Love and Lust* are dozens of examples that unmask the hidden agenda of our hormones. No doubt you'll recognize scenarios in your own life reflected in these pages. What does this mean? Well, for one thing, while environment and culture go a long way in explaining our behavior, our hormones behave the same way whether we are

American or Chinese, Orthodox Jew or Episcopalian. They don't know what your mom told you to do; they just follow their own chemical imperatives.

This book may shake you up a bit, even provoke you; it's unsettling to think that our most private personal problems have underlying chemical causes that we didn't even realize existed, and that molecules can manipulate our minds, even our mates. But it's also liberating—the idea that a long-standing roadblock may very well be your hormones acting up means that you can take stock and consider how to manipulate them in return.

The Alchemy of Love and Lust gives scientific substance to the age-old discussion about the chemistry between the sexes. In that respect, it will make you a better, more knowledgeable alchemist. In so doing you will stop being the unwitting slave of your hormones and become instead their master.

1.

SEXUAL CYCLES AND PEAKS

Oh, the pleasure of it! The exquisite wonder of his hot wet mouth moving over the tip of her breast, drawing in the distended nipple, suckling it like a babe. She felt a shaft of excitement shoot down between her thighs, where his thigh had taken up residence once again. As he kissed and suckled and nibbled she arched her back, pressing her breasts against him with wanton abandon, clutching his head with both hands in his hair as she rubbed herself against that marvelous thigh. . . .

Then one of his hands was sliding down from its play with her breasts, stroking her stomach, a finger burrowing playfully into her navel before moving lower, hovering just above the soft triangle of hair that ached for his touch.

When still he hesitated her hips lifted in instinctive supplication, inviting his touch in a wordless gesture that was as old as woman.

(Karen Robards, Night Magic*)*

Check your breathing. Do these words move you, provoke you, perhaps transport you? Do you detect any subtle physical changes? A faster pulse, some deeper breaths? An involuntary sigh? Publishers of romance novels have built a multibillion dollar industry by stirring up your hormones.

Romantic arousal, no matter what the source, does things to your body. Your skin flushes, heat spreads, your heartbeat quickens, your state of consciousness alters ever so subtly. When a steamy sex scene arouses you or a love story melts you, when your insides do somersaults the moment you encounter that certain someone, it's chemical warfare.

The army that has engaged your every move is made up of hormones and other chemicals that have been deployed in your bloodstream, each patrolling and expanding its territory.

Suzette experienced this chemical charge some years ago—when she first met her husband, Stacy. His look, his touch, even his scent persistently and gradually engaged her to him. As they became increasingly intimate, Suzette realized that Stacy was the man she wanted to marry. It wasn't any logical decision. It was rather some sort of physical instinct, a yearning to be with this man that she couldn't articulate.

But just as our hormones can bring us together to love, marry, and bear children, they can trick us into doing some pretty wild things.

Who hasn't made a bad sexual choice, behaved in the most uncharacteristic, perhaps bizarre, way, made a fool of themselves over someone who wasn't worth the trouble? Actions you wouldn't even confide to your closest, most trusted friend. Behavior you don't want to be reminded of: a phone call you promised yourself not to make, an irresistible affair, a foolish marriage, a one-night stand with a complete stranger, irrational jealousy, spying on someone? The most normal people sometimes behave in the most abnormal way. It couldn't be you, or the person you once believed yourself to be.

One look and she was lost. His turquoise eyes penetrated her out of the blue. She leaned against the wall for support, afraid of losing her balance, not yet fully aware that she already had. It was hard to catch her breath.

Stacy was across the room talking football with a friend. Suzette considered herself happily married, deeply in love with him. Since they had become a couple, she had never entertained the idea of an affair, hadn't even been tempted. But all at once a hunger such as she had never known before overwhelmed her—and for this complete stranger! Earlier, she had overheard this man explaining to their hostess that his wife was resting at home, recovering from some vague illness. She had barely noticed him then, but now, locked in his gaze, she was aware of nothing else. He weaved his way past her friends, who were huddled together in groups sipping drinks and eating hors d'oeuvres. His eyes never left her face. When he finally reached her—everything seemed to be happening in slow motion—he drew so close she could see nothing else. Without thinking, she took his hand, turned, and he followed. She was leading the way, and yet it seemed to be happening against her will. She was asking herself, "What are you doing? Who is this person? Stop. This is crazy." But the argument was lost before it began.

They wound their way down an unlit path in the dark, stopping against a thick tree near the water. Braced against the rough bark, not caring, their bodies took control. In one fluid motion, they pressed through their clothes somehow without effort, and he slipped inside her.

She could hardly believe what was happening. A part of her was standing aside like a dispassionate researcher observing two curious specimens writhing in passion. The rest of her was fully engaged, matching his tempo, racing ahead. He was slow and deliberate, savoring every move. Then intense and relentless, absorbed in his own urgency. They reflected each other's passion, their rhythm responding to a universal beat.

Slightly disheveled, they returned to the party after a hasty repair, praying no one had noticed their absence. There were no words spoken. No hearts broken. Suzette can not comprehend what came over her. To this day, she is still married to Stacy, whom she loves dearly. As for this passionate stranger? She never knew his name.

Now and then, otherwise reasonable men and women get ambushed by forces so powerful they submit to impulses they never dreamed they would have, much less act on. Intoxicated by mysterious forces, would you succumb to your chemistry? Not you? Don't be so sure. Much like alcohol, sexual arousal impairs your judgment. In fact, the mix of the two can often override your common sense. If you happen to be one of the precious few who have escaped such compelling experiences, surely you have heard such stories about a friend who, swept away by hormonal surprise, turned into someone that neither of you recognized.

The fact is this: When you fall in love or in lust it isn't merely an emotional event. Your various hormones, each with unique features to contribute, get in bed with you too. In a way, it's as if there were a corporate decision going on here, with each chemical casting its vote. You may just find yourself outnumbered.

THE LOVE BRIGADE

Let me introduce you to some of the members of this battalion—these hormones and other substances that flow through our arteries and veins, wash over our brains, all the while manipulating our romantic sexual emotions.

DHEA (dehydroepiandrosterone):   There is more DHEA in your body, whether you are male or female, than any other hormone. It bosses us around without reservation. I call it the mother of all hormones because most of our other sex hormones are derived from it, produced by various enzymes acting on the DHEA molecule. In a sense, it tells you when you can and can't have sex. If animal studies hold true for us, then DHEA is involved in your sex drive, your orgasms, and your sex appeal. Oral contraceptives lower DHEA, which might cause you to wonder why they really work.

Pheromones: Pheromones are derived from DHEA. These are sexual signals transmitted from one individual to another through scent. In the animal world, sexual pheromones dictate courting and mating. There is no conscious choice involved. In humans, this hormone may influence who you choose as your mate through its subliminal effect on your sexual scents.

DHEA has a particular maternal role: through the combination of smell (DHEA) and touch (oxytocin—more about this in a minute) babies bond to their mothers and others after birth. DHEA's favorable association with sex drive in women was demonstrated in a Crenshaw Clinic research protocol in 1984. Increased levels of DHEA were associated with increased sexual desire.

Oxytocin: Oxytocin is a marvelous molecule, influencing our life through touch. It is a crucial bonding agent for relationships—think of it as hormonal superglue. If someone holds your hand, your oxytocin levels will rise. If that someone happens to be a person you care for, just thinking about him or her will cause oxytocin levels in your bloodstream to go up. Actually touching will make them surge even higher. Oxytocin bonds and attaches us to those we love, or perhaps causes us to love those it bonds us to—mates, family, friends, babies. It is deeply involved in parenting behaviors, causes contractions of the uterus during childbirth and orgasm, reduces stress, and, most importantly, keeps us "in touch" with each other. Curiously enough, it also makes us forgetful and diminishes our capacity to think and reason.

PEA (phenylethylamine):   Better known as "the molecule of love," PEA is the romantic in you. When you are walking on air, singing out loud and euphorically in love, PEA is probably at work. PEA is a naturally occurring amphetaminelike substance that makes you feel as though you were in a mind-

4

altered state. It is found in chocolates, the bloodstreams of lovers, and diet soft drinks. Its similarity to diet pills is probably why some people lose their appetite when they fall in love. As a special treat, it spikes during orgasm.

Some so savor the rush of PEA, they become love junkies—addicted to the high. Low levels, or a precipitous drop, have been proposed as an explanation for lovesickness. There is even a certain kind of depression caused by fluctuations in PEA that can be corrected with antidepressants that regulate PEA levels.

Estrogen: Estrogen is the Marilyn Monroe in you. It is responsible for a certain softness, both physically and emotionally. Attractiveness to men is enhanced by it. Breasts develop in response to estrogen, which basically endows women with the bodily aspects of sexual appeal. The way a female smells and a woman's sense of smell depends on it. With estrogen flowing, a woman welcomes you into her arms and wants to be penetrated. It governs her *receptive* sex drive and makes her acquiescent to the man moved to pursue her by testosterone.

Testosterone: Testosterone is the young Marlon Brando—sexual, sensual, alluring, dark, with a dangerous undertone. Testosterone is responsible for our *aggressive* sex drive. It makes you want to pursue sex, initiate, dominate. It also stimulates your desire directly, perhaps because it enhances dopamine —a neurotransmitter that is well known to increase sex drive. Interestingly, testosterone seems to have more influence on sex drive than on sexual potency or frequency.

It is also your "warmone," triggering aggression, competitiveness, and even violence. *Testy* is a fitting term; *testing* is too. Without testosterone, wolves don't mark their territory, or confront intruders. With it, they fight to protect their turf, take ownership of their mates, and want to be by themselves, howling in the night.

As a potent aphrodisiac for both sexes, testosterone promotes a drive for specific genital sex and orgasm. It comes with some built-in contradictions. Although full of lust, you may become overbearing or irritable and unattractive to the opposite sex. At the least, it makes you want sex, but it also makes you want to be alone, or thoroughly in control of sexual situations—so it

THE ALCHEMY OF LOVE AND LUST

specifically promotes masturbation or one-night stands (which is as close to being alone as possible with another person).

Testosterone's motto: No emotional entanglements, please. It is fair to say that it causes a compelling sexual urge that spurns relationships, unless they represent a conquest or acquisition of power. Women, having considerably less testosterone than men, are more receptive to emotional intimacy and less reluctant to commit.

Testosterone doubles as an antidepressant, but also makes us (men, especially) angry and irritable when it spikes.

Serotonin: Serotonin is your resident schizophrenic—a two-faced friend that can make you sexually demure or sexually aggressive and indiscriminate, depending on whether its levels are high or low. High levels cool your sex drive; low levels intensify it. It qualifies as a neurotransmitter, which means that it helps to transmit signals in your brain from one nerve ending to another.

At high levels, serotonin has a sensitive side, with a peaceful nature. That's why Prozac makes people feel so good—it boosts your serotonin. (It also takes away your sex drive and delays your orgasms. It is so good at doing so, it has become the first drug used to treat premature ejaculation successfully.)

Aggressiveness toward others and self seems to be restrained by increased serotonin levels. In animals, high levels of serotonin promote selectivity in mates—same species, basic heterosexual pattern. When serotonin levels are low, however, the effect can be macabre. You may become indiscriminately sexual, violent, aggressive, and mean. Low levels in animals induce, along with indiscriminate choice of partner and gender, impulsive behavior and the compulsion for immediate gratification. In fact, when a drug is given to animals to artificially lower their serotonin levels, they will participate in group sex, violently and frenzied, mounting the same sex or the opposite, often harming, sometimes killing others in the process.

Low serotonin levels in humans promotes our responsiveness in addition to our aggressive sex drive. Females tend to have orgasms more quickly, and males ejaculate almost with the speed of light. Given the impact of its fluctuating levels in animals, serotonin may also be involved in our selection of sexual partners and perhaps play a role in our sexual orientation.

Serotonin diminishes with dieting, and may add to that sexy feeling you get as your body image improves.

Dopamine: Dopamine is desire personified—not just for sex, but for the pursuit of any pleasure. It is a neurotransmitter best known for giving us pleasure—of all sorts. Without enough dopamine, we flatten out—feel no joy, no anticipation of pleasure, enthusiasm, excitement, exuberance. Dopamine is the common denominator of most, if not all, addictions, from cocaine to alcohol. It may also be what addicts us to each other. Since its key role is to promote the anticipation of pleasure, it generally increases our sex drive. In the process, it intensifies our experiences and reinforces our desire to have them again. Most importantly, dopamine moves us, literally and figuratively. When we want something, it is dopamine that gets us up into the car or into bed, so we get it, instead of just sitting around thinking about it.

Progesterone: Progesterone is your natural "sex offender" drug. It kills off your sex drive. It does so primarily by reducing testosterone in both sexes. The synthetic version of progesterone, Provera, is such a potent "sex offender" that it has been used to chemically castrate child molesters and other sex offenders. Synthetic progesterone also happens to be the major ingredient of the Norplant implant and many other contraceptives. (Consider the implications for women's libido: Progesterone turns out to have some contraceptive properties not originally intended, like ending your interest in sex altogether.) In addition, progesterone decreases positive sexual scents (pheromones) in animals and may even make women smell bad to men, reducing the likelihood of attracting a man that Saturday night.

Progesterone is a paradoxical hormone. On the one hand, it can make women irritable and aggressive—irritable toward men and aggressive in protecting their young. In this regard, it has a testosteronelike effect. It can direct its force against a husband or mate as well as unknown intruders.

In the animal kingdom, males often attack and/or eat their young. It is the maternal reflex—thanks largely to progesterone—that protects them against all dangers, including their fathers. Yet while causing females to act aggressively, progesterone also makes women nurturing, especially to their offspring. It is known to have mild sedative, anaesthetic properties and a calming effect. These apparently contradictory effects may be explained at

7

some future point by deciphering the presence of several different forms of progesterone, each, perhaps, with its own special features.

Prolactin: Prolactin is a gentle hormone associated with nursing. When it increases, as it does with pregnancy and nursing, sex drive diminishes. When men develop abnormally high prolactin levels, they lose their desire and become impotent. When levels are brought back into the normal range, sexual feelings and response return. Dopamine inhibits prolactin, consequently boosting sex drive indirectly. Estrogen gradually increases prolactin secretion, thus diminishing the aggressive sex drive by default and leaving the receptive sex drive as a woman's primary force.

It continues to surprise me that doctors so rarely explain the influence on sex drive of pregnancy and nursing. Yet the situation perpetuates itself. In February of 1995 in the prestigious *Journal of the American Medical Association,* a seemingly exhaustive study was reported titled "National Assessment of Physicians' Breast-feeding Knowledge, Attitudes, Training, and Experience." There was not one question or comment regarding sex drive. An Australian study concluded that sex drive rebounded after stopping nursing, but didn't make the connection with prolactin—a correlation that is most well established in other medical conditions.

Other than nursing, prolactin rises in response to certain endocrine disorders, prolactin-secreting brain tumors and upset stomachs (particularly nausea and vomiting). In both men and women, prolactin secretion is increased by exercise (hmmm), surgical or psychological stress (explaining in part why sex drive often decreases due to stress), stimulation of the nipples, and sleep. Amenorrhea (absent menstrual bleeding) is also associated with high prolactin levels.

Vasopressin: Vasopressin has been called the "monogamy molecule," and may well deserve this label. It works closely with testosterone, modulating male sexual behavior, keeping it from reaching extremes or becoming "too hot." Levelheaded, orderly, perhaps even a little dull would describe this hormone's personality.

Vasopressin has a "tempering" influence. It is involved in temperature regulation in humans, and somehow governs or influences hibernation in animals, and has some sort of thermoregulatory influence on sex. It also keeps one's temperament from going to extremes, and might even be responsible in some sense for levelheadedness. In the same respect, it may mute the intensity of certain feelings, making your emotional range somewhat more narrow.

Vasopressin turns your attention away from the abstract to the concrete, and away from the past and the future to the here and now, appearing to improve our memory, cognitive powers, and concentration. Consequently, it is usually an advantage to increased focus and pleasure in lovemaking.

These are the key players of this star-studded attraction. You will learn much more intriguing detail about each one, chapter by chapter. There is a support-ing cast as well, over thirty different characters we know of, like LHRH and growth hormone, that influence sex in some way, and whom you will meet from time to time. While each of these substances has its own distinct person-ality, the magic occurs when they interact with each other, creating the cycles and peaks that alternately distract and engage us.

CYCLING THROUGH LIFE

Like the rotation of the earth and the ebb and flow of the tides, many of our hormones rise and fall within our bodies in cycles. Depending on the sub-stance and the mechanism that triggers its release, a cycle might last a few minutes, a day, a week, a month, a season, a year, or a lifetime. There are cycles within cycles. Testosterone levels, for example, oscillate every fifteen to twenty minutes in men, and also follow daily, seasonal, and annual rhythms. Consider what this roller coaster does to a man's mood—and thus his behavior:

Alan has a tight agenda today—lots of things to accomplish around the house, so he wants to be left alone to get the job done—not a guarantee with his family at home. He is brusque and focused—in a no-nonsense mood. His wife, Laura, is looking forward to a relaxing day at home, doing some reading and spending some time with the kids. She is in a particularly warm mood and wanders into the garage to visit. Alan snaps. She asks what's wrong with him. He says, "Nothing, just leave me alone here to get my work done." She replies, "Surely you have a little time for us today, it is so beautiful outside and the kids hardly ever get to see you." He loses his temper and screams that he never has any time for himself. After all, he is only doing all this work for them, and they won't even give him any peace.

Laura decides to leave the snake to his own company, but there is a distinct chill in the air. He finishes his projects, and comes in for dinner, feeling much better, only to discover that the rest of the family has gone out to a movie without him.

If Laura had been sophisticated about testosterone cycles, she could have left the grump alone for fifteen or twenty minutes, caught him in between surges and had a more sensible discussion, preserving the day and her spirits. Fortunately, her estrogens were high at the time, and she didn't handle it badly. (Imagine if she had been premenstrual, instead.) And if Alan had understood his hormonal shifts, he could have told his wife, "I'm in a foul mood. Give me a break."

Testosterone levels seem to be influenced by just about everything: the seasons, the environment, competition, the military, stress, a D cup, just to name a few. The morning highs, daily fluctuations, and seasonal cycles whip men around. Think about the moment-to-moment impact of testosterone levels firing and spiking all over the place during the day, and what this must be doing to a man's temperament. Men who so strongly need to feel in control are in fact in much less control than they realize. No wonder they can be so, well, testy!

The power and influence of hormones like testosterone is exerted in several ways. First, you have to have enough of each hormone to begin with; then you need a sufficient circulating amount of the hormone in its form as a free, or active, molecule. It can't all be bound to a carrier molecule; the *speed* or rate at which your blood hormone levels change can be as important or more important than the amount of change. For example, the man's rapid quarter-hourly surges of testosterone can have more emotional and physical impact than the greater but more gradual changes of some of the hormones that fluctuate during the menstrual cycle.

Other chemicals are secreted in surges as well, pulsing into the bloodstream at intervals dictated by our biological clock, emotional experiences, and outside influences. Some, like oxytocin, do not obey regular patterns at all, but rise and fall according to the needs of the body, the environment, and who touches whom, like air surging into a room in response to feedback from a thermostat.

DHEA, our primary bonding hormone, cycles throughout the day. It is a most volatile hormone with unpredictable peaks and valleys that respond not only to the environment, but to your emotions. Indeed, DHEA may increase up to a hundredfold at a given moment. It drops drastically under stress, which is one of the reasons that sex drive may decrease with acute or chronic stress, and why men may lose erections when they worry about performance. By contrast, its fraternal twin, DHEAS, is consistently stable on a daily basis. This steady pattern is unusual among hormones and certainly makes DHEAS one of the easiest to study. Because both DHEA and DHEAS are metabolically identical (except for the stability factor), for simplicity, I will use "DHEA" to refer to both.

DHEA also cycles on a life scale in both men and women. Of all our hormones, it is the only one that peaks early in life and starts to fall significantly thereafter. It is highest during years of top physical condition—teens and twenties—rising rapidly just before puberty, and spiking between ages twenty-five and thirty. It is the only hormone to reach a peak during this decade, declining progressively from that point forward, reaching a low by the age of sixty, eventually descending to less than 5 percent of peak adult levels, and often reaching undetectable levels after seventy. The clear drop in men past forty is in sharp contrast to testosterone, which declines noticeably in men only after age sixty.

Ironically, because it does not drop precipitously at menopause, DHEA is thought to sustain a woman's sex drive while her estrogens are falling. Although this sounds contradictory at first glance, what it means is that since DHEA doesn't drop as abruptly as estrogens, it buffers some of the sexual shock in women caused by shifting hormones. Nonetheless it gradually but steadily diminishes with aging and may account for some of the changes of menopause in men as well as women as time goes on. It is just not as cataclysmic or pronounced as the better known estrogen withdrawal in women.

Unlike DHEA, which behaves much the same in both sexes, the patterns, rhythms, and concentrations of other hormonal substances are dramatically different in men and women. In most cases, hormone patterns also vary according to age. The differences that we have typically attributed to our upbringing, as reflected in the way men and women behave, actually have deeper roots in our biology. Indeed, these hormonal variations might be said to *define* the differences between the sexes. The resulting features are not right or wrong. They are, perhaps, politically incorrect. I suggest that these sex-linked distinctions between us not only balance one another, but make men and women more interdependent than we would like to believe.

Making the picture even more complex, hormones also respond to the actions of other hormones. The ebb and flow of one often affects the secretion of others. And all these cycles are influenced to some degree by environmental cues—everything from what we eat and drink, to emotional stress, to the company we keep and the games we play. Vasopressin goes up during stress. Testosterone goes down. One study found that when a group of men live together over a period of time, their average testosterone level drops. I'll wager that testosterone plummets when the stock market falls and increases during arguments (verbal battles).

The most well known and most feared hormonal pattern, of course, is the female menstrual cycle. In fact, day by day, a woman's body and brain are awash in a different solution, and this brew alters her behavior and her view of the world around her. Twenty-eight different shades of the same woman. While the relative amounts of key hormones circulating in the body vary daily and usually gradually, in some cases they change abruptly. From one phase of a woman's cycle to another, the changes are always dynamic: estrogen, testosterone, progesterone, and other substances come and go like dancers in a ballet.

We are all familiar with the physical signs and mood swings of this hormonal dance, but remarkably, the more subtle effects often elude us. Most of us realize that many women become irritable premenstrually, but not that more relationship conflicts occur at that time than at any other, and that it is probably the most common time for an argument to trigger a separation or divorce. There are no studies to prove this statement beyond a doubt, but common sense tells you that this is the most potentially explosive time of the month for relationships.

Doesn't that just make you determined not to let a hormone get away with dictating your destiny? Wouldn't you have more power if you tabled a thorny issue for a week or two when your chemistry was more to your advantage?

Beyond the established monthly cycles, women also live what amounts to three lives after puberty; the first is between puberty and menopause, lasting between thirty and forty years; the second and third depend on how she deals with menopause, a condition that may last even longer than her first phase— from forty to fifty years, perhaps half of her life. Women experience this period in one of two different ways, leading to dramatically different lives depending upon whether they take hormone replacement therapy or not. Some do both: hormone replacement therapy for a few years, then nothing, or vice versa.

There is also a substantial group of women who are abruptly forced into premature menopause surgically, through hysterectomy (removal of the uterus) and oophorectomy (removal of the ovaries), or chemically, through chemotherapy. (We'll discuss these subjects in great depth in later chapters.)

While the best known cycles occur in women—the menstrual cycle, and menopause—we are now beginning to appreciate cycles in men: their testosterone fluctuations—an hourly form of PMS—and their viropause—the masculine version of menopause. Today men live only two lives—the first, like the woman, is between puberty and viropause; the second is after viropause. In the future, science may provide them with a third option, just as it has for women: the hormonal treatment of viropause (replete with pros and some cons—another example of the skill of modern alchemists), aimed at prolonging the quality of life and preserving virility.

In the context of these cycles of ours—whether they fluctuate from moment to moment or over the course of a lifetime—come our sexual peaks.

SEXUAL PEAKS

In addition to our various hormonal cycles, we experience sexual peaks on a daily, monthly, and yearly basis. These are times when our interest, receptiveness, or responsiveness is at its height. Until very recently, researchers had not identified any variations in sexual peaks in humans or their correlation to hormones. Now many have been identified, and there are more yet to be discovered.

Orgasm is a peak experience that some enjoy more often than others. Each time it occurs, the changes in our chemistry that helped to trigger this climax change again in response to it. Do we have the sophistication or data available to identify each and every molecular move? No. We have the previews, however, of coming attractions:

During arousal, excitement, and orgasm, oxytocin goes up. DHEA increases in the brain and perhaps in other strategic parts of the body. PEA percolates, making you both nervous and sort of sweet. Testosterone increases sometimes and sometimes not. What other chemicals are intoxicating you during sex?

. . . in the sudden helpless orgasm, there awoke in her new strange thrills rippling inside her. Rippling, rippling, rippling, like a flapping overlapping of soft flames, soft as feathers, running to points of brilliance, exquisite, exquisite, exquisite and melting her all molten inside. It was like bells rippling up and up to a culmination. She lay unconscious of the little cries she uttered at the last. But it was over too soon, too soon, and she could no longer force her own conclusion with her own activity. . . . She could do nothing . . . she felt the soft bud of him within her stirring, and strange rhythms flushing up into her . . . swelling and swelling till it filled all her cleaving consciousness . . . pure deepening whirlpools of sensation swirling deeper and deeper through all her tissue and consciousness, till she was one perfect concentric fluid of feeling. . . ." (D. H. Lawrence, Lady Chatterley's Lover)

Correlating sex drive to fixed biochemical patterns is difficult, because libido varies widely from one individual to another, and because it is constantly modified by outside forces: drug and alcohol intake, illness, fatigue, stress, emotional conditions, relationship conditions, even the weather. Male and female peaks are less distinct or more pronounced, subject to all these variables.

To some extent, sexuality might even run in families, and should this be true, it strengthens the biological interpretations. While this has not been studied formally, I have observed familial patterns among many of my patients: three generations of sexually aversive women, for example, and successive generations of men with premature ejaculation. But couldn't this also be explained as the product of subliminal messages passed from parent to child?

The point is, with something as complex to measure as sex drive it is hard to determine where biology leaves off and environmental influences begin.

We do, however, have a fairly clear picture of a woman's sexual peaks over the course of her cycle in spite of the conditions mentioned above that can cloud the issue. Based on what we already know about how our hormones function, we can also speculate, and the possibilities are intriguing.

Women report three sexual peaks related to changes in their monthly cycles: midcycle, premenstrual, and menstrual.

There have been numerous studies identifying these peaks, some of which contradict each other. Some are flawed, others are well designed. When analyzed as a group, the data points to the following conclusions. Some women, up to 50 percent, notice no sexual cycle. Of those who do, here is a sample analysis: Of thirty-two separate studies, seventeen recorded a premenstrual peak, and eighteen a post menopausal peak. Only eight studies identified a peak during ovulation, and four revealed increased desire during menstruation.

Midcycle Peak: This is the time of the month often thought to encompass the strongest sexual peak, but this isn't quite accurate. In midcycle, during ovulation, testosterone and estrogen both spike. Testosterone boosts the aggressive sex drive slightly, stimulating a woman's interest, and estrogen surges, making her predominantly sexually attractive and receptive. So while there is a mild sexual peak (testosterone dependent) at midcycle, it is almost subliminal. This dynamic makes sense from nature's point of view, because this period is when a woman is also the most fertile. And for conception's sake, *receptive* is the operative word because willingness is more important than the urge to initiate sex. In fact, the aggressive sex drive in midcycle is less intense than later on during the premenstrual and menstrual sexual peaks.

Premenstrual Peak: In fact, many a woman's most intense sexual peak is just before her period, which at first glance doesn't make much reproductive sense, but it is easily explained when you understand the hormones involved. We'll examine this in detail in a later chapter.

Menstrual Peak: As noted, there are women who experience heightened desire during menstruation. This can create a problem for those men and women who feel that the period of bleeding is a taboo time or simply too unaesthetic for sex to be enjoyable. It also creates confusion in scientists and researchers who try to explain human nature. What biological sense is there in high sexual desire during the lowest fertility point?

If we have failed to appreciate the full impact of a phenomenon as familiar as the menstrual cycle, what of the other hormonal connections we are just beginning to uncover? As mentioned earlier, testosterone levels in men oscillate every fifteen to twenty minutes or so throughout the day. Since testosterone has been correlated with irritability, aggression, sex drive, the urge to masturbate, and the need to be alone, this raises a few provocative questions.

For example, are male mood swings related to the peaks and valleys of their testosterone cycles? Hour by hour, some men seem to be considerably more erratic and temperamental than women, who, although they operate primarily on a monthly cycle, fluctuate day by day. Is this constantly surging testosterone responsible for the sexual thoughts and impulses men are reputed to have every twenty minutes throughout the day? It would make sense. Are men most sexual in the morning, when their testosterone levels spike? Many men confirm this, especially in their later decades, but other men's experience is the reverse. Does it follow that the male sex drive has its seasonal peak in autumn when testosterone levels are highest? We have no evidence to confirm seasonal fluctuations in sex drive, although I suspect they exist. In fact, spring, when a young man's fancy is supposed to turn to love, is when testosterone is lowest of all.

While we are still in the early stages of understanding the full scope of how and to what degree fluctuations in hormone levels determine when, why, and how we mate. However, extensive research along with clinical observations have clearly already taught us a great deal about the forces behind the general patterns most of us experience as we wend our way through the sexual stages of life.

LOOKING AHEAD

In forthcoming chapters we will take a closer look at the lifelong sexual drama of man and woman, and get to know intimately the hormonal actors who play the biggest roles. As each becomes more familiar, you will come to understand why men and women think and behave so differently. You will learn, for example, how and when biological forces cause men to shun commitment, act territorial, and avoid talking about their feelings; why women tend to hold grudges, crave touching, and can enjoy sex without orgasm. In fact, behavior that once drove you crazy will come to seem natural, and perhaps even amusing. Once you appreciate male-female differences as the product of strong biological forces, in addition to psychological and social ones, you will also be much better equipped to handle them.

But in order to maneuver effectively, you must understand these differences within another context—our sexual stages. In addition to our cycles and peaks, and the perpetually changing sex soup that runs through our veins, men and women undergo sexual transformations throughout their lives. And these sexual stages are, in large part, hormone driven. What follows in the next chapter is a map to these stages. The hormones affecting them, how they operate, and how to handle them will be introduced to you one by one in greater detail after the stage has been set.

2.

PASSAGES: SEXUAL STAGES

It was just after eleven on a Tuesday night. Pat and Chris, wiped out from another hectic day, collapsed into bed. But instead of going straight to sleep, Pat turned to Chris with a kiss and said, "Come closer. No, I mean really close," clearly in a loving mood.

Chris turned away, saying gently but firmly, "I'm just too tired tonight, hon."

"I can't remember the last time we had sex," Pat said in a voice filled with hurt. "You're working all the time. I never see you anymore, and when I do, you're always too tired."

"You're not being fair, Pat. There's nothing wrong," Chris responded. "You expect too much. I just wish sex wasn't your only yardstick for gauging a relationship. You don't notice the hundreds of little things I do for you. I think of you all the time. Besides, you've been so critical lately, I don't feel much like getting intimate."

"What do you mean critical? If you weren't so preoccupied with everything else in your life, and noticed me occasionally, I wouldn't be so unhappy."

"Maybe if you'd treat me better," said Chris, "I'd want to be around you more often."

"Well, who are you spending your time with then? I obviously don't turn you on," Pat retorted.

"What do you mean?" Chris asked.

"Hey, I know your sex drive, Chris. You must be having it with somebody."

"Look, hon, with me sex is more than just a physical thing," said Chris. "If everything else feels wrong, sex just doesn't work right."

18

A thirtysomething couple in the middle of a fight, right? She's upset because her husband isn't as attentive as he used to be. He's irritated because she's complaining. Well, that's one possible scenario. Then again, this could be a couple in their fifties with the sexes reversed. It depends on the particular sexual stage of life each partner happens to be in. In fact, the same couple having this conversation in their twenties may actually trade places a decade or two later. Same people, same relationship, different perspective.

Men and women experience distinct and predictable sexual stages, which change their nature, on average, each decade.

There is a misguided but prevailing notion that sexual patterns are firmly established during childhood, sometime before age five, and remain fixed—or broken, as the case may be—for life. Many experts still cling to this view, and the majority of the adult population mistakenly resign themselves to it as well. In truth, the sexual patterns of our lives are much more exciting and versatile than most of us realize. They change, with or without our help, for better or for worse, many times throughout our lives.

The problem is, because we think our sexual patterns are immutable, it catches us off guard when, inevitably, something changes. When we, or worse, our sexual partner, shift sexual gears, we feel ambushed, surprised or betrayed—even when it's an improvement: "Why do you want sex now, all of a sudden? You never did before."

When you don't know what to expect you can't be very well equipped to cope. So, let me introduce you to the sexual stages of life before you run into them on your own.

Sexual stages, or passages, are much like the different bands of a rainbow, each separate, but related, and always colorful. One blends into the other although each color is distinctive. Although you may prefer one to another, each band adds depth, dimension, and color to the whole.

Sexual stages involve more than a person's erotic function and feeling. These stages include changes that occur in the emotional and psychological components of an individual as well as a relationship decade by decade. Sexually, some of the changes, like speed of ejaculation and ease of orgasm, result from both physiological and psychological forces. Others, like willingness and ability to communicate meaningfully, are primarily a function of experience and maturity, along with the ability to learn from the past.

Our sexual stages, peaks, and cycles overlap, interact, and influence one another. With stages occurring every decade, sexual peaks interject hourly, daily, or monthly, depending on gender and individual patterns. In addition, most men peak *physiologically* in their teens and *psychologically* after fifty. Women peak sexually in their thirties or forties and psychologically in their fifties.

Cycles, as we've seen, can refer to hourly, daily, monthly, yearly, or even lifetime patterns. Sometimes the defining lines between them blur. For example, do we call a lifetime of sexual highs a peak or a cycle? It could technically take on either label, so some of the terms applied are arbitrarily assigned.

Each time we enter a new sexual stage, a different and usually better relationship forms—either with the same partner or a new one. The transition from one stage to another, however, can be painful. Divorce is a common consequence of mismanaging these passages, which is one of the many reasons for learning how to recognize and handle them.

While there is a general flow to the process that most people experience, there is considerable individual variation. Some stages occur out of sequence, as with women who bear children late in life. Others may skip one or more stages altogether, as in the case of a man who never commits. When sexual dysfunction occurs and persists, one phase may endure indefinitely. Then there are those who, for various reasons, fail to evolve from one color of the sexual rainbow to another, thus perpetuating a monochromatic pattern.

Ironically, as men and women of the same age go through their sexual passages they are not necessarily well suited to one another. Indeed, they typically find themselves sexually out of sync, leading to sexual tensions that can complement or complicate a relationship. To confuse matters further, there are always two levels of compatibility to consider: the sexual and the emotional. Two people of the same age can also be at very different levels of maturity. Consequently, certain age combinations are often more compatible than others. Men and women in distant decades can be quite well suited for one another, at least for a while. An example of sexual similarities between a teenage boy and a woman three decades older is captured in the relationship between Carolyn and Tony:

Carolyn, a fifty-year-old woman living in St. Louis, is streetwise and sexually experienced. Married and divorced twice—the first time for love and the second for money—she has decided that an enduring partnership with a man is an unnatural condition. She loves sex, but doesn't want the emotional entanglements. What she does want are orgasms. Her estrogens are vanishing, and her relative testosterone levels are exerting more influence. She keeps herself fit and is attractive to men of all ages but prefers boys under twenty. There is little or no risk of emotional attachment because she finds young men intellectually deprived. However, they are so delightfully depraved she can have all the raw physical sex she craves. Sexually, she is choosing men in their prime. This is not necessarily synonymous with men at their best.

Nineteen-year-old Tony is her flavor of the month. With testosterone to spare, he has resilient erections, ejaculates often, and participates enthusiastically. Sex is physical, exhausting, athletic. What he lacks in tenderness and technique, she corrects by knowing what she wants and telling him.

Sexually and situationally, this couple is well matched. Emotionally, they are in different time zones, but it doesn't seem to matter to either one of them at the moment. While most people would not consider this arrangement as their idea of a healthy or fulfilling relationship, Carolyn is satisfied, at least for now.

Let's look at this picture in reverse:

Don is in his early sixties. To his surprise, Antonia, one of the young women on his sales force, has taken a romantic interest in him. Although she is only twenty-five, in spite of the generation gap, they have a great time—and not just because of sex. They don't enjoy the same music, but she is fascinated with his wisdom and experience, and he is generous in sharing it. Don has become the envy of his peers, who view his ability to attract a young, beautiful girl as a reflection of unusual virility and personal power—the usual interpretation of older man/young beauty (other than money).

In actual fact, sexually, they are quite well matched, but for the opposite reasons most people think. Don has passed his sexual prime, physically speaking, and isn't interested in sex as often as he once was. He is also not perpetually orgasm driven. Instead, he has learned to savor the journey, and is not as aggressive as he was as a young man. His testosterone levels are dwindling.

Antonia has not yet come into her sexual prime, yet she is at the height of her estrogen receptiveness—which, as you will hear more about later, makes her soft, acquiescent, and sexually receptive, but not demanding or insistent. Beautiful, seductive, and erotic, she offers more intense visual stimulation than most women his own age, and Don is charmed by her warmth and eagerness to please. She enjoys sex with or without orgasm, for women in their twenties are more fulfilled by being penetrated, held, and touched than by fast-paced, hard-thrusting, orgasm-driven sex.

Unlike the average forty- or fifty-year-old, who knows what she wants, when she wants it, and how often, Antonia, at this sexual stage, is comforting to Don, not threatening. She may cuddle up, loving to be touched, but is not inclined to initiate sex or complain if he is not up to a performance. He is more willing to hold her, talk to her, teach her, and give her his full attention than younger men. Instead of unusual virility, it is his increasing sexual insecurity and fading drive for orgasm that makes this match work. Imagine Don trying to get along with Carolyn or vice versa. Consider Antonia trying to develop emotional bonds with Tony or one of Carolyn's other lusty young studs.

Perhaps because Don's testosterone is diminishing, some of his drive for power and control is expressed in more subtle, manipulative ways. He chooses young women and dominates them easily by virtue of age, not ability, while a younger man like Tony, in his testosterone-driven desire to dominate, would be more openly aggressive with a woman his own age. This same-age match could lead to clashes and conflict, perhaps even abusive verbal and physical behavior from him intended to force her into submission.

Clearly then, while emotionally and intellectually men and women closer in age usually make a better fit, strictly from the sexual point of view, they are often quite unsuitable for each other, more so the younger they are. This disparity, when in full force, represents quite a dilemma in cultures where polygamy is not the accepted norm. But add love to the equation and the formula changes exponentially as you will shortly see. Oddly enough, love can either eliminate enormous obstacles or create new ones.

As men and women mature, they become more sexually and emotionally compatible—unless they get derailed along the way—but they don't ever become the same. As a woman gets older she usually manifests more traditionally "masculine" traits—like decisiveness, assertiveness, physical sexuality, and independence. Men expand their "female" dimension of touching, tenderness, insight, patience, and understanding. The net effect is that couples of the same age, if they manage to weather and master the physical and emotional stages of youth, become better and better together over time—especially sexually.

Sexuality is a dynamic, changing process that goes through natural and predictable phases over the course of a lifetime. Each sexual stage you are about to encounter, vicariously or otherwise, is in a sense a rite of passage.

I would like to emphasize that these sexual passages represent typical, but by no means absolute, patterns experienced by the majority of men and women in our society. Social and psychological forces conspire to upset the pattern, often causing stages to occur out of sequence. Major traumas—divorce, the death of a spouse or child, or financial crisis—can also cause wide deviations from the usual patterns. So can early traumatic experiences: a child who is raped or abused or a youth whose first sexual encounters are humiliating take on baggage that may affect all their subsequent passages. Add in differences in appearance, family upbringing, peer influence and culture, and you can see how dramatically patterns can vary from one person to another.

For purposes of convenience, I have divided the adult stages by decades, but understand that they do not begin and end neatly when we pass our landmark birthdays. Sometimes, the characteristics of our sexual stages blend together smoothly, as one phase gradually evolves into the next. But just as the color yellow abruptly changes to green in a rainbow, the shift can be sudden and distinct.

So what do these sexual stages have to do with hormones? Absolutely everything! Culture, environment, and personality all affect the nature and the timing of the stages we go through but not independently of the chemicals that drive us. Hormonal forces influence these stages with the power of an undertow. You can see a tidal wave coming—a divorce or birth, for instance —and defend against it or prepare for it to the best of your ability. Undertows, however—unless someone maps them out for you—will catch you by surprise and do you in. But even then, you can be taught how to maneuver out of their grasp.

As we travel the typical sexual stages from decade to decade, with special attention to the different patterns, needs, and attitudes of men and women, it will become apparent which ages are best suited to form relationships and why. The catch to the deal is that no one stays the same age forever, and the best match at one point in time may become the worst in a decade or so. Enduring relationships require an understanding of this evolutionary process, along with techniques and skills for overcoming the challenges they will encounter along the road. You can actually change your stage instead of your partner, if you know how to do it, avoiding a lot of problems along the way.

By seeing what stages you've already been through, identifying where you are now, and looking ahead to where you want to go, you will be better able to anticipate change, overcome the hurdles, and create the relationship you want. During the process, you will need to recognize corresponding patterns in your mate and how best to deal with them.

CHILDHOOD (BIRTH TO PUBERTY)

Sexuality is an integral part of the human experience from the moment of birth and, surprisingly, even before. While still in the uterus, male babies actually have frequent erections you can see and document with ultrasound. About half of all boys enter the world with a full erection before the cord has even been cut. Females lubricate with a regular rhythm in the womb. Some infant girls actually have minimenstrual periods in the newborn nursery, the result of sudden estrogen withdrawal when separated from their mothers at birth. It is as if nature wants to ensure that all the key equipment is in place and tuned up right from the start. Babies are born with all their sexual hardware intact; it lies somewhat dormant, waiting for instructions from their hormonal software, which makes its dramatic debut at puberty.

But, long before then, boys and girls are sexual beings—and they would be even more so if adults didn't intervene. Childhood sexual exploration is as natural as teething, but more painful to parents. It comes as such a shock to grown-ups in this country that they rush to restrain it. By contrast, certain cultures not only tolerate sexuality in children, they use it as a child-rearing technique. The mothers in one primitive South American society—the Waika Indians of the jungles of the upper Orinoco—quiet fussy baby boys by fondling their penis or stroking their scrotum. This is done to soothe and pacify, not to give perverse pleasure to adults or to exploit children. We, instead, stick a rubber nipple in babies' mouths, which this ancient tribe might find just as incredible.

24

In our culture, playing with your genitals or anyone else's in public is not considered socially acceptable. It is, however, something that young kids just naturally, consciously or unconsciously, do.

Mark and Julie noticed their four-year-old son absently fiddling with his penis while watching television with them in the living room.

"What are you doing?" the father asked.

"I'm playing with my penis, Daddy," answered the child matter-of-factly.

"Why are you doing that?"

"Because it feels good. Would you like to play with it too?" Obviously, this polite child had been taught to share his toys.

The parents were at a momentary loss, but responded with an equally courteous, "No, thank you," and explained that this was an exception to the general rule of sharing—and added that he should do it in private.

Parents tend not to understand that children are very much sexual beings from birth; and even those who see the sexual dynamics clearly usually don't know how to deal with them constructively. Both sexes masturbate, sometimes in the most embarrassing places—in the grocery cart in full view or in church alongside Grandma. Little boys have erections on and off throughout the night, and more often than not, each time you change their diaper. The original Kinsey Report, written in 1948, describes men who recall experiencing orgasms before the age of five, even though they could not yet ejaculate. Young girls discover sexual pleasures sliding down poles on jungle gyms, or from strategic maneuvering around a Jacuzzi jet, like Jenny, who generously shared the news with all her friends.

One young mother noticed her seven-year-old daughter, Jenny, hovering around the Jacuzzi jet increasingly often with a glazed expression on her face. The next time she had a playmate over, she overheard her daughter shout to her friend, "Grab a bullet and jump on." Curious, Mom found her daughter giving detailed instructions to her friend on just how to reproduce the Jacuzzi experience, showing her exactly where to sit and how to find the best angle, all the while asking, "Can you feel it yet? Doesn't it feel good?"

Another couple, in therapy because the wife had an aversion to sex, wanted to protect their children from the strict, puritanical upbringing that had contributed to their own difficulties. I advised them to begin by being candid and responsive to their children's natural curiosity, taking advantage of everyday events as catalysts for discussion. My suggestion backfired, or so it seemed at first blush when Eric and Erin innocently let their parents in on some of their sexual research.

Late one night I received a frantic phone call from Heather regarding her children. They had all been watching a movie on television in the family room. Their son, Eric, was three, and their daughter, Erin, was five. When Heather realized that a rape scene was about to begin, she looked at her husband for guidance. They didn't know what to do: turn it off, as they would have been quick to do in the past, or let it play and see what happened? While they were deliberating over the best course of action, the scene came and went. The children asked what that bad man was doing to the screaming woman. At that point, they turned off the TV and explained that he was putting his penis in the woman's vagina against her will—something called rape. Then, using terms their kids understood and introducing new ones as necessary, Heather and her husband told the children about sexual intercourse, what it was and why people did it, explaining that usually grown-ups have intercourse after they have developed a loving relationship, but sometimes a man forces himself on a woman, committing a crime against her called rape. They elaborated somewhat, trying to make clear that depending on the circumstances, it could be either a very wonderful experience or a dreadful one. (To many adults this may have seemed to be too much information too soon, but if they had not opened the subject they would never have heard what came next.)

Their son's response stunned them. He announced, "Oh, Erin and I have tried to do that lots of times, but my penis is too small and it keeps falling out."

I assured panic-stricken Heather that such experimentation is more common than people would like to believe, and told her how to put a stop to it without making their children think they had done something evil.

After the shock wore off and the discussions, which went quite well, were over, these parents were most grateful that they had let the rape scene play. Without it they may never have discovered the sexual mischief going on right under their noses, and would not have been able to nip the problem in the bud. The consequences of this behavior progressing unchecked could have been quite serious.

Eric and Erin were not knowingly doing anything wrong. As with most young children, they were following their sexual curiosity.

It is important to note that sexuality among children is different from rummaging in the trash or shoving a stick in a gopher hole. It is sexual. They become aroused. In fact, it feels so good they would do it a lot more often if they weren't discouraged. But don't confuse a normal child's quest for arousal with sexual exploitation. The sexual abuse of children by children is not uncommon, particularly as young boys approach puberty. Both sexes can be terrorized by those intent on exploiting them.

To ignore the sexual developments of childhood is to exile a most dynamic aspect of your child's personality, creating a gulf in your relationship and eliminating tremendous opportunities for you to play a constructive, protective role in your child's sexual growth. To ignore these sexual emotions and impulses is also to deny those you yourself experienced as a child that have contributed to your sexual nature as an adult. What you have just read, then, is not only applicable to how you deal with these issues as a parent, but also how you put together the sum total of your life experiences.

In reflecting on your own past, or thinking about your children's future, consider the picture that has been sketched for you here: that children are bombarded with sexuality from the beginning, both from within and without. Their curiosity provokes them, while their sex hormones are working, manipulating these sexual urges and surges. Testosterone and estrogen are at play, keeping a sexual pilot light burning.

The characteristic features of childhood sexuality, then, involve curiosity, arousability, masturbation, "show me" or "doctor" games, secret meetings and private experiments.

TEENAGERS

As childhood progresses into puberty, it becomes more and more difficult for adults to influence sexual development constructively. This is the familiar period of hormonal madness, when nature prepares both sexes to reproduce. The body's chemicals start issuing orders at a dizzying pace. Girls bloom and bleed as they begin menstruating and developing breasts and curvaceous hips. Boys' voices crack and their penises swell at the most unwelcome times. At the height of the testosterone onslaught they swagger aggressively, obsess about girls, and masturbate relentlessly, regardless of the repercussions.

Then came adolescence—half of my waking life spent locked behind the bathroom door, firing my wad down the toilet bowl, or into the soiled clothes of the laundry hamper, or splat, up against the medicine-chest mirror, before which I stood in my dropped drawers so I could see how it looked coming out. Through a world of matted handkerchiefs and crumpled Kleenex and stained pajamas, I moved my raw and swollen penis, perpetually in dread that my loathesomeness would be discovered by someone stealing upon me just as I was in the frenzy of dropping my load. Nevertheless, I was wholly incapable of keeping my paws from my dong once it started the climb up my belly. . . . On an outing of our family association, I once cored an apple, saw to my astonishment (and with the aid of my obsession) what it looked like, and ran off into the woods to fall upon the orifice of the fruit, pretending that the cool and mealy hole was actually between the legs of that mythical being who always called me Big Boy when she pleaded for what no girl in all recorded history had ever had. "Oh shove it in me, Big Boy," cried the cored apple that I banged silly on that picnic. "Big Boy, Big Boy, oh give me all you've got," begged the empty milk bottle that I kept hidden in our storage bin in the basement, to drive wild after school with my Vaselined upright. "Come, Big Boy, come," screamed the maddened piece of liver that, in my own insanity, I bought one afternoon at a butcher shop and, believe it or not, violated behind a billboard on the way to a bar mitzvah lesson. (Philip Roth, Portnoy's Complaint)

Hormones and Hismoans: With the onset of puberty, the already apparent differences between the sexes become much more dramatic. Sexually and emotionally, teenage boys are not a very good match for teenage girls. Hormonal influences, aided and abetted by social customs, dictate widely divergent directions.

"Adolescence?" recalled thirty-six-year-old Susan in response to a question I asked about her teenage years. "It was just great. I cried all the way through it. My mother tells me she can't remember me without red puffy eyes until my sister left home when I was seventeen. Now my brother, who was almost my age, didn't cry at all. But he changed, too. Dave became virtually inarticulate, grunting instead of speaking in complete sentences until he was almost into his twenties."

The distinctions that I'm drawing reflect the most typical patterns for teens, but do not represent the only ones during this stage. In fact, this is the most common sexual/emotion stage where both sexes can get derailed, developing unhealthy, aberrant, even pathological patterns that persist throughout adulthood. The catalysts for disaster are the tumultuous hormones that whiplash their emotions, sometimes beyond their capacity to endure. Fortunately, the majority survive, with the potential to thrive.

Testosterone Poisoning: The first time you encountered testosterone you weren't aware of it. You were in the womb as its forces endowed the unsuspecting female fetus with male features and accessories. Neither did your parents notice this dramatic transition until you materialized at birth and declared your sex.

The second time testosterone makes an appearance, everybody notices. It's almost as dramatic as the mythical werewolf. The sweet tender-looking boy-child with the angelic voice develops an odor, sprouts hair here and there —even on his face—changes body form, mood, personality, and doesn't even sound the same. His face erupts with volcanic manifestations. We don't appreciate how dramatic this transformation actually is, because it is such a common event. But contemplate the force of these changes and consider their lifelong consequences.

Let's start with the "inappropriate erections" boys have to contend with at this time. It's a visible manifestation of the testosterone effect, but the implications of this phenomenon have a far greater impact on a man's sexual development than momentary embarrassment. Most boys already have been confronted by outside forces—parents, teachers, and coaches, primarily— who advise them not to cry, not to be a baby or not to act like a sissy when they have expressed spontaneous emotion. Now, for the first time, they face an internal proctor. In an effort, fruitless though it may be, to repress their uncooperative penis, they start burying their feelings as vigorously as they are able, setting patterns of emotional suppression that may persist throughout their lifetime. Girls have no such incentive, no barometer that betrays their thoughts and feelings for all to see. Having no such frame of reference, it's hard for women to comprehend the difficulty so many men ultimately have expressing themselves.

Under the influence of testosterone, boys become even more competitive as they approach puberty, jousting for status and mates everywhere from the basketball court to the Friday night dance. The same testosterone that gives them acne makes them fiercely determined to win at just about everything. Even their developing relationships with girls are secondary to how they are reflected in the eyes of their male friends and where they end up in the pecking order.

Boys start experiencing nocturnal emissions or wet dreams at this stage, and during their waking hours fantasize about explicit sexual scenes, developing an intense interest in lingerie catalogues and *Playboy* magazine.

During this period, boys collude. Since girls are harder to come by, they often experiment with each other—most often with no relevant sexual orientation. Mutual masturbation, "circle jerks," and peeing contests are just a few examples of the behavior their indiscriminate libido inspires.

In essence, testosterone in males first appears as a secret agent, having great influence with no signs on the surface. Its second coming, however, could not be less subtle. Along with the visible physical changes, teenage boys are subjected to deeper forces of testosterone tyranny—their sexual urges, mood swings, aggressiveness, competitiveness, irritability, sense of omnipotence, impaired impulse control, and perhaps, even their sense of humor—all of which you will learn more about in later chapters.

This is not to say that adolescent boys are helpless subjects of testosterone. Their judgment, emotional maturity, choices, and discipline can strongly influence, even overcome, the influence of this hormone in certain circumstances.

The point is, if a teenage boy behaves well, it is in spite of testosterone, not because of it.

Testosterone and the Town "Slut": Testosterone is operating on girls, too, but at lower levels. In fact, when this hormone goes awry, females get offensive and unmanageable and start behaving like males.

A case report from Sweden describes two teenage girls who manifested sexually bizarre and provocative behavior due to a surge in their "male" hormones as they entered puberty. They became aggressive and violent as well. All these symptoms disappeared when they were given medication that reduced these hormones.

Everyone has a story about the town "slut," but these sexually aggressive or indiscriminate girls fall into two very different categories. The first is comprised of very needy girls, probably low on testosterone, with no self-esteem, who trade sex for attention—any kind at all. The second is almost the opposite extreme: testosterone-driven hypersexual girls who really want sex and like boys. The attraction of sex for her is genital and physical. Power is her aphrodisiac. She's not ashamed of her exploits like her more timid sister; she flaunts them, challenges men, and loves to exercise her control over the poor souls. Neither one is the general rule among teenagers, but because they are so flagrant and attract such a following, they come to everyone else's attention.

Essentially, the high-testosterone adolescent female behaves more like the male. The low-testosterone female behaves like the high-testosterone female, but she doesn't enjoy it. The teenage girl with normal levels of testosterone for her age experiences a sufficient surge of it during puberty to stimulate her sexual urges. However, her specific sexual urges are geared toward masturbation, rather than boys. In fact, she masturbates with much greater frequency than we once imagined, thanks primarily to testosterone. She also lubricates copiously at this age, thanks to estrogen.

Estrogen and the Emerging Woman: While testosterone boosts somewhat during a young girl's adolescence, it is nothing compared to estrogen, which literally sculpts her body into the curvaceous woman she is soon to become. With estrogen at play, she redistributes and increases her body fat. Some girls get chubby during these years. Almost all develop breasts, and, of course, begin to menstruate. Tender breasts, menstrual cramps, mood swings, and bloating don't show on the outside like facial hair, but they are wreaking havoc within a young woman's body and mind.

Not only is she feeling differently—and not always too well—there is a sharp change in how others—both men and women—react to her. Her father stops touching her. Boys start. Some adult women withdraw, finding her a threat. Uncles, family friends, neighbors who never noticed her before stake a special interest. Teenage girls, whether withdrawn, well adjusted, or off center, become sexual magnets, an experience that both thrills and terrifies them.

Just as women have much less testosterone than men, men have considerably less estrogen than women. They do, however, have some estrogen surges of their own during puberty. This aspect of their development has not been well explored, so we do not have much insight into these dynamics.

Sometimes young men will get one-sided breast enlargement during or prior to puberty. One breast becomes tender and enlarged, developing a firm lump about the size of a quarter. It's upsetting and worrisome to everyone, and the young man is often rushed to the doctor for fear of it being cancer, when a little advance knowledge and reassurance is all he needs. It's just a temporary estrogen surge, and will disappear on its own.

The contrasts between the sexes are greater in some stages than others, and teenagers reflect the greatest disparity. While their emotional needs remain similar—self-esteem issues and the quest for approval—the basic sexual objectives between the genders differ most dramatically in these years. Boys are preoccupied with orgasm and conquest, while girls crave endless touching, foreplay, affection, and love.

Although teenagers seem to be emotionally and sexually mismatched, they have one very important point of intersection—touch. During adolescence most boys and girls experience more touching and cuddling than they will at any other time of their lives. It doesn't matter that it's for different reasons. She is allowing him to touch her because she loves it; he touches her for as long as he can, hoping she will eventually respond and allow intercourse or some meaningful alternative with him.

Teenage differences, needs, and desires are too great not to clash frequently and intensely, but occasionally love comes to the rescue. Love is the great equalizer at any sexual stage. For reasons not always readily explained, it can bridge differences, transforming conflict into contentment and euphoria. And that is especially valuable to teenagers whose contrasts are so vast.

Tim and Destiny fell in love as teenagers and somehow avoided the typical tumultuousness of adolescent relationships:

Tim and Destiny started dating in the ninth grade. They were both popular kids, but accessible, not arrogant, so they were well liked more than envied. For the first year or so, they rarely spent time alone, but went out instead, usually with a group of their friends. They held hands and weren't shy of showing affection in public. It was obvious that they cared for each other. Tim's attentiveness to Destiny was in such contrast to the insouciant swagger of the other boys his age that Destiny's girlfriends started using Tim as their gold standard, wanting the same kind of treatment for themselves. Against all odds, Tim and Destiny never seemed to go through the roller-coaster pattern of the relationships around them. Tim loved to talk to Destiny as well as touch her. She adored him in return. During their senior year, they began having intercourse. It brought them closer than they ever imagined, even more so because they kept it to themselves. There was no locker-room talk or girlish gossip—speculation, yes, but no one else ever knew for sure. Other kids marveled and wondered how they did it, and if it would last. At graduation, their relationship was still going strong.

When asked what made their relationship work so well, neither Tim nor Destiny could put it into words, other than to say it seemed like the easiest thing in the world. This is the sort of story you could hear at almost any stage of life. When love, whether puppy love, true love, or temporary love, exerts its influence, it smooths out the bumps and erases seemingly insurmountable obstacles.

As teenagers get older, intercourse is not only more socially acceptable, but opportunities for it grow along with personal freedom. When that happens, ironically, in most cases, touching comes to a sudden and jarring halt. Men no longer have the patience or the persistence to touch to the extent that most women would want. They skip rapidly to penetration thinking that women are just as eager as they are to get to "the good stuff."

He is urged by his testosterone toward genital sex, both his and hers, and orgasm, his, at least. But without the intimacy and arousal of general touching, a woman's oxytocin doesn't get moving. Her estrogens don't rise, her testosterone doesn't take over. She has so much less touching when she needs much more to lead her to genital focus. In the process, she becomes frustrated and discontented. What was once a great relationship deteriorates rapidly. That is basically what happened to Linda and Will:

Will and Linda had been dating since the beginning of their junior year in high school. The relationship was pretty intense in every respect. Their passion for each other was so obvious that their parents took great efforts to keep them apart. They touched, it seemed, from morning to night. Their neglected friends thought it was disgusting.

Linda was too afraid of getting caught to go on to intercourse, not to mention having no comfortable or convenient place to do it. It had nothing to do with her sex drive though. Of the two of them, she was the instigator more often than not.

Will and Linda arranged to attend the same college, and eventually planned to marry. Now ensconced in dormitories for their freshman year, for the first time they were on their own and could do as they pleased. It was hard to tell which one wanted whom more—at first. Will felt that now that they could finally have intercourse, it was no longer necessary to do all that touching, or so he thought. He didn't realize that he was depriving Linda of what she loved the most.

Instead of tenderness, she got interminable pounding. "Is this what all the fuss has been about? What a disappointment," she lamented. Meanwhile, Will was doing everything within his power to give her exactly what he believed she wanted, which was enduring, extensive, deep, heavy thrusting. Exercising great restraint and consideration, he checked his watch occasionally to see if he could beat his last record.

In actuality, many teens who segue to intercourse have little valid information to arm them, and more than enough misconceptions to disrupt their enjoyment. Even when they do know what they are doing, they are usually too embarrassed to talk about what they want with their partner.

Hormones are charging all over the place, and the sexes are simply at odds physically. The differences that define the issues that men and women will have to contend with for the rest of their lives have manifested themselves with a vengeance.

We have primarily been discussing estrogen and testosterone because we know the most about how these hormones behave when your teenagers misbehave. The naked truth is that most of the studies we have on hormones, including testosterone, estrogen, and oxytocin, have been done on adults and we can only speculate on their role in adolescents.

I suspect that just like the hormones we are more familiar with, DHEA, vasopressin, PEA, and especially growth hormone (as well as others we haven't yet discovered) are flexing their muscles, coming awake, and contributing to the general physical and emotional chaos that is so prominent during these years.

TWENTYSOMETHING

At this stage men and women are about as emotionally ill suited for one another as they will ever be but, as we all know, that does not discourage them from relentless pursuit of one another.

Physically and anatomically, they are usually at the peak of their desirability. Reproductively, women are in their prime. Physiologically, they are at the height of sexual receptivity. Emotionally, they tend to be compliant and acquiescent, far less sexually demanding than they will become as time goes on. They are sexual magnets for men of all ages, regardless of their looks.

The woman in her twenties strives to please men so they will love her and she'll feel valued—it's an estrogen effect. She has gone one step beyond her teenage sister, who trades sex for approval. For her, sexual attention *becomes* approval—no longer a trade-off but a necessary payoff. (For those liberated souls who insist modern women aren't like that, think of the popularity of romance novels and who reads them.)

Women in their twenties tend to look for love in all the wrong places. She may be drawn to a powerful, controlling man. He makes her feel more secure because like a parent he takes care of her, but he can also make her feel *less* secure by dominating, criticizing, and infantilizing her. Many women end up with exciting men who turn out to be self-centered at best, dangerous at worst.

Most of the psychological treatises written on this subject disparage women who seek out dominating men, regarding them as emotionally dysfunctional. Women are placed in all forms of therapy from psychotherapy to pills, in an effort to "cure" their "problem." Certainly, many of these women need the emotional support of a caring therapist, and certainly some of them are severely disturbed. But what if we are overdiagnosing here? What if it is *healthy*, not *sick*, for an estrogen-laden woman to be attracted to a high-testosterone male—the dominant leader of the pack who maintains his position of authority harshly when necessary? Such is the ordinary case among lower primates.

What if it is the male, not the female, who requires exhaustive therapy when he crosses the boundary of his mate's physical and emotional well-being, becoming a threat to her rather than a protector? Food for thought.

Going His Way

While she is going out of her way to please him, her male counterpart is going his *own* way. Men in their twenties are somewhat less focused on orgasm than when they were teenagers, but not by much. Their sexual ego has emerged and begins to take on the job of satisfying women—not for her sake, particularly, but as validation of his talents.

Most men still believe that it is somehow their responsibility to lead a woman to her sexual self, to teach her what she likes. Consequently, they do whatever is required to *make* her respond no matter what; they are determined to succeed. They set about the project just as if they were going into training. Practice makes perfect, so they practice a lot. Some men treat sex much like any other athletic event—attitudes that inspire terms like "sport fucking." They become sexual predators, often measuring their manliness by the variety and frequency of their conquests.

Inside, however, men are becoming increasingly disconnected emotionally from the sexual experience. In an effort to perform, to measure up to their imagined competition, they detach. Why? In order to thrust longer, indefinitely if necessary, while they chivalrously wait for the woman to come. In order not to explode in orgasm they have to distract themselves, think of other things, tune out. One man told me he imagined his mother sitting on the toilet having a bowel movement. Nothing less would distract him. Erotic, huh?

The effort most men make is an increasingly lonely one because, sadly enough, they are trying very hard to give a woman what she doesn't really want. Yes, most women like intercourse and enjoy thrusting *if* they get the rest of what they want. They want a feeling of intimacy, of connectedness. They want men to stroke their hair, rub their back, suck their breasts, kiss their thighs, lick their most intimate and exquisite parts and, most of all, savor doing it. When they don't experience all these other things, they detach as well. The result is two warm bodies just going through the motions without the emotions.

While she's looking for romance, he is looking for sex. These contrasting and most urgent needs can cause an enormous amount of tension, confusion, pain, and duplicity. Some men learn to say "I love you" for effect, anything to get a woman into bed. Ironically, the ones who really mean it often can't even get the words out. And with the sexual emphasis for twentysomethings shifting from touch to penetration, the dynamics of a relationship are altogether changed.

Let's pick up with Linda and Will. They are now halfway through college, both wondering silently how to get out of marriage plans that seem to have developed a life of their own, becoming a given for them, their family, and friends.

No longer being touched as much as she once was, Linda feels betrayed. She also no longer feels much, except increasing vaginal discomfort and general irritation with Will as he seemingly goes on forever. She even starts faking orgasm to get him to stop! As she becomes less enthusiastic about sex, she begins finding ways to avoid it. Where have all the good feelings gone? Why has he quit doing all the things she used to like so much? She furtively masturbates after a frustrating sexual session, and finds herself doing so increasingly often.

Will, on the other hand, feels tricked. What happened to the woman who used to love sex, couldn't get enough? Had she just been pretending?

Although technically Will may be past his most intense physical sexual peak, he is still predominantly testosterone driven, genitally focused, and preoccupied with orgasm at the expense of touch. Linda, now deprived of the levels of oxytocin—her touch hormone—that she has become accustomed to, is going through withdrawal. Along with her other fluctuating and cycling hormones, a certain sort of sexual disorientation has set in. Whatever testosterone she had working in her favor is affected by the frustrations she is encountering, turning her more toward her vibrator.

Adding to the pressure on relationships in this stage are several other factors. PMS often plagues women and tortures men. No matter how bad her mood swings were in her teenage years, she never had to deal with them while living day to day with a man. It was her parents and siblings who bore the brunt before; now it's her husband or boyfriend. In addition, married women often feel the shock of loneliness in their twenties. Her career- or work-oriented man does not have enough energy or inclination after a long day at the job to give her the time, affection, and attention she needs. He feels her demands are unreasonable, although she too is usually working hard.

Marital and sexual problems are inevitable at this stage if corrective steps are not taken, because the situation will not improve spontaneously. Couples such as Linda and Will have to face the forces of estrogen and testosterone at their best—and their worst. They also often have a serious time-management problem that undermines their opportunity to respond appropriately to the demands of their personal life. Even the best of relationships with no sexual difficulties whatsoever will eventually develop one, simply by mismanaging time. As a consequence, as much as 60 percent of the sex therapy I do involves addressing this issue. The problem usually begins with the couple's sexual chemistry dying of benign neglect. Eventually, anger sets in and more serious sexual problems ensue—think back to Chris and Pat, whom you met at the beginning of this chapter. The competition for time and attention drives a wedge between them. By the time they see each other, they are too tired to enjoy anything, much less sex. As the chemistry fades, the relationship loses color. And touch, the one thing that could turn the tide, disappeared some time ago.

Let's look at this chain reaction in physiological terms. It is normal for chronic fatigue to take the edge off lust. Between overbooking and overwork, a couple's sexual rheostat gets turned down low. PEA, DHEA, and oxytocin are still normal; they're at resting levels. But there is no sexual chemistry to start a reaction. Because touch has all but disappeared except for perfunctory caresses during sex, the sweet euphoria of the oxytocin effect doesn't have an opportunity to improve the situation.

While they have survived the peak stage of testosterone toxicity, men in their twenties are still governed by that domineering hormone, compelling them toward unemotional, uncomplicated couplings, like one-night stands. But both men and women can be highly ambivalent about settling down at this age, even when a relationship is excellent. They are still far apart on the commitment issue, thanks largely to biology. To goal-oriented young men who pride themselves on working hard and playing hard, commitment represents a burden, while for many women, commitment continues to represent security and emotional fulfillment.

This persistent gender difference was cleverly captured in a Jules Feiffer cartoon depicting a phone conversation between a young couple.

"We saw each other day and night," she says. *"We were in love. I wanted to get married. You weren't ready to make a commitment. I wanted to have children. You wanted your freedom. So we haven't seen each other in six months. Are you happy?"*

"My days are empty," he admits. *"My nights are like death. I get into trouble. I start fights. I drink myself into oblivion."*

"Are you ready to reconsider?" she asks.

"What? And give up my freedom?"

Despite the breakdown of tradition and family structure, men and women still have a strong bonding instinct and continue to build families at the same time they build their careers. In fact, the desire for romance and the desire for sex can come together beautifully in the service of commitment: nothing is more romantic than "till death do us part," and nothing provides the opportunity for regular sex more than a wedding band. Men and women who marry in their twenties can, in spite of as well as because of these differences, make excellent lifelong matches.

This conclusion may appear to be a contradiction, but it is not. Bear in mind that differences don't automatically mean conflict. Sometimes they have the opposite effect. In contrast to teenagers who are extremely polarized, the differences between the sexes found in the twenties, although significant, can lead either to chronic conflict or a complementary alliance. There are two major influences that help to determine the course: love and understanding.

As mentioned before, love overrides many obstacles, regardless of the stage. Understanding what is typical feeling and behavior for this stage enables a man and woman in their twenties to better understand one another, and to accept, cooperate, and help when necessary instead of responding in a critical or hostile manner.

THIRTYSOMETHING

As the thirties approach, the touching patterns of the twenties—namely, little to none—crystallize, replete with their full potential for harm. Ironically, though, as men enter their thirties, the need for separateness diminishes. To what extent this is due to hormonal changes is not yet clear. Maybe vasopressin, the "monogamy molecule," has more of a chemical chance now. Social and psychological forces also conspire to make commitment seem more attractive as men get older. If a man is fortunate enough to have fallen in love, he has learned that sex is more fulfilling when the heart is engaged. His maturity gives him more leverage with his hormones, and he can channel testosterone and its relatives to the relationship of his choice.

But just as men are coming round, women in their thirties start heading in the opposite direction. Many women have been divorced, perhaps more than once. Others have had a series of failed relationships that have never culmi-

39

nated in marriage or children. They've been let down by romance and they are determined not to be fooled again. They're going to pay attention to their own needs first, for a change. This can be a healthy attitude if it's not accompanied by bitterness and resentment. Oddly enough, this independence and strength of character are more attractive to many men in their thirties who have tired of clingy, dependent women.

For those who got married in their twenties, the thirtysomething decade is either the best of times together or the worst of times, with little room to maneuver in between. It is one of the highest-risk times for affairs. Discontent and disillusionment are common, since many marriages don't live up to the hopes and dreams we attach to them, and both partners have probably invested too much energy in their work and not enough in nurturing the relationship. She gets touched even less than before, especially if discord has created distance in the relationship. To fill her need for affection and to get her oxytocin fix, she used to turn to her children. It made the void less obvious. But by now, typically, her children have grown to the age where they don't want to be cuddled anymore. All her sources of oxytocin have been cut off, unless she has an affair or a pet. Touching is at its lowest ebb at this stage.

However, even though her touch quotient is low, subtle shifts have begun to occur in her hormonal balance, enabling testosterone to manifest its influence more prominently. In general, her hormones are making her less reticent and more proactive in going after what she wants.

But, as you will see below, a woman doesn't necessarily confront her spouse with her needs, or ask him for what she wants. She might go elsewhere. Orgasm gets the lion's share of her attention now:

Together since college, now enmeshed in demanding careers, Eileen and Howard had a good sex life. Eileen, an editor at a high-profile women's magazine, and Howard, a stockbroker in Manhattan, were heavily engaged in the social scene in New York as a complement to their jobs. While both enjoyed having intercourse two or three times a week, they found less and less time for it. Practical things started getting in the way. Long hours at work and job pressures made Eileen more interested in sleep than in sex. Howard got his feelings hurt on a regular basis. After one too many rejections, he began initiating less often. Sometime later, Eileen began to notice that their frequency had dropped off drastically, and began to miss the nice times that they used to have in bed. She also felt guilty about neglecting him. So, she made up her mind to remedy the situation.

Because she had been preoccupied with work, she hadn't even been paying attention to her own sexual urges. After a few months (too long for Howard), the absence of sexual release finally caught up with her and she took notice. Ready to pick up where she left off, she arranged a seductive evening.

By this time, however, Howard was sulking and wasn't interested in partici-pating. He was still seething with resentment for all those times that she had ignored him. Eileen thought he was being childish, and waited for him to come around. He didn't. She got angry, accusing him of never wanting sex, or perhaps having an affair. He called her a sex maniac and told her to calm down. Long on conviction and short on patience, she decided to have an affair on her own. She wanted her orgasm and she wanted it now! Her testosterone was talking to her in a language she was just beginning to understand.

But it's not just hormones acting up. Contributing to the risk of affairs are changes in female sexual attitudes. If the twenties were a time of *peak receptiveness,* the thirties are a time of *peak responsiveness.* In her twenties, she was motivated to give pleasure to her man and settled for the leftovers. Now she's concerned with her own satisfaction, sometimes with a vengeance, and consequently responds better sexually. She's more comfortable with sex now, more aware of her body and her needs. She's more sure of herself, more assertive, and typically much more demanding in bed. If she hasn't been multiorgasmic before she is likely to become so now. If she doesn't get what she needs from her husband, she sees there are interesting and interested men at every turn. In the past, she was flattered by the attention but unreceptive. Now she is reconsidering. One man stands out. For Eileen it is Will, a handsome bachelor and a business acquaintance of her husband. Word is he's good in bed. He is on her mind more and more. She maneuvers into position, knowingly or unknowingly, and . . .

Most men in this age bracket have been married at least once. Some, once or twice divorced, have become bitter, angry, and mistrustful of women because they lost custody of their children and are doling out money to an ex-wife they now hate. But no matter how miserable their past experience, it doesn't seem to protect them from jumping headlong into a new commitment. Generally less able to share in close friendships outside marriage than women, men find it harder to be alone. Perhaps this is why most men can't bring themselves to leave a marriage unless they have a substitute woman in the wings, whereas a woman who is miserable with her husband is more likely to leave him even if she has no replacement.

Because this is often the make-it-or-break-it time in careers, many married men fail to do justice to their personal lives. Less accessible than ever, they feel torn and guilt ridden because they want to be more involved with their families. They try to work and play as hard as they did when they were younger, but they enjoy it less and often end up exhausted and depressed. With all this going on, if the passion has chilled, he too might find a willing woman down the hall awfully appealing.

Then there are the thirtysomething men who have never wed. While they are the envy of their married friends, this doesn't usually reflect most bachelors' true feelings. Perceived as having no responsibilities, no attachments, no baggage, no financial burdens, and an endless supply of women, they wonder why they aren't as happy as everyone else thinks they are. By now, being single has lost a lot of its glitter. He's tired of dating, the constant games, the cost. The pursuit of sex doesn't seem to be worth the price anymore. His own behavior sometimes puzzles and upsets him, although he would never admit it to anyone else, especially his envious married friends. A typical pitfall of this period is illustrated by the way Will tries to cope:

Embittered by his sexual experience with Linda during college and angry over being dumped at the altar, Will was determined never to trust a woman again. That wasn't to say, however, that he was going to give up sex. In fact, his hostile feelings toward women made it easier for him to "love them and leave them."

Now in his thirties, Will goes to ridiculous lengths to get a woman into bed. An investment banker on Wall Street who only just survived the crash of the eighties, he might take her to dinner, spend money he can't afford on concerts, theater tickets, and presents. But once he gets her into bed and sex is over, he often can't wait to get rid of her.

Then he'll do it all over again with someone new the next week. As a matter of fact, his last fling was with a married woman named Eileen. Will thought he was cleverly devising a cost-effective method for recreational sex with no strings attached. Eileen couldn't afford to be seen in public; he couldn't afford the wining and dining. But once he was ready to move on, she wouldn't let go and things got messy. She fell in love and started talking about leaving her husband, Howard.

Will eventually found a way to extricate himself from Eileen's unexpected and unwanted attachment to him. Unfortunately for Eileen, Howard, having discovered the affair, did too.

Our thirtysomething man is sick of game playing and tired of disappointment. Besides, he's lonely.

But he's still wary. And so our single man tries to have his cake and eat it too: he looks for a terrific woman to move in with, not realizing that by doing this he may well be setting the stage for the oxytocin effect and *involuntary commitment*.

Let's catch up with Tim and Destiny, the teenagers who fell in love and got along so well together. They made the best of their thirties.

During their twenties, both pursued their education. Tim became an engineer. Destiny studied law. They had two children, both girls. While it was a demanding time financially and academically, they continued to touch, albeit somewhat less frequently. The thirties ushered their relationship into an even better place. They talked, they made love, they had romance—all these resources were there to help them to weather the occasional storm. Their relationship was strong and resilient. Individually and together, they represent the best of the thirties.

FORTYSOMETHING

This is the first time our biological forces are no longer driving us apart, quite the opposite. Men and women reaching this stage have usually achieved a certain measure of sexual security and emotional maturity. They start becoming more compatible. It is no longer such a struggle to complement one another. Yet while the tide has turned, most people don't know it yet, and many are still stuck in the patterns of the past.

She is more orgasm driven now, having learned how to help herself become orgasmic with just about any man. He is less orgasm driven, and has come to value touching as an end in itself. He lasts longer before ejaculating without reciting baseball scores or biting his tongue, thanks to physiological changes and the benefit of experience. He also knows better than to try to thrust forever. He is emotionally engaged during sex, and enjoys the journey at least as much as the destination. He has become more open, more willing to be intimate, to talk. If the nest has emptied, there is even more time to enjoy each other.

The end result of these changes can be a revival of romance and better sex than ever . . . if the man has come to terms with his midlife evaluation and the woman has lost the chip on her shoulder that she may have acquired in her twenties or thirties.

Remember Will and his narrow escape from Eileen? Life has changed. At thirty-six, he made a pragmatic choice to move in with thirty-four-year-old Andrea, one of the women he had been seeing. His thinking was that this way he could avoid married women, angry husbands, disease, and financial ruin—along with commitment—while enjoying readily available sex without having to scrounge around for a date every week.

What he didn't anticipate was the power of oxytocin.

As you will see later on, oxytocin is the superglue of relationships—a bonding agent beyond all others. Since it is produced in *you,* in response to someone you touch frequently (not a casual friend), just living together and sleeping side by side starts the attachment process and you can become involuntarily committed (physically and emotionally) before you even realize it.

After two years, Will started thinking marriage. Once he and Andrea had settled in together as an official couple, however, many of the things that were a problem for him in earlier years began to resurface. Despite Andrea's tendency to overlook his shortcomings, she wasn't willing to tolerate his affairs, and finally left him. His world fell apart when he lost her. He didn't realize until after she was gone how much he needed her, wanted her—and how profoundly he loved her. This hurled him into a midlife crisis, but Andrea returned to help him pick up the pieces when she realized how serious he was about making a solid commitment.

Now in their late forties, Will's a pretty mellow fellow and Andrea has exclusive rights to all the sex he has to offer. She enjoys every minute of it. In fact, she's starting to wear him out. She no longer worries about other women in his life. They have more fun together in and out of bed than either of them ever had alone.

While this isn't their final destination and the road was rough up to this point, Will and Andrea are on the same track, finally headed in the same direction.

For women, as for men, midlife can be a transforming time. They want to spread their wings, express their talents, and have new experiences. They start to turn *away* from the home just when the men in their lives are turning *toward* it. He comes to value the pleasures of domestic warmth and wants to spend more time with his family. She may want to kick up her heels and stretch herself professionally. However, having been there, he can understand these desires. Having treasured her nest, she can comprehend his needs as well.

After Linda left both college and Will behind, she got a job teaching first grade in her hometown of San Diego. Within a year, she met Rick, a warm and loving man whom she wasn't wildly attracted to at first. But he overwhelmed her with attention and determination, and she responded.

After they were married and had their first child, they mutually decided that she would work at home and turn her talents toward their family. Within five years, they had four children. She supplemented their income by taking in neighborhood kids for day care.

Now Linda is forty-five. Her youngest child is fifteen and soon to have a driver's license. One career is coming to a close—that of raising her children—and she's beginning to think about what to do next. Watching the neighborhood change and thinking of relocating herself once her last child is out of the house, Linda decides to go into real estate.

Her husband, Rick, a local architect, found his business shrinking with the recession. To cut down on overhead, he moved his office to the house. He was quite willing to take over what few family responsibilities remained while Linda went out to slay dragons. After all, they had four kids to put through college.

This stage is not without its hazards, but perhaps for the first time, men and women have the same frame of reference—biologically and psychologically speaking. For a change, the challenges couples face are on common ground. If they are empathetic about what the other is going through, rather than threatened, these changing conditions can create a splendid and dynamic life-force. If not, they can certainly create discord, conflict, and even divorce.

Psychologically, the woman has had the opportunity to outgrow her sexual inhibitions and reservations. Sexually she feels and behaves more like a man. Sex doesn't scare her anymore. Psychologically, a man has had the opportunity to overcome his emotional reservations and inhibitions. Feelings don't scare him anymore. (But for those with dwindling potency, sex sometimes does.) It is not surprising, then, that this man and this woman *can interrelate, not just communicate.* Both struggle with the realization that irreversible metamorphoses are gradually moving them from youthful to "mature." They both worry whether they are losing their sexual attractiveness, and what will happen biologically to their ability to function and enjoy sex over time. Physiologically, subtle changes that began in their bloodstream years before begin to manifest themselves. They will not become distinct or definable until the menopausal syndrome appears. The scope of these changes is so profound that I have devoted several chapters to the subject.

While their hormonal chemistry is not alike, it has become closer to the same than it has ever been. Her estrogens are gradually dropping so that her relative amount of testosterone has more impact. His testosterone levels have dropped slightly, making him less aggressive and less sexually demanding just as she becomes more aggressive than she once was. The net effect can produce two most compatible, sexy soulmates.

With one exception—approaching menopause. This is the most drastic and sudden hormonal shift since puberty, and it begins to affect women long before their periods actually cease. So transformative is menopause that we will spend much time discussing it. For now, suffice it to say that, in addition to the purely physical changes it entails, menopause also creates emotional upheavals. Depending on whether or not she has estrogen replacement, how well she adapts psychologically and the quality of support she receives from her mate (and her physician), a woman will find the transition somewhere between smooth sailing and an absolute nightmare.

For men, midlife is also a time of emotional, physical, and sexual adjustment. Physiologically, there are no sudden changes comparable to female menopause—no hot flashes, except for sweet young things, sports cars, and gold chains. If he has not achieved his professional aspirations, he has to deal with feelings of failure and inadequacy. If he *has* reached his goals, he may wonder why he's not as happy as he expected to be. In either case, he has to rethink his future and reevaluate his past. It's a period of life that is increasingly referred to as male menopause, viropause, or climacteric.

Male menopause can occur anywhere between the early thirties and the seventies. Depending upon what sexual stage he is in, the impact on emotions, sexual performance, and relationships will vary. The same is true for female menopause. Although natural menopause is most probable between forty and sixty, surgical or medical menopause, due to removal of the ovaries or chemotherapy, can take place at any stage. Therefore, menopause—of both sexual varieties—will be addressed in more depth in later chapters when the relevant hormones have been discussed.

For both men and women, the sooner they face their midlife crises the better chance they have of becoming centered in themselves and accessible to each other.

FIFTYSOMETHING

For the first time in their lives men and women are perfectly matched sexually and emotionally—assuming they stay healthy and hold on to each other. A fiftysomething couple can have all the romance and sex they've ever imagined, and more. They usually have more time, less pressure, and fewer worries—pregnancy is irrelevant, the children are gone, and the work world is being managed better. Even more significantly, biology and maturity collude to bring men and women this age even closer.

If he has handled the midlife transition effectively, a man in his fifties has usually mellowed and matured even more sexually and emotionally. He has come to comfortable terms with himself. He might also have accepted all his spouse's female attributes, and learned to learn from her. His appreciation makes him a better lover and a more compelling companion.

At this age, younger women often tend to lose their appeal for a well-adjusted fifties man. More appealing than the lure of youth is the excitement and intimacy possible with a woman with whom he has a sense of shared experience—a common history, if not a personal one, then at least a generational one. On a sexual level, a warm, self-assured, sexually aware woman in her fifties projects sensuality and an inner power that a man who is on her level finds immensely erotic. He sees in her a measure of depth, beauty, wisdom, and strength that makes younger women seem one-dimensional by comparison.

And the feeling is mutual. A fiftysomething man who has evolved and has retained his lust for life can be the epitome of sex appeal to a fiftysomething woman. He is forceful, but not predatory: a kinder, more considerate lover. They both get the benefit of his lower drive for orgasm; now he can savor the fun and intimacy of foreplay with her—something he might have missed in the past—and he loves lingering in the afterglow to caress, cuddle, and communicate. For the first time, he may realize how much he enjoys touching and being touched. At the same time, she is more inclined to initiate sex now. She drives toward orgasm, which is extremely attractive to a man who feels sexually secure. Technique, enthusiasm, and experience have all come together. They haven't exchanged places, they have just moved toward one another sexually, and become more alike.

The fiftysomething woman is a piece of work. She has developed a sense of humor about things that used to make her cry. She knows how to run her life without a mate, but often prefers to have one in her world. She has come to terms with most of the physical changes of aging that may have troubled her during her thirties or forties, and reflects a new security that shows in her appearance, her talk, and her walk. If she has kept herself in shape, she carries a look that no younger woman can imitate, a combination of sexuality, sensuality, and self-confidence. She knows her own mind, but knows how to be kind. The men who were once critical of her in earlier years now admire and enjoy her. Sexually, she is athletic and exuberant, innovative and rambunctious—never a dull moment, full of surprises. She meets her man on his own territory and makes it theirs. He can relax and join her there.

The fifties, then, hold great opportunities for relationships, whether two people have been together for decades or have just recently met. Of course, this promising picture assumes that both partners have maintained good health, and are not taking medications that can disrupt one's sex life. Health problems are more likely to arise with each passing decade so it becomes increasingly important to break the toxic habits that youth seemed to forgive, like alcohol, drugs, junk food, and couch addictions, to give your health a chance.

SIXTYSOMETHING

The trends that begin in the fifties now begin to intensify. For people with good health, good attitudes, and good partners, these years can be fantastic— full of romance. Believe it or not, they can be the sexiest years of your life. Men and women this age enjoy other advantages. One is retirement—a hazard, of course, if a terrible relationship is lurking, but wonderful when the relationship is working.

During the sixties, there are no dramatic hormonal changes, but a gradual progression in both sexes. DHEA continues to drop. Estrogens diminish in women unless they are on estrogen replacement therapy. Testosterone shows up more and more. Some women develop a slight mustache and a voice that can't be distinguished from their husband's. So too, men in their sixties become more female. Some develop breasts—an estrogen effect, especially pronounced in, but not limited to, alcoholics.

He notices some changes in his sexual performance, but if he doesn't get overly concerned, they won't interfere with quality sex. In fact, some of them will add to his exquisite pleasure—and hers as well. He doesn't get "fantasy erections" the way he once did—a testosterone-dependent reaction that fades with age. Not realizing that it is normal for men his age to require mechanical stimulation for an erection to develop, he might think there is something wrong with him. But once she touches him, he'll discover that he responds just fine. And she will touch him more often if she wants him to stand up and take notice. His erection might not be quite as firm as it used to be, but it's firm enough and feels just as good; and he won't feel the urgency to ejaculate that once governed his sexual patterns—for most men, a nice relief.

Meanwhile, women are having multiple orgasms (assuming they ever learned how). These orgasms may be somewhat less intense than before, but they are easier to come by. She may lubricate a little less—the easiest thing to fix—but she turns on with the rush of a gas stove instead of an electric burner.

The biggest cause of romantic retirement is widowhood. Women who have lost their husbands in this decade, through death or divorce, have a short supply of men to choose from compared to women in younger stages, since women statistically outlive "the weaker sex." Increasingly, perhaps because of this, many a widow chooses to have an affair with a married man. Other women adapt by taking younger lovers, a practical and productive move. There are more men to choose from and they can avoid the complication and duplicity of affairs. And, speaking of adaptive affairs, some women settle into a sexual liaison with another woman.

The majority, however, follow a more traditional path, spending more time with their grandchildren or adopting a pet. These are not foolish choices.

She needs the physiological impact of touch even more than before to preserve her health. As we will see, oxytocin, the pleasure-giving peptide that increases when we touch, also goes up when we stroke a pet or caress a child. But too many older women write off their love life too soon.

I hear women past sixty complain about the unfavorable gender odds all the time. It's funny but they sound to me just like younger women who are always complaining about the poor quality of the large numbers of men surrounding them. The odds haven't really changed. Smaller numbers of men are available, but they're of much higher quality—after all, they've had women to train them along the way. And with the help of men, most women have become far more manageable themselves.

Remember Carolyn? Last time we checked in, she was having multiple orgasms with Tony, her nineteen-year-old sex machine. (Doesn't she remind you a little bit of Will in his thirties?) The problem was, after a while all her lovers started looking, feeling, and sounding alike. And, just like Will, she tired of the sport. Besides, constant exposure to Pearl Jam was lowering her IQ.

Carolyn started spending more time with her female friends, finding their companionship and common interests considerably more rewarding. When Ruth, one of her girlfriends, wanted to set her up with a man, she declined. But Ruth kept after her and she finally gave in. Why not?

Carolyn, by this time, was sixty-one years old. Her blind date was eight years older. Newly retired, Don had recently relocated to St. Louis to be near his family. No longer needing the reassurances and stimulation of a younger woman like Antonia, in his own way, he had traveled down a path parallel to Carolyn's and arrived at pretty much the same destination. He was settled within himself, open to relationships but not actively looking. He was, however, being distributed from woman to woman by his closest friends like the endangered species he was.

Don took Carolyn to a small restaurant in the Italian section of St. Louis, near the Catholic church. Pepe, the owner, seated them in a quiet corner table. After dinner they took a leisurely walk around the central west end, finishing the evening at Ave's over a cup of decaf.

Something happened. Fate, chemistry, and common experience brought them together in a way neither one had ever known before with anyone else, or, for that matter, even thought was possible. They talked for hours and understood one another seemingly without effort. No matter that they were in their sixties, they had never been so fully turned on before, in every sense of the word—emotionally, intellectually, and sexually. They were both surprised, even shocked, by the intensity of this experience. Instinctively, somewhat awed by the force of their attraction, they proceeded slowly although their emotions were racing ahead. Love at first sight? Absolutely. But what a long look their lives had already taken at the sights.

Isn't it a shame that they had to wait until now to find each other, but isn't it nice that they finally did? I think they could have shortened their learning curve by about two decades had they had a full understanding of how to influence their sexual stages.

BEYOND SIXTY

Health questions aside, the quality of a person's love life after sixty depends largely on how he or she managed it in earlier stages.

I must add, however, that it is never too late to unlearn bad habits and solve old problems. But it's safe to say that if you haven't done it on your own by now, odds are you won't do it at all without some professional help. I have seen dozens of elderly patients who have struggled with sexual dysfunction their whole lives. With short-term treatment, most are able to overcome their problem in time to enjoy a hot old age. In some cases, it requires medication or surgical procedures. In most, it doesn't. A little education and light therapy is sufficient. I recall a woman in her midsixties who had never had an orgasm in her life. In one session we cleared up some misconceptions that had somehow crept into her mind half a century earlier. To her great delight— and that of her new lover—she enjoyed nearly two orgasmic decades before she died.

The bottom line on aging is this: Barring debilitating disease, there is no reason a man and woman can't enjoy love, romance, intimacy, and sex as long as they live. I believe this so strongly that I have devoted the last chapter in this book to a thorough discussion of the subject.

Now that you have been introduced to your potential sexual stages, and those of your mate, you can place the information about hormones that follows into the context of your personal experience. These stages also serve as relation-ship landmarks in another sense. While the transition from one to another often goes smoothly, it not infrequently precipitates a crisis in a relationship. Transition periods are particularly vulnerable times. Relationships either dissolve or weather the storm. Those that survive, thrive. In a sense, whether you stay with the same partner or not, approximately every ten years or so, you have a new and different relationship as you pass from sexual stage to sexual stage. Doing it with the same partner is just a lot less costly than changing him or her—both financially and emotionally.

When you consider that various hormones common to both sexes are at dramatically different levels in men and women, that they are also different for each sex at different times of life and also different times of the day, it makes for pretty intriguing chemistry. If you add to that sexual peaks—which seem synchronized only to make sure that when one of you isn't thinking of sex, the other one is—it becomes quite clear that sexual stages can't be fully understood outside the context of these other related, but independent, hormonal forces.

3.

ROMANCE: LOVE, LIMERANCE,

AND LUST

For thousands of years poets have rhapsodized about falling in love. Songs glorify it, great thinkers have written about it. But still it retains its mystery. What magical magnet attracts us to that one special person instead of another? What happens when that electric jolt overwhelms us? Logic and reason dissolve like smoke and we can think of nothing else. Forget concentration. Destinations don't matter. Your love becomes the atmosphere. You breathe it wherever you are.

We try to explain it. We say things like, "It was her gorgeous eyes," or, "His smile weakened my knees." We single out qualities like warmth and kindness or a great sense of humor. We say, "We have so much in common."

But all these features, compelling as they may be, are just accessories—our mind's way of trying to make sense of the inexplicable. There are millions of men and women out there with gorgeous eyes and great smiles, and we all know people with whom we share common interests. That doesn't mean they turn us on. Not only that, someone with pathetically few of the qualities we think we're looking for will suddenly hit us like a lightning bolt. The Africans call it a *mojo*. One moment you feel just fine, the next moment, your ankles swell up and you fall in love with someone you hate.

Sexual whiplash is what it is. The early impulses can be so gentle you don't even notice them, or relentlessly intense, like a tidal wave. The word we call on to describe this particular magic is *chemistry*—a perfect choice to describe a feeling that combines the giddy rush of a forbidden encounter, the euphoria of an endorphin attack and the sexual surge of a street drug. All of this occurs, of course, without doing anything illegal, costing any money, or subjecting yourself to any drugs—other than those you produce naturally, of course.

The essence of this chapter, then, is how the sights, scents and signals we send and receive influence love, limerance (the romantic feeling of walking on air, a pervading shimmering sense of excitement throughout the day, punctuated by butterflies when your lover comes near), and lust; how the hormones we produce determine who we "choose" to love and what we can—and can't—do about it.

APHRODISIACS AND LOVE POTIONS

Sometimes would-be lovers try to produce this chemistry in the unwilling or indifferent targets of their affection through what we've come to know as aphrodisiacs. Aphrodisiacs are intended to make you sexually irresistible and potent beyond your wildest dreams. Love potions also fall into this category. Their objective, as portrayed by books and movies, is a little different—to cause you to fall madly and uncontrollably in love—Cupid's catnip. While love potions and aphrodisiacs are conceptually distinct—one promotes lust, the other love—the lines between them tend to blur. Both usually contain sedatives, urinary tract irritants (like Spanish fly), disinhibitors (like alcohol), or hallucinogens.

While love potions have been enthusiastically marketed to both sexes, women have pursued them more avidly than men, who historically have preferred aphrodisiacs—anything to make their genitals bigger, better, busier. Indeed, they have been known to smear hot pepper on their penis to wake it up and rub anaesthetic ointments on it to make it last when they actually put it to sleep. Today some are injecting fat into the poor thing to make it long.

In olden days, women were not above using any witches' brew to help their cause of snagging a man. Today, they torture themselves in spiked heels with pointed toes and skin-tight clothes. Her love potions are perfume, makeup, a new hairdo, nail polish, black lingerie, and jewelry. Just like a fishing lure, she sparkles, dazzles, and flashes, trolling for men, hoping to hook a world-class trophy.

But who needs external aphrodisiacs when you have an entire internal arsenal at your disposal? If you were to look for passion inducers within the confines of your body, the first lineup would have to include PEA, pheromones, and DHEA. Let's become more intimate with them now.

PHENYLETHYLAMINE (PEA): THE MOLECULE OF LOVE

Disguised in a box of chocolates is a favorite of mine, PEA (phenylethylamine), also known as the "molecule of love." PEA is a natural form of amphetamine that our own system produces. Romance is thought to cause it to flood areas of the brain that are ordinarily activated during sexual excitement, further stimulating sexual desire and pleasure. Not surprisingly, high PEA levels have been found in the bloodstreams of lovers, probably accounting for the limerance that consumes them both. Chocolate also contains high levels of phenylethylamine.

Since ancient times, chocolate has had a reputation as a quiet, gentle, albeit fattening, aphrodisiac. Nuns were forbidden to enjoy it because of its sexual reputation. Priests, however, could consume it without restraint. Since way back when, suitors have brought chocolates along with bouquets of roses to the woman of their dreams. Perhaps it is no accident that the giving of a box of chocolates has become a traditional part of courtship rituals around the world.

What is PEA?

Synthetic forms of amphetamine derivatives similar to PEA are sold over the counter in drugstores, or by prescription, as appetite suppressants. That is probably why those intoxicated by romance often comment that they have lost their appetite for everything but each other. PEA is churning in their bloodstreams. Chemistry at work and at play.

PEA's role in the chemistry of love, however, extends beyond its amphetaminelike qualities. PEA is a formidable substance with a distinct personality and considerable influence over us. It occurs naturally in our systems and fluctuates according to our thoughts, feelings, and experiences, especially romance.

PEA PROFILE

MOST PEOPLE DON'T KNOW THAT PEA:

- rises with romance
- can cause depression, at low levels
- has been associated with psychosis, at high levels

PEA:

- works as an antidepressant in both sexes
- is similar to an amphetamine and works like a diet pill

AS TO SEXUAL ROLES, PEA:

- is a stimulant
- spikes at orgasm and ovulation

AS TO BEHAVIOR, PEA:

- causes giddiness and excitement
- could be involved in love at first sight
- could be a cause of "love addiction"

TOTAL PEA CAN BE DIVIDED INTO:

- natural
- synthetic

HOW WE CAN INFLUENCE PEA:

Increases PEA:

- deprenyl (Eldepryl)
- L-phenylalanine
- MAO inhibitors
- chocolates
- diet soft drinks
- artificial sweeteners
- romance
- marijuana

Lowers PEA:

- lovesickness
- cause or effect of depression

Sexual Sightings

As important as scent is in the early dynamics of love, lust, and romance, when the potential object of your affections is out of the range of scent, sight has a special impact all its own. The term *eye contact,* touching with your eyes, takes on new meaning. Humans can determine whether or not they want someone with just a glance—even if there is a glass partition making scent impossible. Lower animals can't.

In lower animals, sexual impulses, like those triggered by pheromones, pass directly to the primitive reptilian brain. No thought is involved. But as the human brain evolved, higher centers became engaged in sexual response, including the cognitive areas of the cortex and the limbic area. At the same time, the nature of the key sexual impulses changed from smell and touch, which are direct and reflexive, to sight, which allows decisions to be made at a distance. Further, such desires and attractions can be remembered, stored, and acted upon in the future. Perhaps that is why appearance—how you look, the clothes you wear—has become so important to modern man and woman.

So, is there such a thing as love at first sight? When I give lectures on love, I often ask those who believe in love at first sight to raise their hands and then those who don't to raise theirs. Afterward, I ask for a count of those who have actually been in love at first sight. It comes as no surprise that the only ones who believe in it are the same ones who have lived through it. It's such a ridiculous concept for sane people to entertain. But anyone who has been smitten in so abrupt a fashion knows differently.

PEA could well be the visual component of the chemistry of love at first sight. It's the look across the room that mesmerizes you and casts you under its spell. We do not know how sight (or what particular sight) can cause this response, or how it is processed through our body and brain. We do know the response causes a circulatory surge of PEA. We also know how it feels, the thrill when that happens.

Could love at first sight boil down to this chain reaction? Together with pheromones—the secret scent that draws you close—and touch, which binds you, PEA completes the picture and triggers "that magic moment."

That look! She had been walking to the table carrying a tray of egg-yolk candies when she first felt his hot gaze burning her skin. She turned her head, and her eyes met Pedro's. It was then she understood how dough feels when it is plunged into boiling oil. The heat that invaded her body was so real she was afraid she would start to bubble—her face, her stomach, her heart, her breasts—like batter, and unable to endure his gaze she lowered her eyes and hastily crossed the room . . . But even that distance between herself and Pedro was not enough; she felt her blood pulsing, searing her veins. A deep flush suffused her face and no matter how she tried she could not find a place for her eyes to rest. (Laura Esquivel, Like Water for Chocolate)

Such is the impact of these often unwelcome forces. A burning—and most reasonable—question is, "Does love at first sight last?" The most usual response is, "No, of course not." But I have seen many relationships that began with a quick chemical jolt transform into deep, long-lasting, loving commitments. Either way, there is certainly a chemical underpinning.

Researchers speculate that PEA rises to exaggerated levels during these romantic periods. In fact, it is known to be a potent antidepressant. Conversely, the love-struck soul gets acutely lovesick when romance ends. They don't have the symptoms of garden variety blues: loss of appetite, sleeplessness, and lethargy. Instead, PEA junkies who lose their love eat too much, sleep too much, and go into emotional hyperdrive. It's a chain reaction similar to amphetamine withdrawal as described by Columbia psychiatrist Dr. Michael Liebowitz in his 1983 book, *The Chemistry of Love.* Deprived of the lover (a fix), he or she goes into molecular withdrawal as PEA levels plummet. Certain antidepressants can return phenylethylamine to normal levels, helping the individual to function emotionally again.

Because of the abundance of PEA found in the bloodstreams of lovers, Liebowitz has suggested chocolate as an antidote to lovesickness. There is value in the concept, but there is a catch to the practice. While PEA is present in chocolate, it is metabolized so quickly that it doesn't have time to have much effect. (So eating Hershey's Kisses probably won't raise your PEA levels significantly.)

Another interesting fact is that abnormally high PEA levels appear more often in women than in men. They typically occur at or near ovulation. This would seem to indicate a role for PEA in our desire to mate and procreate.

Yet while we readily accept that sex drive in other animals is governed exclusively by hormones and peptides, not emotion (or so it seems), many of us resist the concept that the same could be held true for us.

Does this complex, intense feeling we call love boil down to a sudden exchange of molecules between strangers who might not even like each other? Is there a molecule of love that kicks in on impact? Do we really want to know?

Love or Lunacy?

Is love a state of temporary insanity? We don't really know. Occasionally, high levels of PEA have been found associated with states of mania and schizophrenia, causing anxiety, disturbed sleep, and possibly psychotic behavior. The genetically identical Genain quadruplets, who were all schizophrenic, had extremely high levels of PEA. It follows that there is a real cause for concern in patients treated with stimulant drugs similar to PEA for diet control or sexual dysfunction: they might become susceptible to "crazy behavior" requiring a different sort of commitment.

Another twist—while high levels of PEA have been linked with love and romance, they can be equally high during divorce—perhaps because many of us get a little crazy while we are going through it. Confusing, isn't it? There is clearly more research to be done here.

Influencing PEA Levels

Looking at "the right stuff" somehow manages to trigger PEA, whether it's a person, a picture, art, or something you read. Just looking doesn't elevate the level of any other romantic molecule that we know of. Since men are much more renowned for being visually stimulated sexually, I suspect that men have a stronger PEA response than women (although there is no firm evidence to support this conclusion). But reading romance novels and fantasizing can also stir up your PEA, while stimulating testosterone and/or estrogen. You are primed. The itch is obsessive. Here is a man's view:

> *Well, I have a whole system if I'm reading. . . .*
>
> *. . . what I do is I read a little of it, whatever it is, the story or the letter or the novel, to see whether it's something I do want to masturbate to or not. If it's something that looks promising, I read it all through very fast, to find out exactly what happens and locate the spot in it where I'm going to want to be coming, and what spots I'll want to skip because they're whatever—violent or boring or somehow irrelevant. Then I go back, not always to the beginning, but I backtrack, and the distance I backtrack from the point where I've scheduled my orgasm I have to gauge exactly, depending on how close to coming I think I am. . . . (Nicholson Baker, Vox)*

Romance Novels: One method of raising your PEA levels may be to read romance novels. This is not just entertainment. They're a cheap PEA ride primarily for women—a quick fix. Indeed, they are not just a read, but a physiological thrill, a borrowed romance, not unlike men using prostitutes (but without the risk, of course).

Romantic movies, X-rated films, TV shows, and even music may trigger the same response. Pornography, erotic art, and topless dancing also stimulate this chemistry, albeit more often in men.

Many women actually object to men's enjoyment of such explicit sexual material. Yet they would be appalled if anyone suggested that romance novels be banned. (Regardless, Danielle Steele is basically a woman's *Playboy*.) A reader complained to Dear Abby:

> *Dear Abby: On a recent trip, while I was in the tub at a motel, my husband got a porno movie on the pay TV channel. On returning to bed, when I saw what he was watching, I refused to watch it with him. I thought it was degrading to women, and I told him so.*
>
> *When the movie was over, my husband wanted to get affectionate, but I told him to buzz off. He turned over and went to sleep.*
>
> *He hasn't spoken to me since. This is the third day of silence. How does one handle a situation like this?*
> *—Oregon Wife*

She responded:

> *Dear Wife: Break the silence and tell him that if he's old enough to watch porno movies, he should be old enough to quit pouting when he doesn't get his own way.* (Los Angeles Times, *March 20, 1995*)

I think they were both acting pretty bratty, put out at what did or didn't turn the other on. Does it make a difference? Whichever your preference—explicit pictures or seductive words—reading or viewing can set your molecules in motion.

Romantic Replays and Fantasies:

> *The physical images were inscribed in her mind so clearly that they might have been razor-edged photographs of his. She remembered the dream-like sequence of clothes coming off and the two of them naked in bed. She remembered how he held himself just above her and moved his chest slowly against her belly and across her breasts. How he did this again and again, like some animal courting rite in an old zoology text. As he moved over her, he alternately kissed her lips or ears or ran his tongue along her neck, licking her as some fine leopard might do in long grass out on the veld.*
>
> *He was an animal. A graceful, hard, male animal who did nothing overtly to dominate her yet dominated her completely, in the exact way she wanted that to happen at this moment. (Robert James Waller,* The Bridges of Madison County)

A second way to raise your PEA levels is to revive a romantic memory and savor it until you recreate the experience and feel the rush. This is an effective method for many women who devour books insatiably, but can't come up with a fantasy on their own to save their lives. Fantasizing is a completely different skill, much more natural to the man. As noted in Chapter 1, it is testosterone driven, so you can begin to understand the disparity.

I have treated numerous women over the years who think they are incapable of imagining specific sexual thoughts. Some even have trouble conjuring up a romantic short story. They want someone else to do it for them (i.e., an author) so they read a steamy book instead. This pattern is most typical of women over forty, who feel more inhibited than their daughters' generation, but it still prevails among some younger women. In any event, if you can't fantasize, don't worry. Most women can be taught to romanticize and fantasize without too much difficulty.

Men are another story altogether. Even though it comes more easily for them, their fantasies are different in nature from a woman's—much more genital and sexually specific. (When they read, they want genital details—action—not feelings and romance.) No plot necessary. The images that arouse them are often considered gross and disgusting by women—nameless, faceless, genital visions—yet they stir their PEA.

Diet Drinks, Chocolate, and Artificial Sweeteners: Diet soda pop contains the artificial sweetener Nutrasweet, which can be converted into PEA in your bloodstream. One should be able to get a sustained PEA "high" more easily from diet soda pop than from chocolates, whose effect is transient. Those "addicted" to diet soft drinks may be attracted by the PEA, not only the caffeine. But this "high" can also transform into migraine or anxiety, so I don't recommend this approach. As you might imagine, no one has invested in seriously researching these questions, so I am speculating shamelessly.

Medications: As mentioned earlier, over-the-counter diet pills and most prescription diet pills have stimulating actions similar to PEA. Untold numbers of women are clandestinely addicted to them, partially because of the PEA-like effect. This is not a desirable way to raise your PEA levels. PEA in your body reacts naturally to sexual thoughts and feelings. Synthetic stimulants have no such sexual trigger. Indeed, with continued use, tolerance develops so that more drug is necessary for the same effect, perhaps overwhelming and making you insensitive to the subtle reactive oscillations of PEA from your own body in response to romance. In fact, the risk of triggering not just anxiety but actual psychotic behavior through abuse of these substances is quite real.

For those with abnormal PEA metabolism, certain antidepressants called MAO inhibitors have been used because they actually modulate wildly fluctuating PEA levels and can treat the mood swings generated by them. In fact, MAO inhibitors have been used with some success in treating the symptoms of "lovesickness," suggesting that PEA is responsible for the emotional upheaval experienced in lovesickness. Marijuana also stimulates PEA, which may partly account for the glow this high creates.

If just a glimpse or a thought of someone you love can stir up your PEA, imagine what adding their scent to the equation can do.

PHEROMONES: SENSUAL SCENTS

Her sweat was pink, and it smelled like roses, a lovely strong smell. In desperate need of a shower, she ran to get it ready. . . .

The only thing that kept her going was the image of the refreshing shower ahead of her, but unfortunately she was never able to enjoy it, because the drops that fell from the shower never made it to her body: they evaporated before they reached her. Her body was giving off so much heat that the wooden walls began to split and burst into flame. Terrified, she thought she would be burnt to death and she ran out of the little enclosure just as she was, completely naked.

By then the scent of roses given off by her body had traveled a long, long way. All the way to town, where the rebel forces and the federal troops were engaged in a fierce battle. One man stood head and shoulders above the others for his valor; it was the rebel who Gertrudis had seen in the plaza in Piedras Negras the week before.

A pink cloud floated toward him, wrapped itself around him, and made him set out at a gallop toward Mama Elena's ranch. Juan—for that was the soldier's name—abandoned the field of battle, leaving an enemy soldier not quite dead, without knowing why he did so. A higher power was controlling his actions. He was moved by a powerful urge to arrive as quickly as possible at a meeting with someone unknown in some undetermined place. But it wasn't hard to find. The aroma from Gertrudis' body guided him. . . . (Laura Esquivel, Like Water for Chocolate)

Well before experiencing an erotic moment, there are sexual forces at work on you—all too often without your knowledge or consent. Scents and touch—in addition to sights—manipulate your choices and actions in ways that are hard to believe. Perhaps not as fancifully as with Juan and Gertrudis, but just as compellingly. Pheromones affect us in particularly intriguing and somewhat disturbing ways.

What Are Pheromones?

A pheromone is a chemical substance produced by one animal that causes a specific reaction in another, usually of the same species, through smell. We call this *chemical contact communication*.

You are no stranger, by now, to the concept that molecules within *our own* bodies can and do affect our mood, mind, and behavior. Pheromones, however, are unique in the respect that molecules produced by *someone else* have the power not only to influence, but indeed to dictate, our own behavior, without our even realizing it. This is staggering stuff!

Our ancestors made some pretty ingenious attempts to master the power of these pheromones, even though some of the substances they used to attract one another were pretty unappealing. Body secretions like urine, sweat, menstrual blood, hair, and nail parings (whether burned, buried, or consumed) were intended to addict the object of desire to the scent and taste of the hopeful suitor. It is not always clear from history whether these items were to be worn, eaten, admired, or slipped secretly into the food or drink of the target (victim) of their affections.

Folklore has it that to excite love in a woman, a man would slip into her drink, without her knowledge, a sample of his semen mixed in sugar. Or, he could burn one of her undergarments stained with menstrual blood and urinate on the ashes. To trap a man, a woman would wipe his penis with an article of her clothing, collecting as much semen as possible and then bury it in the earth beneath her threshold. This would be hard to do if they hadn't been formally introduced! However, from that time on, his erection belonged to her alone, and was guaranteed to wilt in front of another.

There are numerous types of pheromones—sexual, behavioral, territorial, and so forth. While modern science has discovered some pheromones in sweat and vaginal secretions, it has yet to concretely identify sexual pheromones in humans. We already understand their power in animals and insects. Consider the cockroach: A cockroach sexual pheromone has been synthesized, at government expense, no less. It is intended to drive male cockroaches mad with desire, luring them to a preselected sex spot. But instead of an orgy, they get sprayed with a pesticide. This method of executing large numbers of sex-crazed single male cockroaches protects humans and the environment from the hazards of "crop dusting." It is through these same types of molecular scents that a bitch in heat signals all the male dogs in the neighborhood, and that chimpanzees "fall in love."

Since some animal pheromones are derived from DHEA (the "mother of all hormones," which we'll discuss next), we believe the same to be true for humans. The pheromones produced from DHEA seem to be as unique to each individual as fingerprints. Each and every one of us has a "smell print." When we move around, we leave a cloud of scent molecules behind. Our skin is virtually awash in these molecules. In fact, every square centimeter of skin sloughs off *one thousand cells per hour,* leaving our individual smell print behind.

Law enforcement agencies around the world have long understood the concept of personal scent. Criminal suspects in Holland are routinely smell printed and their "prints" added to police archives. An article of worn clothing or something removed from the scene of the crime is sealed in a container and catalogued for future reference. While the police smell-print files contain predominantly criminal fumes, in time, they will probably be expanded to represent the entire population.

Researchers in Utah may have found the first hard evidence of sexual pheromones in humans. Dr. David Berliner, a researcher for Erox Corporation, a company concentrating on developing commercial pheromones, has isolated substances from human skin that behave like sexual pheromones. Erox markets these substances as his-and-her perfumes under the name of Realm.

These newly discovered human pheromones affect men and women differently. Those found to influence women do not seem to affect men, and vice versa. Further, instead of causing irresistible sexual compulsions as they do in other animals, human sexual pheromones appear to influence sensuality rather than sexuality, creating a sense of well-being and inexplicable intimacy with a relative stranger instead of raw lust.

Human pheromones may act on the brain and nervous system through a small cavity inside each nostril, called the vomero nasal organ (VMO). It is lined with a cell type that, according to Dr. Berliner, is "unlike any other cell in the human body, and is of unknown function." The VMO is far less prominent in humans than in lower animals who depend more heavily on smell for guidance.

We appear to have two methods of detecting odor: the VMO described above, which seems specialized to detect pheromones without our conscious knowledge, and a second method that processes odors that we register consciously. Our subliminal receptors (VMO) are located in a region called the pigmented part of the nasal mucosa, the nasal septum, and along the roof of the nose. Since these receptors have constantly renewing cells, they can be affected by what we eat, drugs, age, radiation, and—of course—hormones.

As a rule, a woman's sense of smell is more acute than a man's. It depends on estrogen, and since women have more of this hormone, their perception of odors is more refined. Men are not as sensitive to subtle scents, and seem to handle downright rank ones considerably better than women, which may be a valuable adaptation. Adolescent boys seem to revel in smelling bad, perhaps because they genetically just don't notice when they rot.

What this means regarding each sex's ability to perceive pheromones is hard to tell, but it does explain why men and women react to aromas like perfume and cologne—and pungent footwear—so differently.

While the Utah researchers have come closest to identifying sexual pheromones specifically in humans, the unconscious impact of scent on humans is already well documented. We know that there are pheromones in human sweat that influence the menstrual cycle, and we suspect that there are specific *sexual* pheromones present in human sweat as well.

Women who live together bleed together. They all get cranky and miserable at the same time. Coeds in dormitories and women who work together every day start cycling in synchrony. The group mood swings must be formidable. Sheiks thought so way back when. They kept a handful of wives in their harems in separate quarters at all times, in self-defense.

Several studies have proven that this phenomenon of synchronized cycling is being caused primarily by perspiration odor. In a study of one hundred thirty-five residents of a women's college dormitory, for example, experimenters placed cotton pads under the arms of one group of women and rubbed the pads every day against the upper lips of women in a second group. Gradually the periods of women in the second group meshed with those in the first. This study illustrates that a woman's cycles are profoundly influenced by the environment in particular, and by whom she has constant contact with (whether face to face or through scent therapy).

Male sweat also contains pheromones. In one study, it was found that by dipping a Q-Tip into the male sweat contributed from willing volunteers, and placing the saturated Q-Tip under the nose of irregularly menstruating females three times a day, their menses became regular within just a few months. The placebo group who was not exposed to male sweat but an inert substitute experienced no change in their cycles. Just imagine, he could have the charisma of a Q-Tip, and as long as you were in regular contact with his sweat your body would respond to him!

Although most research scientists have been unable to detect sexual pheromones in humans, apparently grizzly bears, horses, and wild pigs can. They respond dramatically differently to menstruating women than nonmenstruating women. Here are just a few cases in point:

On one of my radio shows some years ago, a grandmother from Oklahoma told me that when she was a young girl, all the women knew it was dangerous to walk in the woods during their menses because they could count on attracting wild pigs and wolves. While camping in Montana, several Girl Scouts had been killed during their sleep by grizzly bears. At any other time of the month, girls could wander the woods without risk to their safety. The girls were all menstruating and it is speculated that a pheromone attracted the bears.

Stallions react to a menstruating woman just the way they do to a mare in heat, becoming so unmanageable and dangerous that a young girl cannot control them. I have witnessed this response myself and it is quite remarkable.

Perhaps these animals are just attracted by the scent of blood, but I don't think that's all there is to it. I suspect there are pheromones sending insistent chemical, sexual messages that humans are too evolved or too "civilized" to detect.

In 1992, the esteemed British medical journal *Lancet* reported a curious medical phenomenon:

> *A woman from Leeds (no, this is not a limerick) went on hormone pills to rid herself of facial hair. She lost the unwanted whiskers all right, but she experienced an unsettling side effect. Her pet Rottweiler fell in*

lust with her. This ordinarily well-behaved animal began making a pest of himself.

That's right, the dog became sexually obsessed with her. The doctor's diagnosis was that the hormone pills caused subtle changes in her body odor, driving her dog wild with desire. He "would not leave her alone." The new scent was imperceptible to humans, but presumably, not to dogs.

Having to choose between a mustache and sexual harassment at home, she solved the problem by castrating the poor fellow. (J. A. Cotter-ill, "Dog days and antiandrogen" [letter], Lancet, *v. 340, Oct. 17, 1992, p. 986.)*

We know that in many animals, the sense of smell is paramount. We assume, perhaps incorrectly, that animals are aware they are being led around by the nose. But perhaps their attraction is just a reflex like our pheromone response. We don't "smell" pheromones in the same sense that we smell freshly baked bread or perfume, but we register the scent at some brain level and respond to it emotionally and/or physically—without any conscious realization of "smell" being a catalyst. This raises a provocative question: What other aspects of scent do human beings unknowingly respond to?

The systems that form our smell mechanism, and the area of the brain where we register certain strong emotions, are located so close together that the stimulation of the nose has a direct pathway to the centers of our brain that control behavior, emotions, and choices. I suspect that the influence of various odors on our feelings and behavior, whether or not we notice them, is far more extensive and powerful than anyone yet realizes.

New and Desirable Uses for Old and Undesirable Odors

Male sweat contains a substance called androsterone, which is commonly found in wild pigs and their relatives. Jovan, an aftershave lotion for men, actually contains a sexual pheromone consisting of alpha-androsterone. It comes from the male boar. The company that manufactures Jovan obviously thinks this pheromone will work for men as well as it does for pigs, which I think the consumer should find insulting. Jovan *will* attract female pigs and probably flies, so I do not recommend wearing it on a farm. Whether or not it can make the male "bore" more attractive to women remains to be seen.

Other commercial enterprises are in hot pursuit of sexual fragrances, too. The International Foundation of Fragrances and Flavors has invested a fortune in hopes of bottling human sexual pheromones as a perfume. With fantasies of controlling the world, the perfume industry persists in its quest. I heard on CNN that one enterprising marketer is collecting celebrity sweat to sell in cologne, just in case.

As discussed, the new perfume Realm has been released to the public by the Erox Corporation. It claims to include a pheromone with the power to provoke a favorable emotional and/or sexual reaction in the opposite sex. (As noted, the composition of the man's version is naturally different than that of the woman's.) If this potion is genuine and effective—and it is theoretically possible—we now have a substance on the market that may elicit an involuntary response from another person. Where does informed consent fit in here?

But contrary to what many people believe, ordinary perfumes (those without pheromones) and aromatic oils have no known pheromonal impact. In general, they fall into the category of sexual attractants along with cleanliness, cosmetics, music, conversation, a full moon, candlelight, and lingerie. Furthermore, if one considers our devotion to bubble baths, deodorants, douches, mouthwashes, perfumes, aftershave, dusting powders, and air fresheners, the actual potency of any olfactory sexual signals we transmit may be obscured.

In Russia and elsewhere, researchers have been experimenting with aroma therapy for some time, according to Robert Henkin, a biologist at Georgetown University Medical Center. Reports have indicated that the former Soviet Union saturated hospitals with various scents and pheromones to improve the mood and attitude of the patients, promoting immune systems, hence healing and recovery. On a more sinister note, years before the fall, the Soviets were rumored to pump scents into prisons and camps as a form of mind control—sedation through smell.

Capitalists of the Western world have pursued a more pragmatic strategy. Forget politics. Never mind love. Make more money! Bodywise, Ltd., a British firm, produces a substance called Aeolus 7+, which contains pig pheromones. They suggest saturating your mail with the stuff for better debt collection. According to David Craddock, the managing director of

Bodywise, Ltd., a mail order cosmetics company in Australia sent out one thousand bills threatening court action if not paid. Half were treated with Aeolus 7+, half were not. Those receiving the scented mail (smells a bit like stale urine) paid up more than the odor-free group (by 17 percent). Pretty aggressive mail!

But don't lose heart. Perhaps there is an antidote. Among certain snakes a substance has been identified that stops male courtship behavior in its tracks. It is a lipid called squalene—not the most marketable term for a perfume, but perhaps the first hint of a defense strategy against all the mail and males that may be assaulting our poor VMOs in the not too distant future.

Don't Lose That Scent

There are numerous questions about our noses and their sense of smell that are tantalizing, but few clear answers at present: Why do some elderly lose their sense of smell two to three years before their death? Will we be able to predict a person's time of death relatively accurately by documenting when their sense of smell disappears? Since estrogens protect against the loss of smell in postmenopausal women could this mean that estrogen replacement therapy ensures a longer life? Why is nasal congestion often related to sexual activity? What happens to sexual interest and desire in patients with chronic allergy conditions like postnasal drip?

Which raises the question: What happens to your love life if you have a cold or can't smell? There are multiple possibilities, some more obvious than others. One patient of mine, who had a chronic sinus condition, couldn't perform oral sex without his nose stuffing up so badly he was unable to breathe. In another example, a group of researchers observed that some men experience nasal congestion during sexual excitement. These same researchers also identified impotence due to the chronic use of nasal decongestants in sixteen patients. These studies suggest a parallel, a relationship between nasal and penile blood flow, meaning that as the penis engorges with blood, so does the nose. (Yes, sexual arousal also makes your nose bigger!)

In another study, physicians at the Vallabhbhai Patel Chest Institute in Delhi, India, reported four cases (three men and one woman) of asthma and/ or postnasal drip triggered by sexual intercourse, some severe enough to require hospitalization. Sexual excitement rather than exertion appeared to be the problem because none of these patients had similar symptoms from climbing a comparable number of flights of stairs.

Clearly if sexual activity selectively increases blood flow to the nose, it implies that certain scents, perhaps subliminal ones, continue to play an important role perhaps beyond sexual attraction, to sexual performance as well as responsiveness.

However, confusing matters somewhat, the reverse effect can also be true. Some men with stuffy noses can experience relief of nasal congestion during arousal and get congested again after ejaculation. Paradoxical reactions are common with hormones and neurotransmitters. In this case, it could explain these unpredictable sexual effects—sometimes triggering asthma and postnasal drip, sometimes curing it.

Two million Americans have a smell disorder and it is reasonable to wonder what impact the absence of their olfactory sense has. It was once thought that humans, like dogs and cats deprived of their sense of smell, would become sexually dysfunctional.

A tiny but valuable study has challenged that idea. Doctors at Georgetown University evaluated the sex lives of four patients (one male, three female) with the congenital absence of smell and found that good sexual functioning remained in spite of the absence of smell. What does this really mean? Actually, it raises a second question rather than resolving the first: When the conscious sense of smell is absent does that automatically mean your VMO and hence your sexual pheromone response is also gone? This small study may instead demonstrate that it is possible to lose one's sense of ordinary smell without losing the ability to detect and enjoy pheromones. Until a similar study is performed that simultaneously evaluates the functionality of the VMO, we won't have the answer.

Also, because other sex hormones are involved in smell, your attraction and attractiveness to others may be affected by a smell disorder. For example, a French scientist back in 1952 showed that women were sensitive to a substance secreted in men's urine. Called exaltolide, this musky fragrance went on to be included in some perfumes. More recent studies showed that women who have their ovaries removed (hence, their major source of untold estrogen production) lose their ability to detect exaltolide. Once they are treated with estrogen, however, they quickly regain this ability. The researchers also found that a woman's sensitivity to exaltolide peaks during ovulation, suggesting that it is, indeed, involved in mating. Further, since smell sensitivity to pheromonal substances spikes at orgasm, the acuity of a woman's senses may influence whether a woman is orgasmic or not. The point is that a woman deprived of estrogens would be less likely to detect and/or respond to the sexual pheromone of another until she is treated with estrogen.

While it is clear that the nose and sex drive are related, and that the nose and orgasm are also connected, there is as yet no comprehensive explanation of this increasingly complex dynamic. Consider the impact of nose sprays. Nose sprays can be purchased in any drug store. As we've seen, some men become impotent after repeated use of nasal decongestants. But certain sprays are also capable of making you a premature ejaculator, even if you are a woman. We know this because some nasal sprays are used to reverse the orgasmic depression of various antidepressants. For example, nasal sprays with phenylephrine (Neo-Synephrine) can and are being used to trigger orgasm—to counteract the adverse sexual consequences of "sex offender" drugs like Elavil and Prozac. (Men with enlarged prostates, however, should not use phenylephrine nasal spray because it could cause urinary retention.) Isn't it striking that what you spray into your nose can cause both impotence and rapid orgasm? Imagine, if a few squirts of nose spray can trigger orgasms —a massive body response—what do you think a few molecules of sexual pheromones could do for your sex life?

Nasal sprays will very probably become the preferred drug delivery system of the future, along with eye drops, patches, and pumps, superseding pills, shots, and suppositories. They work fast, move right into your bloodstream, bypassing the digestive process and, best of all, they are painless. As a result, nose sprays are already being used and developed to deliver a wide range of hormones we will encounter—calcitonin, vasopressin, oxytocin, and LHRH to name just a few. For instance, menopausal women are squirting calcitonin up their noses in thirty countries around the world. Calcitonin is a peptide hormone secreted by the thyroid gland. It prevents osteoporosis by inhibiting bone reabsorption and works for both men and women. It can also be delivered by an eye drop!

There is at this time no noticeable attention in the medical community to the potential side effects on adults of medications that are prescribed via nasal sprays to children and adolescents. For example, clomipramine (Anafranil) is commonly prescribed to youngsters for obsessive-compulsive disorders. Yet it can cause spontaneous orgasms, impotence, and anorgasmia. Imagine adding this dimension to the dementia of puberty.

Vasopressin, as we know, (and will come to understand more fully) has potential sexual and emotional effects in adults, promoting commitment perhaps, and potentiating testosterone. Some children are currently using a vasopressin nasal spray to stop bed-wetting. How could this influence puberty in both sexes? We are far from having dependable answers because this is the first time to my knowledge that these questions have been raised. But, regardless of the medication or age involved, nasal spray administration is the wave of the future.

It won't be long before hormones of all sorts—peptides, neurotransmitters, and other drugs that prove to be sexually valuable—will take their place beside the vibrator and other sex toys, regardless of their method of delivery. Will decongestant sprays be replaced by his-and-her pheromones?

As you can see, smell, courtesy of pheromones, has a pronounced effect on who we connect with; who we are attracted—and attractive—to. Our sense of smell represents a new medical frontier yet to be more fully explored. It also presents an intriguing sexual mystery that we are just beginning to unravel.

DEHYDROEPIANDROSTERONE (DHEA): THE MOTHER OF ALL HORMONES

Now as Debbie has already mentioned, not only is a woman's natural essence nothing to be ashamed of, the truth of the matter is it's a positive thing that works in our favor. Here's a little self celebration I bet you ladies never thought of. What you do is reach down with your fingers and get them wet with your juices. Then you rub it in behind your ears. . . ."

"Behind your ears???"

This brought the class to full attention. It even brought the fat lady back from marshmallow land. It brought Miss Adrian to the edge of a dead faint.

"Yeah, behind your ears. And a dab on your throat if you want.

73

When it dries, there's no whiff of low tide about it at all. It's a wonderful perfume. Very subtle and very mischievous. Men are attracted, I guarantee you. Why in Europe women have been using it for centuries. That's why Neapolitan girls are so seductive. (Tom Robbins, Even Cowgirls Get the Blues)

If you invest a lot of money on expensive perfumes, colognes, or aftershaves, it may be time to get to know the executive producer of your own personal aroma, DHEA, a most versatile hormone that first and foremost has the potential to manipulate our sexual selections through smell.

For much too long, DHEA's value and significance have been underestimated. In the past, it has primarily been of interest as a presumed cause of stubborn cases of acne or excessive hair growth in women. In recent years, however, it has emerged as perhaps the most fascinating, versatile, and dynamic of all our sex hormones.

In fact, DHEA may be the most powerful chemical in our personal world. It not only produces the pheromones that emit our scent through the skin, it acts on our brain to receive the scent of the opposite sex, in this way perhaps influencing our choice of mate. In addition, it may influence whether or not we become pregnant (certainly by whom), the strength of our bond with the babies that we have, and so much more. Since DHEA levels peak around age twenty-five, it may promote courtship and mating at the height of our physical appeal to the opposite sex.

We are introduced to DHEA's power even before birth. During pregnancy, it positively saturates the uterus, placenta, and fetus, and plays an important part in the birth and bonding process that we do not yet fully comprehend. DHEA is the chief hormone produced by the fetus, in amounts some two hundred to over four hundred times greater than concentrations of progesterone and testosterone, and over eight hundred times higher than the levels of estrogen. This tremendous amount present during fetal development and youth has intrigued longevity researchers for obvious reasons. As previously discussed, DHEA, through metabolism to pheromones, may regulate our fertility by manipulating our menstrual cycles to ensure that we ovulate regularly when we are exposed to men (through their sweat). This phenomenon suggests that the intimate companionship of men (or at least their body fluids) has a favorable effect on women's health and fertility.

DHEA PROFILE

MOST PEOPLE DON'T KNOW THAT DHEA:

- is the most abundant hormone in the human body—both sexes
- can transform into almost any other hormone
- improves cognition, protects the immune system, protects against certain forms of cancer
- serves as an antidepressant for both sexes
- promotes a "futile fatty acid" cycle causing weight loss without decrease in food intake
- is actively synthesized and broken down in the human skin
- decreases cholesterol
- promotes bone growth

DHEA:

- is a steroid hormone, regulated by ACTH, manufactured mainly in the adrenals, but also by the ovaries, testicles, and brain
- is a hormone that can transform into almost any other hormone
- blood levels in adult men are one hundred to five hundred times greater than testosterone levels

AS TO SEXUAL ROLES, DHEA:

- is the precursor of pheromones in some animals and most probably in humans as well
- may influence who we find attractive and who responds in return
- increases sex drive, more so in women than men
- increases in the brain during orgasm

AS TO BEHAVIOR, DHEA:

- is the only hormone that peaks around age twenty-five and declines steadily thereafter
- may be a key factor in male menopause
- is beneficial to the quality of life and longevity
- may have different effects on males and females

TOTAL DHEA CAN BE DIVIDED INTO:

- DHEA, which cycles throughout the day
- DHEAS (sulfate), which is stable
- free DHEA in the brain, which is higher in females than males
- DHEAS and derivatives are substantially higher in fat from breast and pubic region than in fat from abdomen

DHEA HAS BEEN USED TO TREAT:

- hirsutism (excessive body hair) in women
- male menopause (outside United States)
- the mental and physical problems of aging (outside United States)
- aging, as a longevity drug (outside United States)
- immune deficiencies
- breast cancer
- AIDS
- hereditary angioneurotic edema
- female menopause (outside United States)
- osteoporosis

HOW WE CAN INFLUENCE DHEA:

Increases DHEA:

- puberty
- prolactin
- bupropion
- digoxin
- smoking
- transcendental meditation
- vigorous exercise
- pregnancy
- stress-reduction techniques

Lowers DHEA:

- alcohol
- stress
- chronic illness
- obesity (cause or effect?)
- first pregnancy
- type A behavior
- corticosteroids
- progesterone and other oral contraceptives
- cimetidine (Tagamet)
- ketoconazole (Nizoral)
- lovastatin (Mevacor)
- anticonvulsants

What is DHEA?

As a sex hormone, DHEA does not discriminate. Since it is produced primarily by the adrenals (two glands sitting right on top of the kidneys), both men and women have nearly equal amounts, except in the brain. (The testicles and ovaries produce a little bit of DHEA, but not enough to make a real difference.)

The brain can make its own DHEA. In fact, DHEA concentrations in the brain are higher in humans than any other sex hormone. Further, the brain levels of active DHEA are much higher than inactive DHEAS—the opposite of levels in the bloodstream, where the reverse is true. Free (or active) DHEA is strikingly higher in women's brains than in men's. We do not yet know what this means.

DHEA and testosterone have similarities as male hormones (androgens). As neighbors in the adrenal gland, they seem to be stimulated by the same "individual"—ACTH, a pituitary hormone. They are, nonetheless, distinct and have fascinating differences. Even though ACTH triggers the release of both DHEA and testosterone, DHEA levels change independently of testosterone, meaning that there is another as yet unidentified control factor for DHEA.

Because men and women have approximately equal amounts of DHEA, and humans are unique among mammals in this respect, it has been proposed that DHEA is primarily responsible for a critical evolutionary event: liberating the female sexual drive to a sex hormone state more like the continuously active (as opposed to receptive) male sex drive, by providing a ready reserve of androgen. In this respect DHEA may be one of the most important hormones for the sexes, fostering the potential for biological and sexual common ground and paving the path to equality.

In looking at the sexual features of DHEA, sometimes through the lives of animals, we can follow the natural sequence of its effect on courtship. As a pheromone, it first attracts our attention to a mate through our sense of smell, then it regulates certain cycles and peaks. Next comes its potential aphrodisiac qualities, its role in our "sex skin," and finally, its impact on orgasm.

The sex hormone concentrations in skin and tissue are usually different from that in the blood. So, even with equivalent levels of certain sex hormones among men and/or women circulating in the blood, there can be large differences in skin sensitivity because DHEA or its relatives transforms itself and/or other hormones once it leaves the circulatory system, enters the tissue, and approaches the skin. The point is that even though one person may have similar blood levels their "skin soup" of hormones can vary enormously causing them to emit different scent signals.

DHEA and Desire

Not only does DHEA influence whom we attract and what we feel, it may tell us who in particular we want and just how much we want them. Information from research performed at the Crenshaw Clinic suggests that DHEA increases sexual desire. In that sense, it serves as a natural "aphrodisiac" in both men and women, young and old. One of the first clues for the researcher was that progesterone, that particularly powerful antiaphrodisiac, plays an opposite role to DHEA in most other metabolic respects. Since progesterone kills sexual desire, perhaps DHEA would stimulate it. That reasonable premise proved to be true. But even though men and women have relatively equal amounts of DHEA in their bloodstream, it appears to affect them unequally.

Women: Women seem to get most of the desirable dividends, but more work needs to be done to confirm this theory. Although DHEA rises precipitously prior to and during puberty and declines with aging, its rise and fall has seldom been studied in regard to sexual drive and behavior. Even so, researchers at the Crenshaw Clinic discovered that for girls at puberty, levels of free testosterone and DHEA were significant hormonal predictors of sexual desire and *noncoital* sexual behavior (masturbation, thinking about and desiring sex) but not intercourse. Is this reticence actually due to the physiological effect of the hormones, or to the strong dictates of society and most parents against adolescent girls engaging in intercourse? Some conclude that social and conditioned psychological restraints are responsible. Others, including myself, think that sexual taboos would apply perhaps even more intensely to masturbation than intercourse (a sin in some religions), thus deducing that it is a true physiological response.

It was not until the late fifties that DHEA was identified as a sex hormone in women. A study was conducted at Sloan-Kettering Cancer Research Institute on women whose ovaries were removed as part of cancer surgery. The results revealed that their sexual desire seemed to be intact. If the ovaries, which produce estrogen, were no longer available, researchers wondered, what was accounting for their sustained sex drive? It turned out that their adrenal glands, which secreted DHEA (along with some testosterone and other substances), were responsible.

Cancer patients who had both their ovaries and adrenals removed were also evaluated. With only ovaries missing, some women experienced diminished drive while most reported little or no change. After the adrenalectomy, however, four out of five women experienced total loss of sexual desire and responsiveness. Since the more serious decline in sexual desire occurred only with the removal of the adrenals, this researcher concluded that adrenal hormones, in particular, DHEA, were central to women's libido. By the same reasoning, the continuing presence of DHEA after menopause may account for the fact that sexual desire is not dramatically diminished in women immediately after menopause.

Only in women, in marked contrast with other mammals, can the ovaries be removed without a predictable reduction of sexual desire. From this we have concluded that human female sexual desire has shifted somewhat from estrogens to androgens like DHEA, and from the ovaries to the adrenals.

On the average, DHEA levels are lower in women who have had children than in women who have never had children. We don't know why, or just what this means yet. While this reduction in DHEA would not affect sex drive as severely as the increased prolactin levels associated with nursing, it does give women a second explanation for diminished interest in sex after having a child, and underscores the concept that a multitude of shifting hormonal forces must be considered. Reduced DHEA may also play a role in postpartum depression.

DHEA may influence women more strongly than men where sexual desire and body image intersect. DHEA levels affect your fat metabolism—in a most delightful and unique fashion—at least by our cultural standards. DHEA helps you stay thin. It triggers a "futile fatty acid cycle" that somehow raises your metabolic thermostat, causing you to burn more energy. End result, you lose weight *without* dieting. The lower your DHEA, the more likely you are to be overweight. The less fat in your flesh, the higher your DHEA. A study of rhesus monkeys showed double the amount of DHEA in lean females compared to obese ones, but no difference in DHEA levels between lean and obese males. Which causes which? It seems to be a self-perpetuating cycle. If you are thin, you have more DHEA, which triggers this spontaneous fatty acid cycle. If you are fat, your DHEA is lower and you do not have the advantage of this cycle. So you tend to stay fat unless you exercise (which can elevate DHEA) and/or lower your food intake.

Men: While men have abundant amounts of DHEA, the impact of DHEA on their sex drive is most probably considerably less than in women for one very simple reason: They're so overwhelmed by testosterone that the DHEA effect is relatively insignificant by comparison.

The importance of DHEA in men is not so much the immediate effect on their sex drive, but the indirect effects on other aspects of their health that ultimately influence their sex drive, such as stress, heart disease, midlife changes, their sexual attractiveness, and quality of life.

DHEA's Role in Our Skin and Scent: Of its various sexual properties, DHEA's most intriguing effect is on our skin and our scent. Through a combination of touch and smell, within the first few days of life DHEA actually changes our brains, determining which cells live and which ones die, and in the process bonding mother and child. DHEA can't penetrate directly into skin because it is not well absorbed by certain tissue. Instead, as it approaches the barriers obstructing its progress toward our skin, it transforms into other hormones that are readily absorbed by the skin, and have a specific purpose there. In this way, DHEA gives birth to sexual pheromones. As a precursor to sexual pheromones in other animals, DHEA drives their mating rituals and courtship patterns. In humans, we are discovering much the same thing. But research has been maddeningly slow.

In addition to producing pheromones, DHEA plays a specific sexual role in how our skin feels and looks. Some animals, notably the gibbon, have a glaringly obvious sex skin that they proudly flash. Human females, not to be outdone, paint their lips and rouge their cheeks—some say to imitate this gaudy relative's behind. However, we have sex skin of our own, albeit less obvious, and it seems that again we have DHEA to thank for it. The most salient parts are our breasts and genitals, including the pubic mound, lips, scrotum and inner thigh. DHEA concentrations are substantially higher in fat from the breasts and pubic region than in fat from the abdomen, while estrogens and most androgen derivatives have the same concentrations in the fat of all areas. Because of DHEA's enormous presence in the breast and genital area, it has the opportunity to spike these erogenous zones in two special ways: transmitting erotic fragrances while receiving sizzling sensations. Perhaps it also primes them for orgasm.

While you may not think of them as a conventional erogenous zone, your lips may give away more than you realize. They transmit sexual signals in addition to all their other responsibilities. Desmond Morris, in his books, *The Naked Ape* and *Body Watching,* suggests that during erotic arousal, your lips become swollen, much redder and more protuberant than usual—making them more sensitive to contact and more conspicuous as well.

Morris also has a theory about ear lobes. He speculates that they, too, have evolved into erogenous zones because women need something to suck during foreplay. I think they have an ample selection of other things to choose from.

Orgasms and Seizures: Measuring sexual responses in animals, as I've mentioned earlier, is limited to what we can observe—masturbation, mounting frequency, penetration, ejaculation, and so on. With DHEA, the investigation may have to be far more sophisticated to discover quantifiable behaviors that can be related to sexual desire and orgasm. Some knowledge that we need about DHEA and sex can only come from direct research on humans during the experience itself. Nonetheless, we do have some compelling data from animal studies.

In the brain, DHEA excites the septum and medial preoptic (MPO) areas, which are known to promote active and pleasurable sexual behavior and reactions. When male rats in their home cage are exposed to the scent of receptive female rats for seven days, DHEA increases in two other brain regions as well—the amygdala and hypothalamus. Brain DHEA has similar levels in male and female rats when females are near or in estrus, but decrease in females when they are not sexually active or receptive. Whether as cause or effect, we do not know. Perhaps both. No changes occur in blood concentrations, however. Also, the increases in brain DHEA that occur during sexual excitement usually are not reflected in changed blood levels outside the brain.

In any event, we can deduce that increased DHEA levels in the brain promote sexual excitement and/or increase in response to it. In addition, DHEA triggers seizurelike electrical activity in the septum. The septum, specifically, is where orgasmic activity is generated in the brain, and is coincidentally associated with theta waves (low brain waves generated during attention or focus; also characteristic of REM sleep and advanced TM states). As DHEA stimulates this portion of the brain, it lowers the seizure threshold, perhaps facilitating the natural and desirable total body response of orgasm, making it easier to have one.

Bupropion (Wellbutrin) and DHEA: It should come as no surprise then, that bupropion, a relatively new antidepressant, can increase sex drive, promote orgasm, cause weight loss and seizures. Sound like a familiar theme? Blood levels of DHEA in women are increased by bupropion treatment, and these increased DHEA levels in women in a bupropion study I conducted along with Dr. James Goldberg in 1984 were associated with alleviation of sexual dysfunction: women with low sex drives improved, sometimes dramatically. The reactions to bupropion by two women in particular was unusually intense, illustrating the "sex skin" response as well as the effect on sexual desire.

One young woman, raised religiously with what she called "a lot of hang-ups," for years thought people were simply lying about female orgasms. She had never had one. "I've always felt sexual in my head," she said, "but my body just doesn't like to go along with what my head would like to do." She felt "nothing" sexually, which had frustrated and angered her for years.

Some weeks into the study, while she was sitting at her dining room table studying, a strange and powerful sensation hit her out of the blue. "I was thinking about economics. All of the sudden I got this fantastic sex feeling," she recalled with a smile. "It was warm, throbbing a little bit, achy and nice." I asked her where these feelings were located, and she said, "You know, down THERE!" and then added, "I wanted to have sex right that minute!"

In fact, sex was all that was on her mind for days afterward, a situation that rattled her marriage until her husband got into the new swing of things. She wanted her orgasm and she wanted it now! This woman, who had previously avoided sex, threatened her husband with finding someone else to satisfy her if he wouldn't cooperate immediately. But she didn't. And in time, he did.

Another woman had a similar explosion, but she didn't like it one bit. Her lifelong attitude was reflected by her statement that, "Sex was to be endured and avoided at all costs." Married and in her fifties, this woman hesitated even touching her husband casually for fear of encouraging him. But she felt that after thirty years of patient tolerance of little or no sex, her husband deserved more. In an effort to turn up her sex drive, she signed up for our study. Several weeks into it she suddenly experienced "a very uncontrollable feeling that annoyed me no end. It colored my whole day." She felt throbbing and heat in her vagina, a sensation she compared to "a kid who has to go to the potty and wiggles." Even intercourse didn't sate this feeling and for three days it persisted at high intensity, although it waned and ebbed somewhat. Over the subsequent seven days, it gradually diminished, to her great relief, disappearing altogether by day ten.

Despite the second woman's initial annoyance at this uncontrollable sexual feeling, she eventually came to enjoy sex with her husband.

The impact of bupropion on sexual desire is probably the result of two things: increased dopamine and DHEA at the brain level. This means that the men and women in our study who were overwhelmed by desire beyond anything they had ever experienced before were probably responding to the combined effect of dopamine and DHEA on the brain. Based on what we know about DHEA's effect on our "sex skin," the intense genital feelings they described were probably the result of local DHEA in the tissues of the genitals. When a different target organ (brain or skin) is saturated with the same substance, in this case DHEA, the emotional and/or physical response will vary according to the dictates of the target organ.

The Big Picture

DHEA's influence is not limited to the subtle dictatorship of smell and almost all other aspects of sex. It affects our lifespan and our health as well.

DHEA clearly decreases during aging, chronic stress, and disease. Its decline could actually be a catalyst for illness and aging. We do not have enough data to assign cause and effect, but we do know that DHEA decreases as these conditions worsen.

Influencing DHEA Levels

Certain substances and medications affect DHEA. Pay special attention to those that decrease it and avoid them when possible. Here is a summary:

Alcohol: Alcohol decreases DHEA in both female and male alcoholics. Chronic alcoholic cirrhosis is characterized by lowered DHEA. Studies have shown that alcohol injections can reduce brain levels of DHEA within thirty minutes. Recovery to normal levels takes about four hours. If all this sounds forbidding, stay young. Don't get pickled before your time.

Bupropion (Wellbutrin):   There are surprisingly few drugs known to increase DHEA levels. Bupropion, a relatively new antidepressant mentioned above, is one exception. It increases DHEA, which may explain the intense vaginal sensation experienced by some women taking the drug. Hormone levels didn't increase when men were taking it. More studies are needed.

Exercise: Exercise can boost DHEA temporarily, but a measurable DHEA response requires about thirty minutes of vigorous exercise a day for approximately one month.

Cigarettes: Cigarette smoking also increases DHEA, but any benefits are certainly outweighed by the diseases and death associated with smoking.

DHEA'S INFLUENCE ON OUR HEALTH

High DHEA:

- protects immune system
- inhibits tumors
- promotes bone growth
- causes weight loss without dieting
- generates higher energy utilization
- lowers cholesterol, LDL, and body fat
- reduces deaths from heart disease
- reduces mortality from all causes, especially in men

Low DHEA:

- is associated with chronic and degenerative diseases
- promotes bone loss
- causes weight gain or obesity
- is associated with low energy in 50 percent of ovarian cancer cases
- is a predictor of breast cancer nine years in advance
- may reduce mortality in women

HOW WE CAN INFLUENCE DHEA:

Increases DHEA:

- puberty
- prolactin
- bupropion
- exercise
- transcendental meditation (TM)
- smoking
- digoxin
- sex

Lowers DHEA:

- alcohol
- stress
- chronic illness
- obesity (cause or effect?)
- aging
- Alzheimer's disease
- autoimmune diseases
- hypothyroidisms
- anorexia nervosa
- glucocorticoids
- phenobarbitals
- phenytoin
- carbamazepine

Transcendental meditation: Transcendental meditation seems to elevate DHEA. The studies demonstrating this effect were conducted at the Maharishi University, a center for transcendental meditation. Their research suggests that long-term transcendental meditation increases DHEA perhaps by decreasing the stress that otherwise lowers DHEA.

In one study, DHEA levels were measured in 252 men and 74 women who were experienced regular TM practitioners. They were compared to levels in 799 men and 173 women who did not practice transcendental meditation. In the TM group, DHEA levels were higher in women of all ages, but levels were higher only in men over forty years old. Mean DHEA levels were 47 percent higher in transcendental meditation women over forty-five and 23 percent higher in transcendental meditation men over forty. Adjustments for diet, obesity, and exercise could not account for the significant DHEAS differences.

Type A behavior appears to lower DHEA levels, similarly to chronic stress or illness. However, type A personalities who had meditated for about twelve years scored higher DHEA levels than normal.

Curiously, proficient meditators demonstrate increased EEG theta power during meditation which peaks periodically (during samadhi states). Such a finding would suggest that highly skilled meditation increases both EEG theta and DHEA, indicating limbic activation, and reinforcing the connection between DHEA, theta waves, sex, and the septum already identified in animal studies.

Decreased stress: TM is not the only way to decrease stress. It stands to reason that biofeedback, exercise (as we know), relaxation techniques, massage, and a wide variety of therapies would be beneficial to DHEA levels. But just like exercise, only consistent, long-term efforts would have sustained beneficial results on DHEA.

DHEA supplements: DHEA supplements are available in Europe and other parts of the world, where they are used to treat aging, male climacteric (menopause), obesity, and various other conditions. DHEA is not approved in the United States for these indications.

In several European countries, DHEA is also used as a supplement in postmenopausal women to prevent osteoporosis, since low levels of DHEA are associated with osteoporosis—at least in women. No comparable studies exist to define the degree of osteoporosis faced by men as they age.

For women, the major concern revolving around DHEA treatment is whether it will have masculinizing effects. The scanty data available suggest this will not be a significant issue, but here in America, this drug cannot be used until the relevant studies have been done and FDA approval is achieved.

Hurdles still to be overcome are the concern that DHEA may foster the growth of prostate cancer and contribute to prostatic hypertrophy.

Remember Our Song?

We can't leave love, lust, and limerance behind without a few thoughts about music and romance. Has there ever been a great love without a favorite song or melody? "Tara's Theme" from *Gone with the Wind*, "Lara's Theme" from *Dr. Zhivago*, "Unchained Melody" by the Righteous Brothers, "I'll Always Love You," "Whatever I Do, I Do It for You," and so on. Music brings back memories even many years later when there is someone you would give your kingdom to forget. An unexpected tune will ambush you on the radio or the band will play a certain song when you are out dancing, and like the electronic transport in *Star Trek*, you suddenly find yourself in a different time and place.

Imagine the Pavlovian power, the impact that this specific connection of love and song has over you. Out of the blue, you hear "that tune" and it compels you to think of the love that you linked with it. It is not transferable, exchangeable, or erasable. You can't get rid of it.

And usually you don't want to. That's why "the oldies" are so popular. It's much more than just bringing back precious memories or familiar music. It is the actual reexperiencing, complete with all the chemical stirrings as you relive the moment through the song.

What is this chemistry that musical memories stir? Let me speculate for a while, because it has been little studied. We know more about how music influences shopping patterns (by helping select the Muzak menu of the day for department stores) than we know about the effect of music on sex and romance. Yet there must be "A Whole Lot of Shaking Going On" in your bloodstream when you are screaming and fainting to the Beatles and Elvis, swooning to Frank Sinatra, drooling over Michael Bolton, and caught up in a maddening frenzy of movement and sex at a hard rock or heavy metal concert. Most music triggers the irresistible urge to move, clap, sway, dance. Is this not hypnotic?

So, back to what could be happening to your chemistry: I would think a rise in PEA would be first. Anytime you feel romantic that seems to follow. Also, the right brain is accessed by melody, like a familiar face, and your affections are stirred. Cortisol drops if you like the music, rises if you don't. Cortisol is your stress chemical and the main reason someone yells "Turn that awful noise down!" You can't understand why, because it sounds so good to you.

When cortisol drops, DHEA usually goes up, setting serious sexual molecules in motion. And you are smiling with such pleasure, your dopamine must be flowing too. Dopamine motivates and mobilizes you. In fact, a dopamine boost is probably why people jump for joy and dance when they are happy. Also, perhaps, why they move to music.

Then what? Welcome oxytocin. When you dance, you touch. At the very least, you anticipate touching sooner or later. Your peptides join the party. Oxytocin surges. A few fast dances—exercise—and up go more hormones. You overheat. Vasopressin tells you to take a break. A slow dance comes on when you tire. You are dripping in sweat and exuding pheromones. You touch again, this time with full frontal body contact—oxytocin overload. Your endorphins are rising. You're stoned.

Once oxytocin gets flowing, it inspires estrogen and vice versa. When these two gang up on a woman, look out—you'll find out why in the next chapter. His testosterone is pumping as he watches her dance and feels her move. No wonder Mormons and Mennonites forbid dancing. "Let's go dancing"—so innocent and simple sounding—takes on new meaning. "I'll put on some music to set the mood." Little did you know you were actually adjusting your chemical rheostat!

THE FUTURE OF PEA, PHEROMONES, AND DHEA

While hard data often remains elusive, we can speculate on some of the future treatment possibilities of the sex hormones, and they are indeed intriguing. Sexually, *human* pheromones in perfumes and colognes could serve as a real "love potion number nine," manipulating courting rituals as it does with sex-crazed cockroaches—with happier outcomes, I hope. Nasal sprays and atomizers by the bedside would be within easy reach to intensify desire, sexual sensations and—sniff—an orgasm, with no more noisy vibrators. DHEA supplements could be used as a "tonic" in both men and women from the time levels start to drop in the thirties to keep you in your prime, to prevent middle-age spread and, later, to ward off disease and the body's general deterioration as it ages.

Until now, the emphasis of research has been on specific physical and genital sex, concentrating mostly on the hormones testosterone, estrogen, and progesterone. DHEA's multifaceted profile will force sex research beyond the physical, to integrate critically involved emotional dimensions, such as touch and scent, perhaps even love, intimacy, and commitment. Simplistic preoccupation with "aphrodisiacal" urges from hormones and other chemicals could then yield to more comprehensive, mature, and useful sexual pharmacology research and treatment.

Fortunately, love, however you define it, seems to be the best aphrodisiac of all. Now that we understand that DHEA through pheromones may play a central role in who we love, and PEA in how much we love, perhaps through further research we can learn how best to preserve it.

4.

TOUCHING: ATTACHMENT, BONDING,

AND COMMITMENT

Confused Identities
Whose foot is this, I think it's mine
But then I'm not quite sure
My arm's asleep, or is it yours
This heart is beating on my chest,
Or is it mine or ours
A stomach growled and woke me up
Excuse me, was that you?
The closer I get to you
The harder I am to define

(Theresa Crenshaw, Confused Identities)

Have you ever wondered why you feel so safe and wonderful when someone you love holds you close? It's as though an invisible force field suddenly surrounds you, protecting you from harm. Danger disappears, problems fade away, and you feel absolutely secure in the magic of someone's arms. It isn't logical. It isn't even true. An earthquake or hurricane could get you. But none of it matters at the moment because your sense of well-being is so powerful. Isn't it amazing how a simple hug can suddenly take your cares away?

Mix this powerful feeling with romance, and see what happens to your chemistry:

"I can't keep my hands off him," gushed Robin, thinking of Eric, the man she had fallen in love with. "His skin is like a magnet I can't resist. When he touches me I get the most exquisite feeling—electrifying and comforting at the same time. I could hold him forever."

When her new lover went on a long business trip, Robin felt as if she were going to die, like withdrawal symptoms from a drug addiction. "It's like his touch triggers some chemical and I'm hooked on it," she said.

Chances are, that's exactly what happened. Touch alters the chemical composition of your body. When you caress someone, or they stroke your skin, when you hug or cuddle or hold hands, a chain reaction takes place that signals your brain, "This is good. Pleasurable. It soothes me. I want more." The more you touch, the more you *want* to touch. And the more you touch the same person, the stronger the bond between you grows. The more you want to be together. Touching is, in a very real sense, addictive.

THE CHEMISTRY OF EMOTION

Investigating the chemistry of emotion is a new and courageous venture for modern researchers. In our society, intimacy, caring, relaxation, domestication, and nurturance are taken for granted. Touch is regarded as incidental, elective. Yet in spite of this neglect, during the last twenty years science has been discovering that touch—and whatever inspires it—is far more important than we ever imagined.

Does touching keep you smart, healthy, and happy? I think so. And so do a lot of other experts. While the health benefits of touch are just beginning to be fully appreciated, we have recognized some of the consequences of touch deprivation for ages. Let's take a closer look.

Vitamin "T"

Touching is a basic animal instinct. Nature seems to have designed it that way. We crave the emotional nutrition that comes from touch, just like an essential vitamin. For evidence you need look no further than your pet cat or dog, the newborn nursery, or a convalescent home. Even solitary creatures like the octopus need to be touched; they instinctively find corners or cavities in rock formations to attach themselves to, not just to hide, but to rub against. Several years ago, Koko, the gorilla, graced the cover of *National Geographic* holding her pet kitten gently in her arms. Snakes snuggle; alligators cuddle—

all the better to face the jungle out there and carry out the biological imperative to mate.

As for humans, we know that babies fail to thrive and sometimes die for lack of touch, even when their other basic needs are met, as Dr. Rene Spitz discovered in the early 1930s during his study of orphans. Those who survive usually become physically and mentally retarded. Older people deteriorate when touch deprived. They die sooner and become senile faster. With such profound effects at both extremes of age, how do you suppose touch, or its absence, influences us throughout our lives?

The truth is, we can't live without it. We develop a form of emotional scurvy, although we call it by different terms: depression, stress, anxiety, aggression, and midlife crisis . . . and treat it with drugs that don't work. Lack of touch is just as detrimental to our health as a lack of vitamin C and just as easy to remedy. Yet I imagine those brave souls centuries ago who suffered and watched their comrades die like flies on long sea voyages couldn't believe that the solution could be as simple as a few oranges, either. Although the scientific research of touch is but thirty years old, we know enough today to dramatically improve the quality of life, in general, and our sex lives in particular.

The Chemistry of Touch

Touching feels good, but that is only part of the story. It causes us to secrete endorphins, the natural opiates the body produces to protect us from pain. When someone holds you in his or her arms, your endorphin levels rise, making you glow with contentment. But endorphins are only a small portion of a bigger picture. Perhaps nature's chief weapon, the substance that truly compels us to get close to another person, is oxytocin, a wild and wonderful peptide. It spikes when someone touches you. If you spend time in that person's presence, a more profound lifelong pattern develops. Oxytocin will surge just at the thought of him or her. Then it goes even higher when he or she arrives, and up again at a mere touch.

Involuntary Chemical Commitment: Let's say Robin's boyfriend doesn't want to get married. He enjoys being with her, but she's not the girl of his dreams. She, however, presses, he acquiesces and they start living together.

He feels great. Steady sex, no commitment. Robin's happy. She's got her man, at least for now. For both of them it seems like a step in the right direction.

But guess again. It depends on your point of view. Being together, they touch a lot and have sex often, spending virtually every night side by side. Their oxytocin response becomes increasingly pronounced as they establish bonding patterns that get stronger and stronger over time. He comes to love and depend upon her *being there*. She gets increasingly attached to him.

Eric moved in with Robin to avoid a permanent commitment only to become "chemically committed." After not too long, he finds himself attached to her. Perhaps even in love with her. He feels miserable without her and anxious at the thought of losing her.

If he had not moved in with her, would this have happened? He didn't really change his mind. Proximity and chemistry changed it for him. Had he understood the predictable effect of oxytocin bonding, he would probably not have cohabited as a temporary defense. He would have understood that encouraging continuous physical contact with someone you feel lukewarm about can get you hooked against your will and better judgment. The reverse is also true. Women become involuntarily committed to men they live with and sleep with courtesy of the oxytocin effect—perhaps their risk is even greater due to the incestuous nature of oxytocin and estrogen (which we will discuss in more detail).

But then what determines whether we stay or stray? Vasopressin, perhaps.

The Monogamy Molecule

He's an affable bachelor with lots of friends of both sexes. Then he meets Ms. Right and he changes overnight. Warm and gentle at home, he's now indifferent to other females, and so suspicious of other males that he has a fit if any come near her. When his child is born, this homebody is as affectionate and protective as a mother. The ideal husband? A man of the nineties? No, he's a prairie vole, a tiny rodent found in the Midwest.

When these montane voles mate they stay hooked to one another for life. And guess which chemicals are responsible. Oxytocin for the females, vasopressin for the males. It seems that along with its tendency toward moderation and good sense, vasopressin is also a bonding agent, at least among prairie voles. The first time a male vole mates, his vasopressin soars, and before he knows what hit him, his bachelor days are over.

Like many stable midwestern humans, the prairie vole has a wild cousin on the West Coast. This western vole is polygamous—and has much less vasopressin.

Vasopressin is a dazzling peptide with numerous attractive patterns. However, it is also a "checks-and-balances" chemical that discourages emotional extremes. When oxytocin makes us giddy, vasopressin, it seems, brings us to our senses.

Do differences in vasopressin levels explain why some men have commitment phobia while others are devoted one-woman types? It will take some time before research on humans gives us a definitive answer.

You may also wonder, if vasopressin applies checks and balances to oxytocin's craving for touch and attachment, how can it also foster monogamy? Well, perhaps it helps women avoid the extremes of attachment, while helping men avoid the equally dangerous extremes of independence and multiple partners.

Unfortunately, there is much less research on vasopressin, so there are gaps in its profile. What we know, I will include here, but by default, oxytocin will get the majority of our attention.

Oxytocin and vasopressin responses probably increase or decrease more as a result of our choices and behavior than whatever sexual stage we happen to be in—how much a person touches and is touched in return, for example; whether or not they live with someone; how often they have sex (touching); how many, if any, orgasms they experience; if, by design or by chance, they have children. All these circumstances and many more determine how your oxytocin levels and vasopressin levels fluctuate over a day, a week, a month, a lifetime—and whether these chemicals are working in tandem, or are at odds with one another.

This chapter will introduce you to the yin and yang forces behind attachment, bonding, and commitment. You will learn how touch influences your physical health and state of mind beyond anything you have ever imagined, and how you, in turn, can influence oxytocin and vasopressin.

Complementing estrogen and testosterone, oxytocin and vasopressin define and refine the masculine and feminine traits we manifest. They balance each other with a remarkable synchrony—or clash, depending on how you handle yourself. They also interact, often confoundingly, with a host of other sexual chemicals.

In addition to their impact on commitment, oxytocin and vasopressin are highly sexual peptides, affecting desire, arousal, nipple responses, erection, orgasm, and ejaculation. Strictly speaking, peptides are not hormones, but are active, potent chemicals that interface with them. Moreover, peptides behave like hormones, and most scientists today refer to them as hormones. For our purposes, there are no important distinctions.

The best way to get to know oxytocin and vasopressin is to follow them throughout our lives, relationships, and experiences in their natural sequence, beginning with birth and continuing on as they impact our sexual and emotional world.

This is a global journey with a common denominator—touch—a theme that carries through all aspects of this diverse discussion. It connects our early experiences to later patterns, creating a circle that takes us from how our parents touch us as children to how we touch each other as adults—for, ultimately, one can determine the quality of the other.

Touch is, of all our resources, the most powerful. Withholding a touch at a crucial moment can break a relationship. Maintaining continuity of touch during troubled times can save one. Healthy enjoyment of touch throughout the course of a relationship can prevent numerous problems and dramatically modify the sexual stages we experience. In addition, touch has a direct bearing on our health.

Up until now, we have examined forces like DHEA with its most powerful pheromonal tyranny, along with PEA and other molecular manipulators that dictate our lives. *Oxytocin is the first player that we can influence at least as much as it influences us*—by what we choose to do or not do, whom we touch, and whom we spend out time with. It illustrates the premise that if you don't take charge of your life, your passive choices—choices still—will take charge of you.

OXYTOCIN PROFILE

MOST PEOPLE DON'T KNOW THAT OXYTOCIN:
- promotes touching
- promotes bonding between mates and parents and children
- is involved in the birth process, breast-feeding, and orgasm
- decreases cognition and impairs memory

OXYTOCIN:
- is secreted by the posterior pituitary
- has a synergistic relationship with estrogen
- is pulsatile
- and dopamine modulate each other
- sensitizes skin to touch
- is widely distributed throughout the brain and body
- increases dopamine, estrogen, LHRH prostaglandins, serotonin, testosterone, prolactin, and vasopressin

AS TO SEXUAL ROLES, OXYTOCIN:
- spikes at orgasm
- causes uterine contractions during orgasm and during labor
- increases sexual receptivity
- speeds ejaculation
- increases penile sensitivity

AS TO BEHAVIOR, OXYTOCIN:
- rises in response to touch and promotes touching
- induces parenting behavior
- promotes affectionate behavior
- has been linked to obsessive-compulsive disorder

OXYTOCIN HAS BEEN USED TO:
- induce labor
- reduce postpartum bleeding
- treat schizophrenia, depression, and obsessive-compulsive disorders

HOW WE CAN INFLUENCE OXYTOCIN:

Increases oxytocin:

♦ estrogen

♦ touch

♦ acetylcholine

♦ tactile genital stimulation
  or vaginal stretching

♦ intercourse, both male
  and female

Lowers oxytocin:

♦ estrogen deprivation

♦ touch deprivation

♦ alcohol

OXYTOCIN: OUR BONDING AGENT

Oxytocin is a peptide secreted from the posterior lobe of the pituitary gland. It flows to receptor sites in various parts of the brain and throughout the reproductive tract of both men and women. Without the presence of estrogen, it has almost no power. Together, they do quite marvelous things.

In animals, oxytocin increases sensitivity to touch and encourages mating, grooming, and cuddling in both sexes. When given to female rats around ovulation, oxytocin heightens their efforts to contact males and intensifies lordosis (sexual presenting). Through its association with smell via estrogen sensitization, it orchestrates the body's response to pheromones. I suspect the same is true for humans. The fact that oxytocin spikes at orgasm in women (and seems to in men as well) adds to the complete picture of a molecule that takes you from the first touch to the height of orgasm.

Researcher and author Niles Newton studied oxytocin for over thirty years before her landmark 1978 article, "The Role of the Oxytocin Reflexes in Three Interpersonal Reproductive Acts: Coitus, Birth and Breast Feeding." Subsequently, other researchers have studied sexual bonding through oxytocin in a series of papers during the 1980s. Its main function, as we've seen, appears to be to keep sexual partners and progeny *in touch*.

The Birth of Emotion

Before it begins to bond us, oxytocin cuts the first and most elemental cord. By causing the forceful contractions of labor, it expels us from the womb, separates us from our mothers, and tosses us out into the world alone.

Vaginal stretching in the course of vaginal delivery releases oxytocin, producing pain-relieving, euphoric, endorphinlike effects, as well as triggering the uterine contractions. Of course, the birth process is a much more painful trigger for oxytocin response than intercourse, but remarkably enough, the uterine contractions brought about by orgasm (triggered by oxytocin) are just about as powerful as those of labor, as documented by actual measurements performed in the Masters and Johnson laboratory.

Pitocin, which can be injected to induce or assist labor contractions, is a form of oxytocin; the same effect can be created naturally by nipple stimulation, which also releases oxytocin and has been used as an alternative to these injections to induce labor. To illustrate the point in a most unusual way, I would like to introduce you to a thirty-one-year-old married man with normal male physical characteristics who sought treatment to enlarge his breasts and nipples. Stimulation of his nipples caused him sexual pleasure, and he wanted more. He surmised that feminization of his breasts would enhance this experience, and his wife supported his desires. He found a physician to accommodate his request, and was treated with ethinyl estradiol (a form of estrogen) daily for nine months, which resulted in moderate breast enlargement and *two normal-sized female nipples*. He was delighted in particular with the increased intensity of sexual responsiveness in his nipples: their stimulation caused erection and ejaculation.

As you can see, oxytocin responses can be enhanced in either sex. But let's leave our man and get back to the mundane. In normal women, after birth, oxytocin continues to contract the uterus and reduce bleeding, expediting the healing process.

Oxytocin is necessary for milk let-down in the breasts and sensitizes the mother's body to her child's touch. During nursing more oxytocin surges; it engenders pleasure and relaxation, again reinforcing the bonding between mother and infant. Some women describe exquisite sensations from nursing, with even an occasional orgasm.

PROLACTIN PROFILE

MOST PEOPLE DON'T KNOW THAT PROLACTIN:
- is the cause of low sex drive in nursing mothers
- surges during stressful experiences such as nausea, vomiting, fainting, jet lag

PROLACTIN:
- is secreted by the pituitary
- is involved in sperm production and the maintenance of genital tissue
- in chronically high levels decreases testosterone
- causes milk secretion from the breast
- is pulsatile

AS TO SEXUAL ROLES, PROLACTIN:
- is primarily inhibiting sexually
- spikes during copulation and orgasm
- decreases sex drive/orgasm
- can cause impotence
- increases in pregnancy and in nursing mothers
- decreases testosterone
- peaks midcycle and remains high until menstruation

AS TO BEHAVIOR, PROLACTIN:
- depresses sensation and alertness in general
- can cause mild depression and fatigue

HOW WE CAN INFLUENCE PROLACTIN:

Increases prolactin:
- breast-feeding
- pregnancy
- prolactin-secreting pituitary tumors

Lowers prolactin:
- bromocriptine
- testosterone
- dopamine

- estrogen
- oxytocin
- progesterone
- vigorous exercise
- surgical/psychological stress
- stimulation of nipples
- sleep
- hypothyroidism
- sexual intercourse (in women)
- opiates
- orgasms
- high protein meal

Have you ever seen milk shoot from a mother's breast in response to her baby's cry—even when it's in another room? Pavlov and oxytocin are at work. Bonding of the highest physiological order.

Oxytocin spikes some more when we touch our newborn child, creating skin hunger for our progeny. And the more we touch them, the more attached we become to them, the more we need them and they us. Strong natural bonds, once formed, will be continually reinforced through time and touch.

Love at First Whiff

Newborns also form a powerful bond with their mothers, not on first sight, but based on smell (a DHEA-dependent response). The formation of scent memory is only possible during the first few days of life and actually produces changes in the anatomy of the brain. These memories form in less than ten minutes and stick with you for life. They are much more profound than memories acquired later on.

This smell memory, however, will not form unless a new scent *is accompanied by touch*. That way, the baby doesn't just bond spontaneously with any passing odor, a pizza for instance.

The brain cells that store these learned smells survive, while unstimulated cells die out. Fortunately, we no longer take newborns away to a sterile incubator immediately after birth, depriving mother and child of this early body chemistry, although newborn nurseries sometimes recruit volunteers to come and hold preemies to help them grow.

But what about fathers? Are they afforded the same bonding opportunities with their child during these formative moments?

Parenting

Touch is the first sense to function fully, and it plays a key role in the growth of our other senses. Early touch deprivation can irreversibly damage a baby's personality, social skills, and our ability to express affection as adults. Extreme deprivation of touch has been shown to destroy brain cells. Yet parents are often unaware that how often they touch their children when they are young will influence how much they touch, and how well they bond with their mate when they mature.

Imagine if you could give mothers and fathers a "parenting pill" to make sure they'll take good care of their offspring! Today, it is theoretically possible. And oxytocin is that "parenting hormone."

In animals, oxytocin can create remarkable sudden maternal behavior in otherwise unwilling beasts. Virgin female rats normally avoid pups. Wild mice, hamster, or gerbil virgins may even attack and eat them. However, if oxytocin is injected into virgin rats, full maternal behaviors begin within the hour, including nest building, retrieving and grouping pups, and crouching in a nursing posture.

If virgin rats are put into continuous close contact with pups for a week or more, they will gradually show all these maternal behaviors. Twenty-four to forty-eight hours is required for these rats to become physically comfortable with the pups. The change occurs in response to natural oxytocin levels rising, providing orphaned offspring with virgin rat love.

When oxytocin is given to male rats, papas start acting like mamas. They will build nests for their young and guard them ferociously. When given a drug that blocks oxytocin, they neglect them instead. They may even eat them. Doesn't this give you the impulse to inject deadbeat fathers who abandon their children, don't pay child support or bother to "stay in touch?" Quite a twist! What about mothers who neglect their children? Are they oxytocin deficient or just indifferent?

Oxytocin and other chemical cues can shift sexual behavior from copulating to parenting. And, of course, one tends to lead to the other unless you take measures to prevent it. Since estrogen facilitates oxytocin reactions, it may explain why the effect is less intense in the male, and why some mothers (moving into menopause without estrogen replacement) withdraw from their grown children, although neither of these questions has been studied. Further, since oxytocin depends on the presence of estrogens and disappears in its absence, women who decide to use estrogen replacement therapy will sustain higher oxytocin levels than those who do not. Yet oxytocin deprivation is one of the risk/benefit considerations I never hear discussed regarding hormone replacement therapy. In addition to menopausal women, what about the younger woman whose ovaries are removed without checking or replacing her estrogen levels? What happens to her sexual and emotional responses to her husband or boyfriend(s)? Does her behavior and/or devotion to her children change? Clearly, more research needs to be done.

VASOPRESSIN: THE SENSIBLE AND STABLE ONE

Vasopressin is a peptide much like oxytocin and one of its closest neighbors, secreted from the same general area of the brain—the posterior pituitary. While it resembles oxytocin structurally, its metabolic impact is quite different. In most respects it balances or opposes oxytocin's influence.

Vasopressin is best known in medicine as an antidiuretic, which prevents water and salt depletion. One way it does so is by discouraging urination. For that reason it is used to treat bed-wetting in children, its primary clinical application in the U.S. today. In other parts of the world it is used to treat problems associated with aging, as well as enhancing our immunity to disease.

Vasopressin improves cognition—our ability to think clearly. We'll talk more about this in Chapter 9. Apparently it does this through enhancing attention and alertness while reducing emotionalism. These cognitive benefits are especially captivating. By enhancing learning, attention, and memory, vasopressin provides a nice counterbalance to oxytocin, which makes us forgetful and absentminded. In fact, oxytocin impairs memory, which is one of nature's ways of making us "forget" the pain of childbirth. You can speculate about the implications of this phenomenon.

One of the most valuable assists that this pragmatic peptide provides to relationships is the way it calls our attention to sexual cues: helping a man register eye contact, a come-hither look, an inviting smile.

Vasopressin is also active in the septal (orgasmic) region of the brain, and is somehow involved with setting our emotional and physical thermostats. It seems to synchronize with DHEA in the brain and magnify the effects of DHEA in our bloodstream. Did you ever wake up soaked with sweat and realize you were having a nightmare? That's because you were heating up from the emotional intensity of it all. When you have an excessively emotional dream, the brain cuts off REM sleep—the dream stage—to prevent your body from overheating. How does vasopressin fit in with this? Research has demonstrated that a deficiency of vasopressin causes a reduction in REM, and the deficit can be corrected with vasopressin treatments. Having enough REM sleep is important to our health and emotional stability. All of which is evidence of the peptide's value in moderating emotional extremes. We don't fully understand how it does this, but we know enough to recognize and appreciate the connection.

In an important study, one researcher stimulated the septum of humans with electrodes, and the subjects had ecstatic reactions that resembled orgasm. Now add this fact: During both orgasm and REM sleep, the brain generates theta waves from the limbic area. Vasopressin can strengthen the theta potential for sexual reactions. This means that theta waves, orgasm, and vasopressin are somehow mysteriously related. Incidentally, during REM, men get nocturnal erections and women lubricate.

VASOPRESSIN PROFILE

MOST PEOPLE DON'T KNOW THAT VASOPRESSIN:

- is strongly associated with testosterone
- improves cognition and "sensible behavior"
- is reputed to be the "monogamy molecule"
- regulates body temperature/lowers fever
- is a brain hormone involved with memory, learning, and recall

VASOPRESSIN:

- is secreted by the posterior pituitary
- is widely distributed throughout the brain and body
- is a peptide generated from the medial preoptic nucleus (sex center in the brain)
- is active during REM sleep and theta activity
- is central to controlling hibernation patterns in animals
- is an antidiuretic hormone and prevents bed-wetting
- is a vasoconstrictor

AS TO SEXUAL ROLES, VASOPRESSIN:

- may be a catalyst for orgasm but inhibits lordosis in animals
- may decrease sexual receptivity in females without decreasing proceptivity
- discourages sexual extremes
- facilitates attention to sexual cues

AS TO BEHAVIOR, VASOPRESSIN:

- facilitates testosterone assertive/aggressive behavior
- discourages emotional extremes
- may improve concentration during psychotherapy by focusing attention on the "here and now"
- encourages flank marking and flank grooming in animals, a key marker for social status
- increases with arousal and returns to normal prior to ejaculation

VASOPRESSIN HAS BEEN USED TO TREAT:

- bed-wetting (Diapid nasal spray)
- the symptoms of aging (outside United States)
- male menopause (outside United States)
- sexual dysfunction (outside United States)
- diabetes insipidus—used to elevate blood pressure and retain body fluids
- memory disorders due to alcoholism
- posttraumatic amnesia, amnesia
- lost memory associated with normal aging
- depression (NIMH found some depressed patients had lower levels of vasopressin in cerebrospinal fluid)
- loss of short-term memory

HOW WE CAN INFLUENCE VASOPRESSIN:

Increases vasopressin:

- testosterone
- nicotine (cigarettes)
- estrogen
- acetylcholine
- dopamine (at DA-1 receptor)
- yohimbine

Lowers vasopressin:

- dopamine (at DA-2 receptor)
- progesterone
- serotonin
- opiates
- endorphins
- alcohol

The Testosterone Connection

While oxytocin affects women more profoundly than men, courtesy of estrogen, vasopressin is mainly a man's chemical. It depends on testosterone for its sexual effects. In animals, when testosterone levels fall, vasopressin also puts on the brakes. When testosterone is replaced, vasopressin levels return to normal.

During hamster hibernation in winter, testosterone levels fall, decreasing vasopressin activity in the area of the brain responsible for regulating body temperature. These hamsters sleep soundly with no REM dreams.

This connection to testosterone makes a lot of sense. Testosterone and vasopressin seem to support each other in regard to assertiveness, confidence, cognition, and orderly thinking. With regard to aggression, does vasopressin reinforce it or prevent its extremes? Perhaps both.

Vasopressin delays the sexual consequences of castration, so it seems to complement testosterone's sexual features, except in one respect: monogamy (remember the male vole?). If vasopressin is truly a monogamy molecule in humans, imagine this molecular tug of war—a clash of the Titans, if you please. Testosterone wants to prowl, vasopressin wants to stay home, and oxytocin just wants to hold you and not think about it anymore.

Testosterone: *I'm feeling horny. Think I'll cruise the bars and pick up some chicks.*

Vasopressin: *What's wrong with Robin? She's a wonderful woman who loves you. She's great in bed, too. What more could you ask for?*

Testosterone: *Are you kidding? Too complicated. I don't want all that responsibility. Besides, I like variety.*

Vasopressin: *Settle down. Don't be such a jerk. When are you going to realize what is really important?*

Testosterone: *Look, you're ruining all my fun and trying to run my life. Oh, I give up! I think I'll just go masturbate instead.*

Oxytocin: *Come on over here. Why be alone? Touch me, hold me, squeeze me.*

Testosterone: *Will you all just please leave me alone. I can't stand it anymore. You're crowding me beyond belief! I'm getting out of here.*

Animals have territorial scents that mark their turf—ownership chemicals called territorial pheromones. When people touch regularly, a similar ownership process may occur through scents left by touch that have subliminal effects on ourselves and others. Certainly, people who sleep together begin to smell alike—just as the sheets pick up familiar odors. Are there commitment chemicals that result from regular body contact, signaling to others that this man or woman is forbidden territory?

It is interesting to speculate about the existence and function of ownership chemicals and commitment molecules, the way they influence bonding and affect love. For example, frequent gentle caresses might promote fidelity and discourage affairs. Perhaps as a man or woman touch one another often, each is unknowingly establishing territory and ownership, leaving fingerprints subconsciously detectable by other potential suitors. Perhaps, more simply, the mutual fulfillment experienced when touch is a dependable dimension of a relationship makes someone less apt to look elsewhere.

SKIN HUNGER

Let's take a closer look at the patterns of touching throughout our sexual stages, and how they change, bearing in mind that men and women usually have quite different experiences throughout their life cycle.

Robin was an accident. Born ten years after her youngest brother, she has been the baby of the family for all her life. As the only girl, she was showered with attention by her delighted parents and four doting brothers. There were always arms waiting to hold her, rock her, pick her up when she cried. For her, touch was a way of life. Tagging along behind her brothers, she grew into a tomboy full of rough-and-tumble play. Her brothers would wrestle with her, amused by her spunk. But her dad loved it when she settled down and curled up in his lap.

When she hit puberty, however, everything changed. Not a word was said, but her brothers' roughhousing suddenly stopped and abruptly her father withdrew from his affectionate behavior. She was "too old" now, whatever that meant. The end result was acute oxytocin starvation. By the time she started dating, she had developed tremendous "skin hunger." Her parents shouldn't have been so horrified at the dregs she chose. For Robin, just about anybody would do at that point for an oxytocin fix.

The touching experience boys have up to this point is a completely different story.

Eric was touched growing up about as much as most boys. In our culture, that's not much. He was allowed to cry in his crib much more than his sister—to develop his character, and it did. As a toddler, when he fell down and hurt himself, instead of a comforting hug, he got a pep talk: "Whoops, don't cry, try again. You're a big boy, you can do it."

Those who receive a paucity of touch as children become aggressive and belligerent as adults. In this way, our boys become men. In North American culture, once a young boy reaches age four or five, most tender touch, if it was ever there, dries up, replaced almost entirely by roughhousing and contact sports. From this point forward, almost all body contact he experiences is aggressive. Out of fear of turning their sons into sissies, mothers often back off and dads keep them at arm's length, except for a slap on the back or a handshake greeting. Boys don't touch other boys and, up to a certain age, they don't want to touch girls. They grow into young men as touch deprived as orphans.

As we discussed in Chapter 2, in contrast to teenage girls, teenage boys are not accustomed (or addicted) to the pleasant sensations associated with touch and its companion, oxytocin. But their testosterone-driven goal of orgasm entails a lot of determined hands-on contact. While these boys have a more pressing objective, the touching en route may be all they get the opportunity to enjoy.

Indeed, these years of petting and foreplay, taking advantage of movie theaters, baby-sitting forays, parents' night out, back seats of cars and other dark and precariously private places usually provide—for both sexes—the most touching they will ever get from one another in their lifetimes. From this point forward, touching between men and women dwindles as the man replaces it with the specific sexual satisfaction he has been craving. Women, even though they acquiesce, are acutely aware of their increasing skin hunger. Men usually don't realize what they are missing.

Courtship

Oxytocin appears to mediate this "skin hunger." When primed with estrogen and testosterone, and working together with DHEA, oxytocin spurs us to seek and pursue a mate so we have someone to touch and who touches us in return. Clearly, the presence or absence of these intimate touching patterns has a great influence on the nature and quality of each sexual stage.

In addition to triggering an overwhelming urge to touch, it inspires grooming behavior, contributing not only to sex appeal and bonding but to pecking orders and social status. In this respect oxytocin and vasopressin work together toward the same objective, but exactly how is not clear.

One group of researchers studied the effects of vasopressin on hamsters. The rank of each individual in the social hierarchy is very important to these creatures. By marking and grooming their flanks (flank marking) they communicate their status and preserve the dominant-subordinate relationships in the group. When vasopressin is injected into certain parts of the hamster's brain, flank marking intensifies; when a substance that *blocks* vasopressin is injected, flank marking decreases.

Vasopressin may play a real but as yet undefined role in human status orders, manifesting itself in fashion consciousness: what we wear, how we dress, our jewelry, perhaps the toys we buy (in essence, how we decorate our flanks).

Seduction

In humans, smell and touch are remembered in association with thought. Oxytocin surges and spurts in response to internal thoughts, as well as external events. Through its association with smell via estrogen sensitization, oxytocin orchestrates the body's response to pheromones. What this means is that molecular memory of how someone smells or feels can be brought to the surface just by thinking about that person, recreating the experience in our mind. We can actually look forward to smelling and touching a particular person, thanks to the involvement of another area of the brain called the prefrontal cortex. We take this talent for granted, not realizing how unique we are among animals to be able to recreate and to anticipate like this. When we expect that special someone to arrive, our brain receives a charge from memory structures, telling us to be on the alert (and move toward where he or she is expected to appear), in anticipation of this pleasure. Dopamine contributes to the anticipated pursuit. Also, as you see the object of desire now with your mind's eye, PEA may kick in too, making you feel excited and nervous.

All this is to say that oxytocin is both a response to touch and a cause of it. It feels so good, you go to great effort if necessary to enjoy the experience again—over and over. Your endorphins add to the addiction, and dopamine inspires you to go after it.

Robin has been suffering in Eric's absence. Here's what happens to her, anticipating his return.

Robin is looking forward to Eric's arrival home from a business trip. She has attended to her appearance as if she were her own lady-in-waiting—she's bathed, oiled, manicured, coiffed, and just barely dressed in sheer slinky stuff. While she has been preening and anticipating, her hormones and peptides have been mobiliz-

THE ALCHEMY OF LOVE AND LUST

ing. *Many of our now familiar friends are involved—DHEA, PEA, estrogen, testosterone, dopamine, and, of course, oxytocin.*

DHEA is increasing in her brain and seeping through her skin, sending signals into the atmosphere, tuning up the nerve endings in her erogenous zones. PEA is flowing, making her feel a little giddy and high as she sees Eric's smile and feels his touch in her mind's eye. Zap! Oxytocin fires off a volley. What pleasure she is looking forward to—forgetting that he hasn't called her in two days. Pow! Dopamine pulses in her brain. She can't wait!

But wait a moment. Vasopressin cools her down. She suddenly gets side-tracked, a vasopressin moment: Why won't he marry her? Slap! Her anger pumps out some adrenaline. But then oxytocin takes over. Oh, never mind; stop thinking so much. It doesn't matter. As she puts on the finishing touch of perfume, she hears his footsteps coming up the walk. Primed for action, she goes to meet him at the door, back arched, breasts out.

Robin's brain is awash in sex soup, some would say a witch's brew. The blood in her veins is percolating, spiking and steaming. She is full of skin hunger and compulsive craving.

While Robin was getting ready, Eric was on his way home from the airport. He travels frequently on business and likes to get away. It gives him a chance to see other women without Robin's knowledge. Testosterone rules on the road. He gets to be separate, has his one-night stands, flexes his freedom. He doesn't care about these women. It's not as though he's being unfaithful. After all, he and Robin may be living together but they're not married.

As he drives toward the house they now share, he's feeling irritable and defensive. Testosterone is peaking and vasopressin is giving him some guilt. He is not looking forward to seeing Robin, expecting recriminations for his absence and neglect.

Eric walks into the house with his armor on, a little belligerent, on the defensive. He sees the preparations she has made and how inviting she looks. He wishes he had brought flowers and snaps at her instead. She doesn't bite back, surprising him by putting her arms around him saying, "You must be tired, how about a glass of wine and a back rub?" Oxytocin, conditioned by her touch, washes over him. He loosens his tie and folds himself into her familiar arms as though he were born there, forgetting his anger, suddenly glad to be home, wondering why he was so eager to leave in the first place.

Robin unbuttons Eric's shirt and rubs his neck. Oxytocin and endorphins soothe him, almost to sleep. With his testosterone levels sinking, he feels like a wet rag—poised for war, with no one to fight with.

Then Robin strokes his chest and starts kissing his chest. Hello, dopamine, testosterone, and vasopressin. She teases his nipples. Welcome oxytocin. This is good. He is waking up—below the waist—starting to stir, anticipating more.

Although female breasts have long been acknowledged as erogenous zones, nipple stimulation releases oxytocin in males as well. Remember our man with the enlarged nipples? In fact, the male nipple is much like the woman's clitoris. Its only reason for existence is sexual stimulation. Male nipples have no other function, no dual purpose or other redeeming value. So don't neglect them so!

In an inquiry to *Medical Aspects of Human Sexuality*, a sixty-nine-year-old man asked how to suppress an embarrassing urge to stimulate erection and orgasm/ejaculation through rubbing his nipples. Many men feel ashamed of their desire for nipple stimulation as though it makes them effeminate. This man was told to go ahead and enjoy it.

Masters and Johnson created a whole generation of "nipple watchers" among men who wanted to be sure that their woman wasn't faking orgasm. They popularized nipple erection prior to orgasm when they pointed out that it was a predictable visual signal of orgasm in women. However, you shouldn't disregard the fact that nipples also perk up when it's cold.

Dangerous Memories/Scary Thoughts: Once we are primed by oxytocin and its chemical assistants, it's difficult to hold on to rational thoughts. Not surprisingly, grudges or anger over events—real or imagined—can be dissolved by oxytocin-induced forgetting. The only drive we have, once in this state of selective amnesia, is to keep these good feelings and have more of them.

Eric wants her now, to kiss, to suck, to probe—testosterone and vasopressin have come back to life. He is locked on a flight path toward orgasm. Testosterone pushes him on. Serotonin drops some. Oxytocin continues to pulse with every touch and makes him forget his earlier hostility.

As if this weren't enough, oxytocin simultaneously impairs our judgment and makes us sexually receptive.

Eric sheds the rest of his clothes and slips off Robin's bikini panties, stroking and patting her bottom as he does so. She arches her back in response (estrogen, lordosis) and he grabs her hips with both hands, pulling her to him. Their temperatures rise.

Body Heat: Vasopressin is involved with body heat in relationship to sexual activity. A series of studies in rats within the past five years show that vasopressin controls sexual behavior by keeping the rats from "overheating." Getting "hot" is almost synonymous with a surge of sexuality or aggression in the vernacular. Staying hot, however, can lead to burnout. Vasopressin paces the action and avoids excesses that can lead to exhaustion.

Although Eric has been hard since Robin first stimulated his nipples, Robin hasn't taken much interest in his erection until now. Suddenly it is the focus of her full attention—her hands, her mouth, her breasts.

Eric's penis has gone from firm to swollen. The head has flared to its full size. He is on the verge of orgasm, trying to distract his mind. Slow down now. Where is vasopressin when you need it?

Erection: Basically, oxytocin gets the penis to pay attention. It is present throughout the male genital tract. Oxytocin may facilitate VIP (vasointestinal peptide, another sexual neurotransmitter) secretion in the penis, thus giving a hand to erection. Oxytocin also heightens penile sensitivity. Contraction during orgasm/ejaculation is probably dependent upon its pulsatile release, much as it is in women.

Rat studies show that oxytocin improves erection and speeds ejaculation in both normal and sluggish rats. It enhances contractions of penile tissue and increases ejaculate volume and sperm count. If oxytocin is blocked, sperm count is reduced. Oxytocin is of little help with hopelessly impotent rodents, while it does improve sexual performance in old male rats (both sexually normal and sluggish).

Oxytocin is also involved in one of the most amusing sexual phenomena of all—the penile erection and yawning reflex. A monkey yawns, stretches, and gets a huge erection. We have as yet no explanation for the connection between yawning and sexual arousal, although we have identified some of the chemicals that trigger erection and yawning. Indeed, there are some prescription drugs in use today that have the rare and unusual side effect of causing an orgasm when you yawn. Wouldn't it be marvelous if sexual arousal were as contagious as a yawn? Or if you could just yawn whenever you want an orgasm? Maybe the yawn is some primordial cue that it is time to go to bed together?

Long before science took an interest in this phenomenon, Mennonite elders, enforcing a strict moral code, forbade the children from stretching or yawning, saying it was the devil's doing. How did they know?

Consummation

Robin wants him inside her. Oxytocin and estrogen are demanding penetration. As he enters her vagina, more oxytocin pulses through her system. He thrusts in response. She paces him.

Penetration: Contact between the penis and the vagina releases oxytocin in both female and male primates. Oxytocin, when in league with estrogen, makes a woman want vaginal penetration. The instinctual drive to be filled— with a partner via his penis or a baby—derives from this dynamic duo. The estrogen connection, this "female hormone," may explain why this intense desire for contact and cuddling seems so much stronger in women than in men. The sexual pleasure associated with physical bonding, attachment, and fullness—not just mechanical stimulation—is what most women crave more often than not.

In rats, once the sex act has begun, the female controls the rhythm of thrusting. One researcher calls this control a "vaginal code." For adequate control, the sense of touch must be heightened. Oxytocin enhances spontaneous attention to such touch cues.

Orgasm: *Vasopressin cools him some but not enough. The septum in Eric's brain reverberates with DHEA, dopamine, and electrical current. Theta waves start pulsing. Heat rises. Vasopressin betrays him right about now and changes sides. Oxytocin joins forces with it and triggers a well-controlled localized seizure in the right side of his septum. He jerks, throbs, and ejaculates, whether he wants to or not.*

Both vasopressin and oxytocin are involved in orgasm. Vasopressin increases prior to ejaculation in animals and remains high throughout orgasm. This may mean that vasopressin helps prime a person for orgasm, or it could mean that vasopressin levels increase as a result of arousal. I suspect both are true.

Oxytocin is one of the relatively few chemicals known to change during orgasm. Just before orgasm and ejaculation, oxytocin spikes to levels three to five times higher than usual.

Afterglow: Oxytocin may also be involved in the inertia that typically follows orgasm in men. Perhaps that is the reason why men become so much more romantically sluggish than women, once all is said and done. Not surprisingly, this drowsiness is typical of theta brain states and also occurs during breast-feeding. Before such sedation happens, though, there is an increase in oxytocin levels in both men and women, with a sharp rise during the orgasmic crest. The sedation immediately afterward may be due to the oxytocin-filled orgasmic "flush." Women are used to having high levels of oxytocin—like chronic alcoholics. Men are not, so they get drunk on it.

Robin came close. Eric is half-asleep after ejaculating. She's not done, but can't get his attention. Hovering in the glow of oxytocin, DHEA, and vasopressin, she starts to stimulate herself by hand. He notices and responds in spite of his oxytocin coma. She comes. He enters her again. She has a few more orgasms as he's trying to inspire one of his own. She enjoys the ride. They are both on an oxytocin high. He finally ejaculates again and they collapse into each other's weary arms.

Oxytocin, released during sexual behavior, sensitizes or magnetizes a female's sense of touch. In males, such a sexual direction may be less intense, probably because they have less estrogen than females. Also, men may find the influence of oxytocin more difficult to translate into affectionate relationships since it is more foreign to them. Even the experience of oxytocin during and after orgasm/ejaculation may be undermined by men's functional notion of ejaculation as "the end" rather than part of an emotional/biological process.

INFLUENCING OXYTOCIN AND VASOPRESSIN

There are both artificial (or, at least supplemental) and natural ways to influence these hormones. Some are provocative, some just common sense.

Nasal Sprays

Oxytocin: Some European researchers are investigating an oxytocin nasal spray to treat impotence and orgasmic disorders, such as ejaculatory inhibition. And why not? Women have used an oxytocin spray called Syntocinon to stimulate breast-feeding, to initiate milk let-down, to induce labor, and to control postpartum uterine bleeding. For nursing, the dosage suggested is one spray into each nostril two to three minutes before breast contact.

Years ago, when we mentioned our interest in oxytocin to Syntocinon's local company representatives, they said nothing much was being done with it but they had heard that Las Vegas show girls were sniffing it to make their nipples stand out. In spite of this short-sighted view, it may well turn out to have some tremendous treatment potential. Could it be used to treat touch disorders? Parenting problems perhaps? Orgasmic dysfunction in both sexes? However, it is not without worrisome side effects—amnesia, psychosis, impaired judgment.

Vasopressin: A man can take this peptide by injection or via nasal spray. Imagine spraying your partner with peptides to inspire touch, sensitivity, and orgasm. Is it a useful tool for firmer erections? Durk Pearson, author and promoter of life extension, claims that vasopressin gives him great erections and orgasms. But, beyond the physical, could vasopressin be used to prevent infidelity or promiscuity? Is it of value in helping us make more sensible choices in a mate?

Natural Solutions

While we may be able to get specific clinical effects from vasopressin medications, in your natural state vasopressin is more likely to manipulate you than vice versa. Vasopressin will automatically come to the rescue during times of stress or at extremes of emotion to help you maintain your balance. How this happens, we don't know yet. Oxytocin, however, can be more easily manipulated: it is so simple—just *hold someone.*

It's Sunday morning. You sleep in late. Your toddler crawls in bed with you and cozies up. Or it's a quiet, firelit night. You are curled up with a book, your head in his lap, his arm around your shoulder. Later, going to sleep, spooning. You are promoting oxytocin, spreading it around.

You feel completely content, safe and sound, together alone. There is submission and dependence, born of trust. Allowing yourself to be held requires dropping all pretense, relaxing, and becoming vulnerable. That is why it is harder for some people and easier for others. But holding someone, as opposed to being held, is also a way of receiving pleasure yourself while being protective and comforting for them. Men often find this posture easier psychologically, and the result is just as good.

Physical intimacy helps to shield you from arguments. If you are having a fight with someone whom you usually love, the hardest thing to do is to hold them, but if you manage to hang on tight for a few moments, it helps the conflict dissolve of its own accord.

But what if you are alone? Why do you think men get so attached to their overstuffed chairs? They are being cuddled—indirectly—without having to admit it, like the octopus in its crevasse. Kids attach to blankets, dolls, and teddy bears. Women wrap themselves in fur coats, hold the telephone, pillows, and wear flannel nightgowns. Other surrogates include down comforters, cozy slippers, a cup of tea, a bubble bath. It is no wonder that warm baths are so soothing, or that we prefer to take them at "womb" temperature. If anyone were to look, they would probably find nice high oxytocin levels bubbling up. No studies have been done on oxytocin responses to inanimate objects and I don't think a teddy bear protocol is going to materialize any day soon, but I would love to know the whole story.

Pets count too. Look at their names. Their role in our lives is to be "petted"—touched.

With a little thought and not much effort, you can experience more tenderness, closeness, and intimacy than ever before.

Touch is calming, reassuring, and relaxing. Men and women react differently to the absence of touch. Depression is the most prominent consequence for the woman, who often becomes aversive to sexual touch, while aggression is the most typical consequence for the man, who often becomes aversive to nonsexual touch. If sexual problems stop a couple from touching, symptoms of touch deprivation begin to surface—depression, irritability, increased pulse rate, blood pressure, and other physical malaise.

Other upheavals such as midlife crisis and male menopause can be aggravated or precipitated by unhealthy touching patterns, although the connection is usually not appreciated.

Yet in spite of your upbringing or life experiences, introducing touch to your life as an adult can transform you. It can neutralize anger and depression, fostering contentment and joy. In fact, the association between touch and healing is ancient and worldwide. The "laying on of hands" is a traditional part of medicine. It is also a central feature of mystical healing from the miraculous touch of Jesus to the modern-day faith healer.

Holding close, hugging, snuggling, petting, stroking, touching—it's good for your health, your heart, and your relationships. It's habit forming. It seems a terrible shame that such a wonderful resource is often limited to times of grief and sexual encounters, when it can do so much to improve the quality of life when enjoyed on a daily basis.

In the final analysis, touch is essential to our survival as individuals, and through sex, as a species. Touch is free, and readily available; it can soothe you better than a Valium or a drink, and lower your blood pressure as well as a diuretic—with no adverse side effects. In fact, those who drink alcohol to relax and loosen their inhibitions will be surprised to find out that alcohol actually lowers oxytocin (which also makes it less effective than you would think for seduction purposes).

Touch is your most precious and powerful resource. Don't underestimate it, and make sure to use it to your full advantage.

5.

$\mathcal{T}$HE AGGRESSIVE SEX DRIVE

The twenty-five-foot fiberglass Bertram yacht was a luxury cruiser on a serious mission. Fully equipped for fishing, it was headed for the Coronado Islands to catch the first yellowtail run of the season. Harry was taking the boat out alone, moving slowly but deliberately through the San Diego channel just as dawn was breaking.

He loved the solitude, communing with nature, the danger, the hunt. There was always a rolling surge in the narrow channel, but he was looking forward to skimming the morning glass once he reached the open sea.

As he passed the landmark high-rise hotel, he was riveted by the sight of a tall woman standing stark naked in her room, just barely in view, in the throes of sexual ecstasy. He put the boat on autopilot and grabbed his binoculars for a better look, moving toward the stern to keep her in his line of vision.

She was masturbating against the edge of a door, arching her back as she rubbed herself, her gyrations becoming more intense. He was mesmerized, forgetting where he was, what he was doing. He could almost reach out and touch her through the magnified lens, captivated by the idea of watching without her knowing. He was so immersed in the experience, leaning over the transom for a last look, he tumbled into the water.

The cold shock brought Harry back to his senses and the full impact of his predicament hit him: his transportation was moving on without him, taking a small fortune with it. The bank would own his hide, if he could only save it. It was all he could do to keep his head above water in the swells, watching his boat disappear in the distance. Every time he moved, he got this horrible pain in his chest, and his left arm had gone completely numb. How on earth had he gotten himself into this mess?

If someone didn't come along soon, his boat wouldn't be all he lost. And for what?

Although he didn't know it yet, it turned out to be his lucky day. A trawler came by and dragged him out of the deep like a big tired fish. This rusty old workhorse of a boat caught up with his sleek toy, and one of the crew jumped aboard to get it under control. Harry got away with three cracked ribs, a broken arm, and a bruised ego.

"What happened to you?" his friends asked.

"Well, you won't believe this but . . ."

This true story was told to me by an admittedly embarrassed friend, whose name I changed to protect what little dignity he had left. What on earth had come over him, you might reasonably ask?

Testosterone! It had them both under its spell. Governed by this hormone, she wanted sex—not intercourse, but masturbation. His brain, awash in testosterone, could no longer think straight. Swept away by sexual reflex, he risked life and limb without a second thought.

Testosterone is supposedly the best-known sex hormone of all, but in this chapter I am going to break many conventional assumptions, and expose you to new facets of its character. I will introduce you to testosterone's influence on both sexes—its effect on mood, personality, power, aggression, and sex. We'll also take a look at the use of testosterone to treat certain sexual disorders in women, a controversial practice. And, naturally, we will investigate the role of testosterone use and abuse in men. Without it, women wouldn't like them much. With it we sometimes cannot stand them!

Testosterone cycles daily and seasonally. We'll see how these cycles influence the way men and women get along socially and sexually. In addition to its role in lust, fantasy, masturbation, and intercourse, we'll look at the role testosterone plays in social status, courtship, date rape, affairs, domestic violence, criminal behavior, perversions, competition, sports, commitment, divorce, lust, fantasy, masturbation, and intercourse.

Can we control testosterone or does it control us? Are women getting their fair share? Can they get even more—and if so, is this desirable?

TESTOSTERONE PROFILE

MOST PEOPLE DON'T KNOW THAT TESTOSTERONE:

- is a steroid hormone, regulated by LHRH, manufactured in the testicles, ovaries, and adrenals
- can transform into estrogen
- fluctuates daily and seasonally
- improves cognition
- inhibits serotonin, opioids, prolactin, and MAO
- facilitates dopamine, adrenaline, and vasopressin
- cycles every fifteen to twenty minutes

TESTOSTERONE:

- determines the masculine characteristics in males
- produces and maintains sperm
- works as an antidepressant in both sexes
- increases ratio of lean muscle mass to body fat
- is pulsatile

AS TO SEXUAL ROLES, TESTOSTERONE:

- increases sexual thoughts and fantasies
- responds to novelty, inspires one-night stands and affairs
- increases aggressive sex drive in both men and women, but—
- doesn't have a strong effect on erection except indirectly by increasing desire
- increases the urge to masturbate rather than the desire for intercourse

AS TO BEHAVIOR, TESTOSTERONE:

- is activating
- maintains separateness and promotes aggression
- increases assertiveness and self-confidence

- has been implicated as a cause of certain types of criminal behavior and domestic violence
- can trigger or contribute to psychotic behavior
- rises in response to winning, social status, and pecking orders
- is higher than usual in career women

TOTAL TESTOSTERONE CAN BE DIVIDED INTO:
- free testosterone (metabolically active)
- protein-bound testosterone (metabolically inactive)

TESTOSTERONE HAS BEEN USED TO TREAT:
- low sex drive disorders in men
- low sex drive disorders in women, primarily in postmenopausal women but also less commonly, premenopausal women
- erectile dysfunction in men
- hypogonadism
- menopausal women in hormone replacement therapy (HRT)
- menopausal men in hormone replacement therapy (HRT) (outside United States)

HOW WE CAN INFLUENCE TESTOSTERONE:

Increases testosterone:

- winning competitions/ arguments/battles
- sexual thoughts, activities
- diet containing meat
- exercise

Lowers testosterone:

- losing competitions
- vegetarianism
- progesterone

MALENESS

Testosterone is a predominantly male sex hormone (an androgen) that women have too, although in much smaller amounts. In fact, men have about twenty to forty times more of it than women, which is one reason why our sex drives are so different. This forceful hormone is responsible for the "active" libido—the drive associated with sexual appetite, attention, motivation, and action. It is less involved in sexual responsiveness than we originally thought, and more involved in patterns of aggression than we previously realized.

While testosterone does increase your sex drive, it fosters masturbation over intercourse. Men often feel guilty or immature that they continue to masturbate as adults, even after marriage with a willing, appealing woman available, but adult masturbation is not just a persistent adolescent pattern. It's a forceful impulse from a natural hormone. Almost everybody masturbates and almost everyone feels bad about it. Catholics do it, and then confess to God about it. Orthodox Jews become bed masturbators. They turn on their stomachs and thrust against a pillow or mattress without using their hands. If they don't actually touch themselves, how can it be their fault? Evangelists get prostitutes to lend them a hand, and southern Baptists don't masturbate at all. They get possessed by the devil, who does it for them. It isn't Satan. It's just testosterone.

Testosterone is also extremely sensitive to its environment—not only to internal physical changes, but to outside forces. Sex, a confrontation, competition, stress, and the presence of other men or women are just a few of the externals that whip testosterone around. Even the rhythms of day and night and the changing of the seasons affect its availability and activity.

Testosterone is constantly in flux, characterized by its cycles and peaks. Under normal conditions, it is higher in the morning, lower at night, vacillating up and down in between. It also peaks and cycles over the course of a lifetime, first peaking prenatally in males, then at puberty for both sexes. It then gradually diminishes over the decades, reaching its lowest ebbs in the eighties and beyond. Thus far, most clinicians do not consider these dwindling testosterone levels with age of much importance. I am inclined to disagree and suspect you will too, once you have an opportunity to read what I have to say.

A minor amount of testosterone is produced in the adrenal glands of both sexes, and in women, in the ovaries. But the main factories, the twin power plants that manufacture masculinity, are the testicles.

As the chief testicular stimulant, testosterone makes and pampers sperm. Inside each of the testes of a mature male are about seven million Leydig cells, taking up 5 to 12 percent of the available space. If anyone wants to protest the things men do, that's where they should start, for it's in the Leydig cells that cholesterol is converted to pregnenolone and ultimately to testosterone.

Boys Would Be Girls

Were it not for the Y chromosome instructing testosterone to act at critical moments, all boys would be girls. In fact, the female is the universal sex. Men are an afterthought.

Here's how that happens. Testosterone works its wonders in the uterus, transforming the female embryo into a male with all his ornaments. It usually does a nice job, crafting the penis and its neighbors, the scrotum and testicles, along with the requisite masculine body contouring. Having done its mischief, testosterone then takes a sabbatical.

We do not see much action from this sexist hormone until it has an encore at puberty when adolescent boys are overcome by "testosterone toxicity," and it seems as if all their other hormones have gone fishing. Teenage boys become walking grenades, just waiting for someone to pull their pin.

As production kicks into high gear, the psychological and physical impact of testosterone is overwhelming. More than any other substance, testosterone controls the development and maintenance of masculine characteristics. Facial hair sprouts, competing with crops of acne. The voice cracks and deepens. Shoulders broaden, hips narrow. Muscles become lean and powerful. Body hair and body odor make fine companions. Sperm gets produced and wants release, often.

At about this same time, testosterone takes over a boy's brain. It becomes a live-in tyrant leading him, in general, to behave as though he were thinking with his other head. It has also been known to impair judgment, better than almost any other mind-altering drug.

Once a man survives puberty and enters adulthood, his testosterone levels off, dropping slightly in the thirties and maintaining itself well until his fifties or sixties, after which there is a gradual decline. This decline is truly minimal compared to what most people anticipate, and cannot completely account for the diminished sexual drive and function common to so many men as they age. However, testosterone is not the only hormone affecting sex drive. Others play a role. (More about them later.)

For the most part, sex drive in the young adult male is pretty straightforward and follows a fairly simple repetitive pattern. Sex is on the mind. If single, he cruises and perhaps carouses. He finds a target, initiates the courtship ritual, goes to absurd lengths to have his way, and then falls asleep, preferably alone. Sound familiar? (The married man does almost the same thing, usually with the same partner.)

Not for Men Only

Contrary to what was once believed, testosterone is not for "men only." A woman first encounters its power at puberty when the ovaries start to produce increased amounts of testosterone. Although it is only about 10 percent of the amount circulating through teenage boys, it is this testosterone, not estrogen, that causes the heightened erotic sensitivity of the clitoris, breasts, and nipples. It maintains the fullness, thickness, and health of her genital tissue as well. Testosterone also accounts for her newfound romantic interest in boys.

Thanks to testosterone, a woman will go after a man. She becomes more sexual, more responsive and more assertive. She chooses, instead of being chosen. As she matures, testosterone turns her interest to orgasm—gets her enjoying her genitals as well as her fantasies. Oddly enough, testosterone increases don't necessarily influence the frequency of intercourse. But to the extent it appears to increase the frequency of masturbation, testosterone's masculine effect of "separateness" may play a role even in a woman.

Testosterone can be a source of confusion for both sexes. It compels toward masturbation on the one hand and boosts the aggressive, proactive sex drive on the other, putting both sexes on the prowl. Sometimes one compulsion wins out, sometimes the other, depending on what other hormones are at play and who, if anyone, is available. It increases desire yet decreases tactile sensitivity especially of the penis (vibrotactile sense). It causes irritability and aggression yet is an antidepressant.

In women in particular, it serves as an antidepressant, increases assertiveness, and improves her self-confidence. But when it interacts with progesterone, it also contributes to a woman's irritability and aggressive behavior (more about that later). Interestingly, increased testosterone *in the male* increases the attractiveness of male odors to the *woman*.

Different Drives

The biochemical urge we call the sex drive comes in two basic styles: aggressive and receptive. The aggressive sex drive is controlled not just by testosterone, as most people think, but by vasopressin, DHEA, serotonin, dopamine, and LHRH as well. The receptive sex drive—which we will investigate more thoroughly in the next chapter—has been overlooked altogether in humans by researchers and therapists alike. Receptive doesn't necessarily mean passive, but that is at one end of the spectrum: passive means available, and perhaps willing, but without the initiative to pursue sex. At the other extreme of the receptive sex drive is a proceptive stance. This is not aggressive but assertive: available, willing, and interested, but not dominating. By contrast, the aggressive or active drive means interested and motivated to pursue sex, often against all odds. One is anticipatory and willing, the other is motivational and aggressive. And while estrogen governs the former, testosterone the latter, neither one works alone.

It follows that women are generally more receptive, men more aggressive —although the tables can turn, particularly at certain stages in our lives. At any rate, the dichotomy of feminine and masculine is so fundamental, and so necessary for the continuum of life, that even unisexual species conform to it: when an animal is unclear about its sexual orientation, and a few are, active-passive sexual behavior takes place. For instance, a unisexual lizard leads up to sex by acting out the role of both sexes—alternatingly behaving as a male and female. We do not know why this "bisexual" split is necessary to their courting, but evolution has seen fit to preserve it.

LHRH PROFILE

MOST PEOPLE DON'T KNOW THAT LHRH:

- can be used both as a contraceptive and to treat infertility
- has paradoxical effects depending on whether administered episodically or continually

LHRH:

- is pulsatile (one- to two-hour cycle)
- triggers production of testosterone
- regulates production and release of LH (luteinizing hormone) and FSH (follicle-stimulating hormone)
- regulates the relationship between testosterone and estrogen
- is synthesized in the hypothalamus

AS TO SEXUAL ROLES, LHRH:

- may mildly increase erections
- increases lordosis
- may mildly increase desire

AS TO BEHAVIOR, LHRH:

- is hypersensitive to stress
- fluctuates dramatically in response to environmental, visual, emotional, or sexual cues

LHRH HAS BEEN USED:

- as a male contraceptive
- to treat low libido
- to treat impotence
- to treat sex offenders and certain other sexual behaviors through reversible chemical castration
- to treat prostate cancer and breast cancer
- to treat low testosterone and anovulation
- to treat endometriosis

- to treat uterine fibroids
- to treat infertility
- to treat precocious puberty
- to treat PMS

HOW WE CAN INFLUENCE LHRH:

Increases LHRH:

- low testosterone
- low estrogen
- appearance of attractive woman
- watching sexually explicit videotapes and perhaps other erotic material
- vigorous exercise

Decreases LHRH:

- stress
- danger

LHRH: THE REGULATOR

Testosterone is not allowed to run amok without some necessary regulatory controls. Within the body, testosterone levels are regulated by the feedback loop between the gonads and the brain. When everything is working properly, a decrease in testosterone will stimulate another hormone, LHRH-LH, which in turn inspires the testicles to produce more testosterone. When testosterone levels reach an acceptable high, the brain receives a message to stop LHRH-LH secretion. This continuous feedback marvel ensures that the male "will always be ready" when the woman is available.

Pulsing and Throbbing

Human LHRH secretion cycles approximately every ninety minutes. As the central mechanism in the feedback process, it not only affects but is affected by the production of testosterone. Too much, too little, or inappropriately timed testosterone secretion disrupts the LHRH cycle. In turn, if LHRH is poured nonstop into the human bloodstream, testosterone will actually decline toward castrate levels. Just like the feedback loop with testosterone, LHRH (through its effect on LH) influences the ovaries' production of estrogen. LH secretion, a messenger triggered by LHRH, increases estrogen. Elevated levels of circulating estrogen shut down LH and LHRH production.

In these cases, as with various other hormones, total hormone levels are less important than a change in levels: how often and how much hormone is released and the ratio of one hormone to another. Indeed, momentum, rhythm, and crescendo typifies sexual function. LHRH is a prime example, since it normally behaves with a pulsatile rhythm, much like a flashing light.

LHRH was among the first of the sexual hormones to be appreciated for its pulsatile, rhythmic release nature. Until this chemical cycling was understood, it was impossible to manipulate with LHRH in a meaningful fashion: the effect of giving a constant dose of the hormone is quite different than when it is delivered in a natural pulsatile fashion. Same drug, same dose; different delivery, different result. We are finding the same to be true for many other hormones.

Intermittent peak intensities—such as the LHRH pulses—most effectively charge up your sex life. This is because sexual feedback requires sharp, clear signals generated by predetermined hormone rhythms. These sexual rhythms are very sensitive to stimulation from the environment and are also disrupted by emotional upset and distress.

The LHRH-testosterone production circuit in men is particularly dependent upon external cues. For example, the appearance of an attractive woman startles a male into releasing LHRH, while a threatening intruder can short-circuit his current.

Women's cycles can become irregular due to emotional difficulties, which interfere with LHRH and other critical chemical rhythms. One woman I know who hates to travel gets her period predictably as clockwork on the day she leaves, no matter what day of her otherwise normal cycle she is on. The stress of packing, planning, logistics, decisions, and the conviction that no matter how hard she works she will forget something crucial triggers her to bleed out of sequence.

We can't explain the exact series of chemical steps that lead to this woman's pattern, but it has occurred consistently for too many years to attribute to coincidence. What we can conclude is that for both men and women, our delicate sexual and reproductive circuit depends upon how well we are coping with the outside world.

Sexually Sluggish Stallions

As a companion to testosterone, LHRH is another natural sexual stimulant. Impressive animal research in the 1970s and early 1980s demonstrated that LHRH reliably increases mating and sexual behavior in frogs, lizards, pigeons, mice, rats, horses, monkeys, and man. The fact that it increases both male copulation and female lordosis (sexual presenting behavior—arched back, flipped tail) suggests that it might be a universal sexual stimulant for both men and women.

Sexually sluggish stallions can be stimulated by an LHRH injection, resulting in intense mating behavior. They manifest their interest by spending more time near mares sniffing, licking, nuzzling, and nipping. Stallions injected with LHRH show frequent flehmen responses. (Flehmen responses are lip curls used to channel odors while sniffing urine, feces, or vaginal secretions.) When a stallion does that, he sucks pheromones up his nostrils. Even castrated stallions (geldings) will show flehmen responses in the presence of estrus females if they (the boys) are given testosterone and LHRH together. It won't happen with LHRH alone.

The DHEA connection is evidenced by the horses' flehmen response, triggering that "feeling in the air" that sets everything else in motion. We can speculate that the same occurs with men and women. The production and detection of pheromones provoke feelings of desire and lust. With a look or glance, PEA gives rise to a flutter, giddiness, limerance. Testosterone teams up with DHEA to translate desire into action. The pursuit of sex escalates until the drive to orgasm becomes so compelling it competes with good judgment and common sense.

SEROTONIN PROFILE

MOST PEOPLE DON'T KNOW THAT SEROTONIN:

- is more abundant and influential in women
- can decrease anxiousness and aggressiveness
- can facilitate calm, warm sociability that promotes continued intimacy short of sexual peaks

SEROTONIN:

- is primarily inhibiting
- has a preferential relationship with estrogen
- facilitates opioids and progesterone
- increases prolactin

AS TO SEXUAL ROLES, SEROTONIN:

- causes physical sexual excitation
- inhibits sex drive
- inhibits orgasm in both sexes
- blunts impulsive pleasurable arousal

AS TO BEHAVIOR, SEROTONIN:

- promotes contentment
- has paradoxical unpredictable effects (i.e., can cause spontaneous orgasm)
- causes craving for sweets and carbohydrates

TOTAL SEROTONIN CAN BE DIVIDED INTO:

- multiple receptors with different features and functions

SEROTONIN HAS BEEN USED TO TREAT:

- depression
- obsessive-compulsive disorder
- panic

- anxiety
- phobia
- PMS

HOW WE CAN INFLUENCE SEROTONIN:

Increases serotonin:

- Prozac and other serotonin uptake inhibitors (5HT-UIs)
- obesity
- castration

Lowers serotonin:

- dieting
- PEA
- phenylalanine
- lysine
- PCPA

SEROTONIN: THE MODULATOR

In regard to influencing sexual desire, in addition to pheromones (and other DHEA effects), testosterone, and LHRH, we must consider two additional forces: decreasing serotonin and/or increasing dopamine (which you will hear more about shortly). Both dynamics increase desire in different ways.

Researchers have linked serotonin to our hunger for sweets, appetite for food, eating disorders, aggression, shifts in mood, sexual aberrations, and even arson. Let's focus here on the sexual and aggressive aspects of this most intriguing character.

The frequent, predictable fluctuations in testosterone levels throughout the day, along with the unmapped spikes responding to triggers like sex and aggression, are in marked contrast to serotonin cycles. Serotonin, which mutes sex when high and intensifies it when low, fires at a slow, regular beat, not responsive to the swift changes in excitement characteristic of intense sexual activity.

There is good evidence that serotonin and testosterone are like a seesaw, with "serotonin down, testosterone up" being a natural state for men and the exact opposite, "serotonin up, testosterone down," being the natural state for women. Apparently this inborn formula depresses sexuality and aggression in women, in contrast to men. When either of these ratios is altered in one direction or the other, the result is comparable degrees of hypersexuality or hyposexuality.

With higher serotonin levels, the initial relaxation and decrease in defensiveness may initially improve sexual compliance, desire, and responsiveness by taking away anxiety and resistance. However, once that elevated serotonin set point becomes the established norm, sexual excitement and responsiveness will decrease. This is just what happens with drugs like Prozac, Zoloft, and their relatives. These drugs are serotonin boosters. They depress sexual desire and inhibit orgasm in both sexes so successfully (sometimes to the point that orgasm is impossible) that Prozac is commonly used in the treatment of premature ejaculation.

Conversely, it stands to reason then that women with lower serotonin levels would be more readily arousable and easily orgasmic. This seems to be so. Not only that, they take more initiative. In fact, when serotonin is lowered in female rats, not only do they mount other females and smaller males, they behave more like males. Some females even show thrusting body jerks similar to male ejaculatory behavior. These patterns are eliminated when serotonin levels are brought back to normal.

Meddling with the serotonin levels of women pharmacologically raises the concern that one could precipitate certain sexual perversions and dysfunctions ordinarily found chiefly in men. For example, women who chronically abuse amphetamines (which lower serotonin) have been shown to have an increase in promiscuity, compulsive masturbation, prostitution, and intensification of sadomasochistic fantasies. Milder forms of this type of sexual behavior could be expected with chronic abuse of over-the-counter diet pills, most of which are distant cousins of PEA.

Regardless of the cause, low levels of serotonin, however stable, lead to the most unstable, and even dangerous, behavior in men. In fact, one prominent feature of low serotonin is to decrease impulse control, increasing sex drive to a point where it can be abnormally aggressive. When mixed with alcohol abuse, a condition called "intermittent explosive disorder" can erupt. As with women, with or without alcohol, their masculine sexual behavior is reinforced, but in a manner that exaggerates and perverts it.

PCPA (parachlorophenylalanine): The hardened criminal

PCPA is a drug that reduces serotonin dramatically. It isn't used in humans because the side effects are too toxic, but we have learned enormously from observing its influence on animals. These studies have important implications for human sexual response.

In animals, serotonin depletion causes violent, perverse, aggressive, and homicidal sexual behavior. Animals will torture, wound, kill, even devour their mate during sex. They will cross species and gender barriers, becoming sexually indiscriminate and will even mount dead animals. What does this imply about the biochemical nature of violent sex offenders? Could there be certain biologically governed neurotransmitter disorders controlling their behavior that are correctable with medication that increases serotonin?

If research were able to identify serotonin deficiencies associated with certain antisocial human behaviors, it would not be beyond the scope of our experience to diagnose and to treat them. Psychopharmacology has done just that with regard to depression, anxiety, and schizophrenia—identifying missing chemicals and replacing them. In fact, serotonin-enhancing drugs, such as Prozac, are some of our most promising resources in this respect.

Could it be that certain violent criminals, including spousal abusers, would benefit from the use of Prozac and like drugs? I think so, but the concept is controversial, raising many issues, such as should treatment be voluntary or court imposed? How, in practical terms, can someone who is not incarcerated be persuaded to take their medication reliably? Are these behavior-altering drugs somehow a violation of human rights? Or should it be a human right to have access to such treatment? The issue today is not *if* we can modify violent, antisocial behavior, but should we, and where do we draw the line?

THE PLEASURE PRINCIPLE

Thus far, the chemicals you have been introduced to that play a role in aggressive sex drive have predominantly compulsive features, with testosterone compelling masturbation, variety, competitiveness, and serotonin depletion dictating frantic, indiscriminate, aggressive acts. While gratifying powerful impulses and needs, in neither case is there a sense of joy, delight, fulfillment.

Dopamine, however, is what makes you smile. The aggressive sex drive could not be complete without it. It does not provoke aggression or other nasty behavior. Rather, it is inspirational, motivational, and anticipatory, so it mobilizes you to recognize, seek, and pursue pleasure. In this way it contributes both to the aggressive sex drive, and to the enjoyment of the experience. Dopamine drives libido up and makes you *want the pleasure* of sex, urgently. Without it, sex would be just another bodily function.

Like oxytocin, dopamine fires in relatively rapid bursts. These chemicals pace one another by mutual reinforcement. When dopamine increases, oxytocin recedes, and vice versa. Together, you get the pleasure of dopamine along with the intimacy, comfort, and security of oxytocin. Both facilitate your orgasms.

Dopamine takes you on an emotional high. Whether it keeps you there or drops you, it is especially desirable, and particularly addictive. Given that dopamine is the common denominator in most human addictions from drug abuse to hypersexuality, it most probably plays a major role in healthy human attachments as well—and after oxytocin, DHEA, and PEA have had their way with you, dopamine addicts you to this person you have come to love, helping you to weather the tough times that come. This addiction may also be one of the reasons that losing that love triggers such a painful withdrawal process—you *were* actually addicted: not to a drug, but to a person in the natural evolution of a powerful relationship.

By contrast, there is a concept of "love blindness" that is probably related to the absence of a dopamine response, and perhaps a glitch in the natural endorphin system. These people cannot feel love or become attached to another as much as they might wish to do so. It is a bit like being color blind and wishing to appreciate a painting. We don't know with certainty that low levels of dopamine cause this effect, but appreciating how these hormonal systems can malfunction helps us to better understand how they work under normal circumstances.

And, as always, these molecules work together, interfacing in complex yet sometimes quite predictable ways. Dopamine is increased by a testosterone injection, by the presence of a sexually receptive female, or by sexual

DOPAMINE PROFILE

MOST PEOPLE DON'T KNOW THAT DOPAMINE:

- is the key common denominator in almost every addiction from cigarettes to cocaine
- induces spontaneous activity and movement

DOPAMINE:

- is a neurotransmitter produced by specialized nerve cells located in the arcuate nucleus of the brain
- is pulsatile, with rapid-burst firing (like oxytocin). When dopamine spikes, oxytocin decreases, and vice versa
- is inspirational, motivational, and anticipatory
- mobilizes you to recognize, seek, and pursue pleasure

TOTAL DOPAMINE CAN BE DIVIDED INTO:

- multiple dopamine receptors

AS TO SEXUAL ROLES, DOPAMINE:

- increases sex drive in both men and women
- facilitates orgasm
- can cause premature ejaculation
- is perhaps the "missing link" in the treatment of sexual aversion

AS TO BEHAVIOR, DOPAMINE:

- causes us to perceive and pursue pleasure
- decreases craving for alcohol and other abused substances. When dopamine is given medically, the search for a dopamine high from illegal drugs is diminished (analogous to cigarette smoking/nicotine patch)
- improves alertness and energy
- can cause schizophrenia

DOPAMINE HAS BEEN USED TO TREAT:

- Parkinson's disease
- addictions

- sex drive disorders
- aging (outside United States)

HOW WE CAN INFLUENCE DOPAMINE:

Increases dopamine:

- schizophrenia
- cocaine
- L-dopa
- Eldepryl
- Wellbutrin
- testosterone
- sexually receptive women
- sexual activity

Decreases dopamine:

- antipsychotics
- Parkinson's disease
- castration

activity itself—at least in the case of animals. Castration reduces dopamine and increases serotonin.

High dopamine levels are not always advantageous. As with most other peptides, hormones, and neurotransmitters, their effects can be a mixed blessing, even problematic. One undesirable effect of high levels of dopamine is its tendency to cause or aggravate premature ejaculation. Another problem is dopamine's involvement in schizophrenia, as reflected by the fact that most of the medications used to treat these psychoses are antidopaminergic (dopamine lowering). That may be the main reason why schizophrenic patients so often reject their drug treatment. They must miss the dopamine high.

That's one reason why the dependence or addiction of women to over-the-counter diet pills, amphetamines, or prescription diet pills can be so dangerous. Some may be pushed into a psychotic breakdown. And when you consider my seemingly lighthearted comments about PEA (a dopamine stimulant), love, and madness, it is probably not far from the truth. PEA raises dopamine; high dopamine levels have long been associated with schizophrenia (madness).

Electrochemical reinforcement of dopamine in the brain is essential for the pursuit of happiness. If these reinforcement areas do not charge when they are supposed to, behavior becomes subdued and listless. Sexual overtures and initiatives from your partner will not arouse you. Your favorite dessert or the chance to go out and run and play will also fail to inspire you. When you actually do participate in sex, dessert, or play, you may enjoy yourself, *but only when you are directly involved.* Such a lack of anticipatory or reflective pleasure may underlie bulimic or addictive behaviors, where one is only satisfied while directly involved in the activity or substance. Sexually, the absence of dopamine is especially disruptive to anticipations and sex drive. It is especially characteristic of a problem called sexual aversion: during sex, you want it, like it, respond, and perhaps have orgasms. Afterward, you think, *That was so nice, I wonder why I don't do it more often.* And then you continue to avoid it. No anticipation of pleasure in between—only apprehension.

Jonathan used to pursue sex quite often with numerous different women. After his marriage, his frequency and enjoyment remained high and sustained for years. In spite of the prevailing notion that the chemistry eventually disappears, he and his wife, Amy, rarely had sex less than once or twice a day whenever they were together.

About twelve years into the marriage—when Jonathan was in his midforties —Amy became concerned because there had been a gradual, but dramatic change. They rarely ever had sex anymore. It had dwindled to once or twice a month. Even when she tried to entice him, he was lukewarm about it. When he did cooperate, albeit reluctantly, he had no trouble functioning and the erotic quality of their encounters was much as it used to be.

A year after this distressing development, Amy contacted me for help, and I had the opportunity to speak with both of them. Their relationship was strong in every other way, the love intact. They had ordinary disagreements, but were more compatible with and considerate of one another than most couples. I could find no significant subterranean resentments or hostilities to account for his sexual withdrawal.

As with many men experiencing this pattern, Jonathan commented to me after our first session that he was shocked and surprised at himself. He loved sex, had always enjoyed it, and still did. But he hadn't been unhappy without it—hadn't even noticed it disappear or appreciated how much he had been depriving himself. He wondered how on earth he could have allowed himself to get into this predicament and become so complacent.

This is the kind of apathetic behavior that diminished dopamine can produce. A strong dopamine response not only ensures anticipation, and attention to pleasurable experiences, which Jonathan was missing, but supplies the motivation to pursue them. It actually sets you into motion.

The way dopamine motorizes us can be seen in reverse with patients who have Parkinson's disease. Because the dopamine receptors in their brains have become defective, it is as if they had too little of it. Lack of dopamine is what causes these individuals to walk with the hesitant shuffling gait they develop. Their locomotor system is impaired. In extreme cases of dopamine disorder, they become inert and exhibit an almost stuporous state of suspended animation, as you might have seen in the movie *Awakenings*. Replacing their dopamine does enable these people to function smoothly again. The results can be as dramatic as replacing a battery.

WHAT YOUR DOCTOR HASN'T TOLD YOU ABOUT TESTOSTERONE

Testosterone comes in two forms: attached (bound) and unattached (free). Bound testosterone is free testosterone trapped on a protein molecule (either albumin or globulin)—it is in prison, so to speak, not active or metabolically effective until it escapes and becomes a free, circulating molecule of testosterone. A free testosterone surge causes a man to go on the prowl, get aggressive, compete for females. He is potentially belligerent, triggered to fight another suitor, or to take what is not freely given.

The sum of bound and free testosterone is total testosterone. Until recent years, whenever a man worried about his sex drive and went to his doctor for a blood test, all the doctor measured was total testosterone, a value without much value. A free testosterone count must be specifically taken because it is the most important component from the sexual point of view. It is also usually the smallest component—approximately 5 percent of the total. But in this case, a little goes a long way. Free testosterone is like a grenade with the pin pulled out. Bound testosterone is the grenade with the pin in place. Total testosterone is the combination of, say, ninety-five inactive grenades and four or five with the pins pulled. As you can see, even though there are fewer active grenades, they are the only ones with high impact.

If a man has total testosterone levels above normal with below-normal free, active testosterone available, his sex drive is basically napping. Consequently, the high results reflected by total testosterone measurements could be (and often are) misleading. He is sexually sluggish if his free testosterone is below normal no matter what his total values show. Today only some doctors know to order free testosterone tests, so you need to be prepared to request it specifically.

In addition, research performed in 1991 suggests that albumin-bound testosterone may have some metabolic activity of its own. If this turns out to be true, our evaluation of testosterone levels will have to be completely revised.

Multiple Morning Samplings Needed

The normal range of total testosterone in a man's blood varies between 250 and 1,200 ng/dl (that's nanograms). This a very broad range of "normal."

With a normal range this wide, a man might wonder, why bother to get tested at all. The first and most important reason is to check on how much free testosterone he has (normal range: 1.0–5.0 ng/dl, or a high of 41.0 pg/ml in the 20's to a low of 9.0 in the 80's), regardless what the total value is to make sure he has an adequate amount for normal sexual response. The second is to check his total testosterone to make sure that he has enough. If it is low normal (below 400 ng/dl), it bears watching. If it is below normal (below 250 ng/dl), it usually requires replacement therapy with some form of testosterone. High normal values of total testosterone are nothing to crow about, since the main character is the free molecule. Levels above normal have no known significance unless they result from overly aggressive replacement therapy, in which case, the dosage should be corrected accordingly.

The cycling of testosterone makes it difficult to get an accurate measurement from just one blood sample. The pulsatile secretion of testosterone (as well as seasonal variations) is caused by the fluctuations in LHRH secretion from the brain. Consequently, within any fifteen-minute stretch, testosterone levels will vary dramatically; they could be 100 ng or more apart with a daily variation as great as 50 percent. Therefore, to get a correct reading, the blood must be drawn three different times, at fifteen- to twenty-minute intervals, *in the morning* when levels are highest. Free and bound testosterone levels are recorded for each of the three samples. These three results are then averaged to provide a meaningful reading.

Testosterone fluctuates in both sexes but because women have so much less to begin with, their range of variation is much narrower than the male. As a result, its influence on feelings and behavior is usually more subtle. Consequently, most women do not need to consider having these blood tests unless they are disturbingly hairy, past menopause, or experiencing low sex drive.

Since testosterone levels are routinely higher in the morning, it may be one reason many men feel more vigorous and virile at dawn than at dusk, especially as they age. When well rested, with a morning erection in the vicinity and a naked woman at arm's length, nature obliges.

What About Erection?

How do these numbers affect a man's ability to have an erection?

An erection results from a hydrodynamic process that depends upon the flow of blood to and from the penis. To sustain an erection, more blood must enter than leave, or the blood must be trapped there for a while. This process is mediated chiefly by certain peptides and neurotransmitters. It is somewhat dependent upon testosterone, but not as much as formerly believed. We know this not only from scientific research but from ancient tales of eunuchs. Those castrated men were employed as harem attendants because, lacking sperm (and testosterone), they could not reproduce, making sure that the gene pool was safe for the sultan's seed. While the majority of castrates lost both their drive and their erections, a small percentage still found their plumbing could function fine, although their drive was almost gone. We have concluded that very low levels of testosterone can steal away both your drive and your erections, but testosterone seems to have much more impact on desire than function. It fuels our libido, so to speak.

AGGRESSION: LOVE AND WAR

A friend told me this story of a high-profile, fifty-year-old aviation executive in charge of multimillion-dollar contracts and important worldwide government connections. Here was a most accomplished man with an impeccable reputation, a salary well into the middle six figures, fair, even tempered, and respected. My friend recalled this man once told her in strict confidence that every three or four years he would go into a bar for a singular, premeditated purpose. He arbitrarily chose a man approximately his size and age to pick a fight with, usually with these words: "I don't like the way you looked at the lady over there. You gave her a dirty look."

When the victim protested, he would egg him on until his target became insulting and abusive in return. He would then invite him to step outside to settle the argument physically. He said that if he was "lucky" enough to get a good fistfight out of the provocation, it would give him a rush that topped any he had ever experienced. He explained that this "craving" was nothing he felt that was central to his sexual makeup; he was otherwise a most considerate and gentle lover who had never been abusive toward anyone. He was wrong.

The aggressive sex drive, primarily a masculine trait as we've seen, is closely related to other forms of aggressive behavior—the relentless pursuit of any desire, such as a job, a home, a car, a prize. It also compels men—and women to a lesser degree—to fight and compete for these acquisitions and to go to battle when necessary to protect their possessions.

From Lunge to Lust

The link between aggression and sexual behavior is dramatically illustrated during the courtship ritual of the male stickleback fish, a small fearsome creature with a big sexual dilemma. He usually can't decide in advance whether to attack or to mate, so his romantic ritual begins just like aggressive encounters with other males.

First, the male establishes a territory, which he dominates. His testosterone makes him sensitive to space and boundaries, inherent to the male psyche of all species. The establishment, protection, and defense of these boundaries is characteristic of the male's need for separateness, as defined by the exclusion of other animals except for females, who do not threaten his autonomy.

Enter another stickleback. The female stickleback is indistinguishable physically from the male. She apparently looks and smells just like one—no lipstick, hourglass figure, spiked heels, or perfume to help him decide whether to make love or war. What is he to do? He attacks. If the intruder is another male, he will fight back or retreat. If the intruder is a lady, she remains passive.

For the stickleback male, this passivity under assault is her only appealing

141

feature, in fact, her only feminine feature. But this alone is enough to trigger his hormones to switch from lunge to lust. The key dynamic switching his behavior is his aggression versus her passive receptivity—a complementary action and reaction. Without such reciprocity, stickleback mating would be just another battle. Now the stickleback gets interested and begins tentative circling, trying to get a closer look at his guest without getting hurt. If she holds still he becomes increasingly aroused until copulation is irresistible. Ejaculation is the final act that catapults this male back into his separateness. He is not open to commitment, marriage, or living together. Happy at last, he retires. There is a clear refractory period during which time he will not be enticed by sexual invitations from her and wants no company whatsoever. This is not merely fatigue. At this point, sexual behavior is chemically impossible.

Sound familiar? It's no wonder. You'll find a similar mating dance in every species on earth: humans, naturally, and our close cousins in the primate family, and cats and wolves and birds and snakes—and simple fish like the stickleback.

Consider the human animal: He's a single guy on the prowl. He stakes out a perfect location, a barstool with an empty seat nearby. He sits down, checks his appearance in the mirror, orders a drink. He hears a rustling sound. Someone is sitting down next to him. He feels his blood surge. The chemistry of his body shifts, telling him it's time for action. What kind of action? That depends on who he sees when he turns around. If it is a man, he will firmly assert that he is saving the seat for someone else, and he'll let the newcomer know with body language that he's not welcome to trespass. If it's a woman he will turn on the charm. His body is prepared for either response, aggressively territorial or aggressively sexual.

Home Alone

Now, humans aren't sticklebacks, although some men act like them. After intercourse, he too rolls over and drops dead. Or he decides to get some work done or catch the sports report. Either way, he is "separate" and lost to her. A woman basks, luxuriating in the afterglow, longing to snuggle, nuzzle and cuddle. Instead, she gets his (stickle) back. She is intensely frustrated and "home alone."

Ejaculation is a terminal act that removes the male and returns him to his separateness. In some species, the male even makes ultrasonic sounds that signal the equivalent of "Not now, sweetheart, let's see what's on TV."

Whether a rat, a ram, or a rooster, once a male ejaculates, his body demands a period of rest before he can ejaculate again. (Note that the refractory period *does not* apply to erection, only ejaculation.) When the refractory period ends, he mounts his mate again and once more ejaculates. This next refractory period is longer than the first one. The following one is longer than the second, the fourth is longer than the third and so forth until the animal is exhausted. This should come as no surprise to humans, especially women who have waited impatiently for their men to spring back to life.

This "loner profile" of testosterone is absolutely crucial to understanding what men are all about. Just as testosterone separates the sexes physically at birth, it separates them emotionally as adults. Testosterone motivates the male to strive for separateness in ways a woman is not designed to comprehend. He wants to be alone! But, psychologically, and sometimes physically, she wants to visit. He will fight her intrusion on his privacy and in the process of defending his solitude, his resistance can transform into active aggression, which, ironically, often winds up in sex.

Testosterone is often at odds with itself and others. It wants novelty, shuns commitment but presumes ownership. The female sex requires accommodation for fertilization and mothering to occur. In turn, the male resists "commitment" (accommodation)—another testosterone effect. On the other hand, ownership of females is of critical interest to him because of testosterone "territorialism."

Wolves are well known to mark their territory in the wild. They urinate around the perimeters to signal other males to keep their distance, and fiercely defend their terrain if one should challenge their boundaries. Females are simply their property. Ownership is the issue. But if these same wolves are given a chemical that cancels their testosterone effect, they no longer mark their territory, object to visitors, or protect their females. They share their women without worry. What does that tell you about the origins of jealousy?

Yet look what often happens after long periods of monogamy. Who is usually the first to develop a wandering eye, the husband or wife? Who is more likely to have an affair? Who is the one who needs to spend more time alone or with buddies? Who is the one who has to assert the right to autonomy and independence, often in the most childish of ways? The one with the most testosterone, naturally. How is it that such a possessive creature is typically the first one to stray away?

The Coolidge Effect

Now it gets more interesting, hormonally speaking. If, at any point in the series of sexual trysts, the original female is replaced by a new one, the male's refractory time resets itself to its initial time. To our eyes, one female monkey might look like any other, and so do hens, ewes, and mares. But to the male of the species, who presumably smells the difference, the new gal in town is the fountain of youth. This is called the Coolidge Effect. It was not named after a researcher, but for the thirtieth president of the United States.

Reportedly, while visiting a government farm with his wife, Calvin Coolidge noted the large number of chicks and the high production of eggs. Mrs. Coolidge admired what must have been prodigious efforts by the local roosters. The proud farmer replied that the roosters performed their sexual duties many times a day. "You might point that out to Mr. Coolidge," said the first lady.

The president responded by asking if the roosters were monogamous. "No," the farmer replied. Introducing a fresh hen inspired the roosters to new heights. The president asked if he would kindly point that out to Mrs. Coolidge.

Novelty, therefore, is important to male sexual desire—more so, apparently, than for the female. The aggressive sex drive seems to crave variety. Sexually exhausted males will perk up like dandelions if a new prospect arrives—unless vasopressin takes charge, keeping him home and seeing that he's content to stay there.

The Psychology Behind the Drive

Testosterone has one of its most pronounced effects on mood, but this dimension has received little attention. It has antidepressant properties, probably much more powerful than those possessed by estrogen. For obvious reasons, we can't easily study these forces in humans, and animals won't usually tell you their feelings except through behavior, just like many men. But occasionally one of nature's accidents or medical intervention for another purpose gives us a tantalizing glimpse:

I had a glamorous Las Vegas show girl come to me as a patient. She was gorgeous, with endless legs, rich red hair, and a flawless complexion. She was terribly depressed and had considered suicide, explaining that if she couldn't find help for her problem she didn't wish to live. Her husband of ten years had married her knowing that she had once been a man. He didn't know this until after he had fallen in love with her and by then it made no difference. She had completed the surgeries (testicles and penis removed, artificial vagina created, Adam's apple refined) and hormone treatments, but there was one aspect she could not cope with. She remembered how it felt to be a man—her confidence, her usually happy state of mind.

After a few weeks without testosterone, taking estrogen supplements instead, she became severely depressed and described it as "out of the blue, I develop this dowdy housewife syndrome, lose my confidence, my enthusiasm, and don't have the energy to leave the house." She got to feeling so bad that she went back on testosterone out of desperation. Within weeks, she felt like her old self again, but she also began growing facial hair. Electrolysis wasn't sufficient to counteract the effects. She was afraid that her friends and associates would begin to notice, and couldn't face that. The only workable solution was creating a delicate balance of phases in which she would either go without her estrogen or her testosterone replacements, and then phases on medication that commenced at the first signs of hair or the first signals of depression. This was a high-wire balancing act; she basically had to invent her own cycle, but it worked for her. She lived in both hormonal extremes, but in neither for too long. This is just one person's tale, but it got my attention.

The next story disturbed me further:

Helen was in her late thirties. She had been referred to me by her general practitioner for a second opinion on the testosterone treatments she had been taking for about two years for low sex drive. She and her husband wanted to know whether or not this therapy was harmful to her health long term—to stop it or continue. She explained that before she began on testosterone, she was a different person— withdrawn, shy, without confidence, depressed, and sexless. With the shots, she developed a voracious sexual appetite, felt happy and assertive and got along much better with her husband. He said, somewhat apologetically, hoping that I would not take it wrong, that when she was on testosterone, he could communicate with her like a man. She was "logical and made good sense." Off it, she was hopeless. Indeed, concerned about the side effects, and already noticing that her figure, voice, and hair growth had changed, she had gone off it once. All the old undesirable characteristics returned.

Even after I explained the full scope of the health consequences of remaining on these treatments, she decided to continue. She said she just didn't like the person she was without it.

Once they left my office, I had to sit and catch my breath. You can't tell me that a hormone makes you logical—and a male hormone at that! That just couldn't be. Who said men were logical, other than other men? Testosterone isn't even rational—an antidepressant that makes you irritable and warlike? A libido booster that makes you obnoxious? I wasn't at all happy about this. If it were true, women were missing out naturally, but not completely.

More About Women

It is becoming increasingly clear that testosterone plays a powerful role in female sexuality, and that when a woman's testosterone dwindles, so does her sex life. But what about her social life and work relationships? What about her attitude, her mood, her assertiveness, and her effectiveness? Do men, having so much more, have an advantage that has not been fully appreciated? Are women being expected to compete in a male-dominated world according to their rules without the same equipment?

In particular, how does it alter the playing field when men and women are in professional competition with each other, winning and losing, establishing pecking orders and control? How do their testosterone levels respond—as friend or foe, to mate or to fight? Is this not perhaps part of the problem facing us in the work force and surrounding sexual harassment issues? The stickleback syndrome, perhaps?

A little speculation here. In many animal studies as well as in male humans, testosterone levels rise in those who are victorious in combat. "King-of-the-hill" states of mind include confidence, self-esteem, optimal coping, and well-being. Some women may be able to boost their own testosterone levels naturally, achieving these favorable psychological states by being highly competitive. Small studies proposing that women who are successful in business and sports have higher testosterone levels than other women have merely whetted our appetite for more information.

Curiously enough, while the dominant female in primate society is the most sexually attractive to the male, gets more food, reaches maturity faster, and reproduces at an earlier age, "boss" baboons pay a price in fertility. They have a much higher rate of miscarriage than their more submissive sisters, and in some cases they are infertile—a very rare condition among female mammals, caused perhaps by her higher level of testosterone.

Too Much of a Good Thing

When society is in danger, men's "warmones" are desirable. However, when peace prevails, these very men tend to disrupt the home they are charged to protect. What happens when men have abnormally high levels of this hormone?

Men and animals with higher testosterone levels are more aggressive than those with lower levels. In fact, men born with a genetic disorder that gives them double male hormones are substantially more aggressive than normal men. These men have XYY chromosomes instead of the usual XY chromosomes. The Y chromosome, as you know, is the male chromosome. When two Y chromosomes trigger testosterone production instead of just one, the hormonal effects skyrocket. That's why a high percentage of these men end up in prisons for the commission of violent crimes. They are lean, mean, and oddly enough infertile.

The trouble is, all that testosterone makes some women weak in the knees. In 1991, scientists at the University of Engee in Turkey found that men with deeper voices—which they linked to the quantity of male hormones— were apparently more attractive to women, making baritones and the disc jockeys who reported this news tidbit particularly proud.

Perhaps tough testosterone macho triggers some deep primal mating instinct in women that goes back to the days when a brute was a good catch because he could protect his wife and children from marauding tribes and saber-toothed tigers. Nowadays that sexy hunk might turn into a temperamental monster with high blood pressure, explosive tantrums, a wandering eye, and a hairless head. These are the "men who can't love," the "dance-away lovers," the men who "can't commit." You can find their wives on Oprah and Sally Jessy every week.

A Pennsylvania State University sociologist named Alan Booth obtained the hormone levels of 4,462 men and then studied the history of their relationships. He found that those men with high testosterone were less likely to get married. More troubling was what they were like when they *did* get married. They were more prone to having affairs, hitting their wives or throwing things at them, and leaving home because they could not get along with their spouses.

One Scandinavian study suggests that the higher men's testosterone levels, the higher the divorce rate. Not surprisingly somehow, a high-testosterone male is more self-centered, selfish, and has a personality profile not unlike a psychotic. Hmmm. But some women think he's irresistible.

In other words, these guys win over a woman with their aggressive pursuit and sex appeal. Once the conquest is made, they might be good at bringing home the bacon, but they resemble pigs themselves.

I suspect that this behavior, which women despise and men treat with pride, is a compulsion, not recreation. These men are swept into a primordial pattern that they participate in with determination, not joy. They may kid themselves and each other, but they are not having a good time. In some respects they are as much victim as predator, lonely in their isolation, alienated from the intimacy they fear, incapable of maintaining the bonds that would enable them to find peace in this life, sentenced to making war on the world.

Let us pursue some of the social disarray that men who are unwittingly colluding with testosterone subject themselves to.

Date Rape: This man is attracted to the passive, reluctant female. Like the stickleback, her passivity makes her more alluring. Like the cat, when she resists, he thinks it is just part of the choreography. He is sexually aroused, flipping back and forth between his sexual urges and his aggressive instincts. He pins her down, has his way, and "knows" she wanted it too; just like those noisy cats, she loved it. If she pouts or complains, he gets angry and defensive. Through his testosterone filter, he can't see very clearly. He has acted in accord with primitive sexual and emotional reflexes. Then he casts her aside ready for the next one and wonders why she gets so furious. It was great. It's over now. What's the matter with you? Leave me alone.

If, in spite of having fractured all civilized sexual etiquette, he comforts her, treats her gently, apologizes or seems at all concerned, many if not most women would be inclined to forgive him, or at least would be less inclined to inflict punishment he distinctly deserves.

Violent Rape: Taking this sexually aggressive behavior one dangerous step further, we must consider violent rape. Not all men, of course, are rapists. However, as you can see from the history and chemistry of testosterone, they must contend with sexual forces for which women have no real frame of reference. Yes, women have a dash of testosterone, but it is diluted and muted by all their other hormones; there is no comparison.

Rape is an act of aggression, without question. It is also (contrary to current political thinking) a sexual act. They are not incompatible acts in nature—among animals. What makes humans unique is our physical and emotional evolution sexually, enabling us to find passion without violence, sex without bloodshed, and intimacy within our sexual relationships. Men who rape are reverting to primitive patterns, incompatible with modern civilization, and unacceptable to modern man and woman. For a man to civilize, socialize, and acculturate himself to nondestructive sexual behavior is no easy feat. (But neither is toilet training. And it can be done.) In fact, *the choices* men make based on their aggressive impulses actually increase or decrease the aggressive chemicals in their systems. In other words, violence begets violence. Peaceful behavior reinforces and rewards itself. Men (and women) need not be helpless victims of their hormones!

Primitive behavior, of course, doesn't have to be as extreme as rape. It erupts in other ways. For example, some professions are notorious for their jungle tactics.

David Margolick in a *New York Times* column described the hypothesis of psychologist James Dabbs, Jr., who is testing the saliva of trial lawyers to see whether or not they have abnormally high testosterone levels:

"Tort-Tosterone: Like juvenile delinquents, substance abusers, rapists, bullies and dropouts, trial lawyers may have too much testosterone but have learned to harness this primordial force for fun and profit." He also found that actors, entertainers, and football players had higher testosterone levels than ministers, whether they were pastoral or missionary, concluding that the hormone variables might affect occupational preference in subtle ways.

The Sporting Man

In addition to sexual aggression, there are other social manifestations of testosterone-dependent aggression all having to do with competition, social status, dominance, and submission.

Competitive situations bring out testosterone-dependent aggression. We see it in office politics, in singles bars, in courtrooms, and, most clearly of all, in sports where teams and individuals compete for positions of great status. Here, winners and losers are clearly defined. In fact, a great many studies have shown that testosterone goes up in victory and down in defeat—and that high testosterone appears to be the *result* of winning, more so than the cause. This was inferred since the lack of a clear-cut victory, or winning by chance (such as in a lottery) results in no change in testosterone levels. But thinking may soon change. Two new studies showed that among college men, losing or winning five dollars (based on a task entirely by chance) caused testosterone levels to fluctuate. Winners had significantly higher testosterone than losers and were in better moods. Let the guys win, gals!!

In groups of male monkeys, those who become dominant show elevated testosterone, while those who are defeated show lower levels. Once the dominance/submission patterns are established and stabilized, differences in testosterone between dominant and submissive males tend to even out. Giving monkeys more testosterone, or even castrating them, doesn't cause any immediate change in this stable hierarchy. From this, we can conclude that testosterone differences among males only appear when animals are actually participating in competition or battle, not when they are resting on their laurels. Testosterone levels do not reliably predict which animal will win, although it would seem to me that the most irritable, fierce, and belligerent ones would have a definite advantage.

These studies demonstrate a most important feature of male psychobiology: men who achieve status through their own efforts feel good as a result and experience a boost in their hormone levels. But once is not enough. To sustain the feeling, they must do it over and over.

One small study out of England suggests that becoming the king of the mountain not only produces a feeling of dominance along with a boost in testosterone, but engenders anxiety as well, by raising prolactin levels. While the implications of this study are not yet completely clear, perhaps, along with success comes stress over how one is going to stay on top.

Naturally, although winning may have some cost, it does have its rewards. In most animal groups, primates included, dominant males not only eat better and claim the best shelters, they get to copulate more often. Maybe nature raises their testosterone so they will be driven to have sex and perpetuate their winning genes.

Compare this monkey business to humans. How do testosterone-triumphant men behave compared to losers? Look at any college campus after a big game. Puffed up with victory and pumped up with testosterone, the winning players celebrate. They cruise, carouse, and cavort. Mainly, these cocky devils are out to keep scoring, this time with girls. Some are so aggressive they get into fights. Add alcohol to the equation and you can count on drunken brawls. Rape and date rape are all-too-familiar side effects.

Today, the association of sports and violence—sexual and otherwise—is politically incorrect. However, that does not change the psychophysiological realities. A new study led by Todd Crosset, a University of Massachusetts sports management professor, studied one hundred seven cases of rape, attempted rape, and fondling at thirty schools in the National Collegiate Athletic Association division I. Figures from 1990 to 1993 involving ten schools showed that while athletes made up only 3.3 percent of all male students, they performed 19 percent of the sexual assaults. In the other twenty schools, athletes scored slightly higher than nonathletes, but the differences were not as significant. Clearly more study is necessary before firm conclusions can be drawn but these results are certainly suggestive.

TESTOPAUSE

Testopause is a term I have coined to represent an aspect of male behavior that has not been described before in the human. It is a self-defeating testosterone cycle—a pattern that can occur at any time in a man's life, but is most pronounced during viropause (male menopause consisting of midlife psychological, physical, and hormonal changes) which you will learn about in Chapter 7.

Testopause results from the *relative* testosterone deprivation that occurs when a man thinks he has been kicked off the top—lost to the competition. Surprisingly, it also happens to the man on the top who has mastered success, but collapses sexually. To all the world, he looks like the king of the mountain. At home, he can't get up the slope.

I call this phenomenon testopause, because testosterone drops (pauses) when he feels defeat at work or in bed. (Stress of any kind will do it too.) It goes up again only when he finds someone to pick on. Desperate for his fix, his feeling of control, dominance, and power, this man abuses someone close to him to get his high back. Remember, testosterone only peaks during the actual fighting or competition, so he must perpetrate discord in order to feel good again.

Most of the scientific community thinks that testosterone has little to do with male menopause or female menopause. I believe they are wrong on both counts. Beginning with male menopause, let us review the facts. Testosterone levels drop just slightly in the thirties and forties, then somewhat more in the fifties and sixties. Because this decline is not precipitous, one usually hears that male menopause can't be related to such a subtle change. However, there are several flaws in this logic. The first is that the free (active) fraction of testosterone has not been considered. Free testosterone, the potent one, drops more than bound testosterone. Secondly, as you will soon see, in the case of male menopause (viropause), absolute levels of testosterone, free or bound, may have little bearing. It is the rhythm and nature of short-term rise and fall of free testosterone that may make all the emotional difference. Specific circumstances, such as winning, success, assertion, and control dramatically affect the rise and fall of testosterone at all ages. Since high levels of testosterone are stimulated and maintained by winning and by dominance, it holds true that the converse also applies.

I propose that the impact of changing social status and pecking orders so pervasive among other species, including primates, is at the heart of the problem for human males as well, triggering them into a hormonal nightmare of life-altering magnitude.

Pecking Orders

A study of soldiers in officer candidate school found that testosterone levels declined in the arduous early stages, but went up at graduation. Clearly, when they were being stressed, oppressed, and dominated by superiors, the trainees' hormones were suppressed. With success—when they became "superiors" themselves, their testosterone surged.

In a similar vein, young, frightened novice parachutists had depressed testosterone levels. Once they became practiced jumpers, their mastery increased their testosterone.

Apparently, testosterone thrives on the stress of competition and ignores the *status quo*. The advantage of "pecking orders" is that a defeated male may recover quickly by bullying another "animal" lower in the pack, returning his testosterone to normal by this "displacement" behavior.

So, the man who is pushed around all day at some miserable job can recover his "masculine dominance" by making his subordinates—or wife and children—miserable later on.

What do defeated athletes and warriors do? They sulk, they drink, they feel sorry for themselves. They are not beating their chests and pouncing on women. Their mates need to cajole and comfort them. The men may get nasty in response to these tender overtures. Why? They're getting their testosterone fix by snarling at someone else. Why do you think men can't stand to let another car pass them on the freeway? They have to get back in front at any cost, furious at a complete stranger. Winning. They must preserve that testosterone high.

SOLUTIONS: INFLUENCING AGGRESSIVE SEX DRIVE

Aggressive sex drive can be influenced in a variety of ways for a number of different reasons, resulting in a spectrum of different outcomes, from hyposexuality to hypersexuality.

Lowering the Aggressive Sex Drive

There are occasions when lowering the aggressive sex drive is desirable—as in hypersexuality and sexually aggressive or violent men. The differences in how these categories can be treated pharmacologically are not that great, and have mostly to do with dosage and strength of drug. For the more violent of the group, both lowering testosterone and raising serotonin would be theoretically desirable (and has worked in practice). There are numerous drugs that can raise serotonin, including Prozac and its relatives. Other drugs can reliably lower testosterone.

Lowering dopamine could have some effect on curbing sexually aggressive behavior, but would diminish the capacity for pleasure across the board, and therefore may not be a sensible approach. The antipsychotic drugs that lower dopamine might be a valuable resource, but have not been tried for this purpose. The main problem with them is the patients' unwillingness to take the medication on their own. They have so many undesirable side effects that it is hard to get patients to continue to take them outside of a hospital setting.

Castrating Chemicals

The most effective methods for reducing aggression involve the use of progesterone, a hormone you will become more familiar with in the next chapter. Progesterones and their relatives are generally so unkind to the male sex that they can be used as a form of chemical castration—a controversial practice employed to treat certain sex offenders. It is also used in the treatment of prostate cancers—if the choice boils down to your life or your sex life.

For the past thirty years, medroxyprogesterone (MPA), a particularly powerful form of synthetic progesterone, has been used in Europe and the United States to treat male sex offenders convicted of rape, exhibitionism, voyeurism, and other paraphilia. Some sex offenders are not violent—many exhibitionists and voyeurs (peeping Toms) do not escalate their behavior. Furthermore, some, perhaps most, child molesters and date rapists use persuasion and intimidation rather than violence. MPA has been used with more success treating such nonviolent offenders rather than violent ones.

Testosterone reduction by progesterone is generally attributed to the inhibition of the LH signal necessary for testosterone production. MPA accomplishes this and also lowers testosterone directly by increasing its destruction (metabolic breakdown). Within the first month of MPA treatment, testosterone levels can fall more than 90 percent, but not always.

Although MPA's power is attributed to its ability to reduce testosterone, it can decrease sexual arousal even when testosterone levels don't drop. In fact, deviant sexual behavior has remained suppressed by MPA even when testosterone levels return to normal, suggesting that there are other forces in play. Interestingly, although MPA treatment effectively knocks out the male sex drive, the capacity for erection is not dramatically reduced. We can't explain this nor do we know what these other forces might be—another paradox we have yet to understand.

Within a few weeks of stopping medication, testosterone levels usually return to normal. The chief nonsexual side effects of progesterone therapy are drowsiness and weight gain.

The Price of Prostate Cancer

Some prostate cancers are testosterone dependent, much like some breast cancers that thrive on estrogen. Progesterone stops this testosterone effect. The progestogen megestrol (Megace) is widely used to treat prostate cancer. Negative sexual effects occur, however, with marked loss of sex drive in up to 70 percent of men.

While there are new drugs that seem to have fewer sexual consequences, what happens in the long run is still an open question.

RAISING THE AGGRESSIVE SEX DRIVE

Testosterone Replacement Therapy (TRT) for Men

Testosterone was recognized as a sex hormone at the turn of the century, but it was not until the last fifteen years that its remarkable role in male sexuality was demonstrated in controlled experiments. Since then, a preponderance of studies has shown that men deficient in testosterone obtain definite sexual benefits by replacing it.

This suggests that testosterone is not only a natural aphrodisiac but an antidepressant as well. And why not? Depression often sinks libido right along with spirits.

However, if a man with a *normal* testosterone level has low sex drive or erection problems, giving him more testosterone won't improve matters and may actually cause harm. We know this for a fact because testosterone has been one of the most abused, misused, and overprescribed medications for male sexual dysfunction in medical history. It is like putting oil in your car. It won't run any faster, better, or differently as long as it has enough oil to keep running smoothly. Adding more can actually cause problems.

Like too much oil in the car, testosterone given to men with normal levels can cause an overflow. Most male patients do not realize that the use of powerful synthetic steroids like testosterone interferes with the natural hormone circuitry, confusing it, rather than just adding to it. For example, when testosterone is given to men who do not really need it, the LHRH-LH mechanism is shut down, upsetting normal hormonal cycles. The cycling period of human LHRH secretion as mentioned earlier is approximately ninety minutes. If this frequency is changed, or if there is too little or too much LHRH available to maintain the optimum pulsatile frequency and amplitude, testosterone production stops. In an effort to correct one problem, you create another.

In addition, testosterone can cause salt retention, which causes fluid retention, secondarily causing or aggravating hypertension. Also, since testosterone levels go hand in hand with irritability and aggression, I propose that giving testosterone to a man with normal levels who does not need it can make an impotent man into a temperamental monster with high blood pressure, explosive tantrums, and less hair. Even if he got his erections back—who would want to get near him?

On the other hand, when sensibly prescribed for men or women whose testosterone levels are low, the results can be remarkable.

A seventy-four-year-old patient of mine, who had been impotent for twenty years, lost his wife to cancer. He met a young, erotic sixty-two-year-old widow a few years after his wife's death. His first wife had not been interested in sex in her later years. This woman was, and, to his great surprise, so was he. A few months after they met, he came to see me, hoping I could help.

I performed a physical examination, laboratory workup, and psychosexual evaluation. His testosterone levels were 150 ng/dl (the normal range is between 250 and 1,200). I referred him to an endocrinologist for further evaluation, and we soon began testosterone replacement injections. Within two weeks he had his erections back and was functioning as well as he had in his fifties. A few months later he remarried.

At present, testosterone therapy is somewhat cumbersome. While pills are available, a man with severe sexual symptoms due to low testosterone usually does better with injections, which require trips to the doctor's office every three weeks or so. The rise and fall of the testosterone effect is obvious to most men who undergo this treatment, since there is an artificial and unnatural testosterone level established that ebbs and flows with the injections.

Given the pulsatile nature of testosterone release that occurs from moment to moment, the ideal delivery system would probably be by testosterone pump. Such a device that the man could wear continuously would eliminate the inconvenience of frequent doctor visits, and the pump feature would allow the drug to be delivered in a fashion that imitates the natural fifteen- to twenty-minute surges that occur throughout the day.

Sometimes improvement can occur before the injections have had time to work. It is because of the placebo effect. Desire and erection are perhaps the most easily affected by the power of suggestion. If a person believes that a token, symbol, or "medicine" has sexual potency, the belief itself can cure the problem, if it was psychological to begin with. There is a tricky way to tell: A placebo effect will show up the next day after the shot is given; a medication effect takes a week to three weeks to develop.

New forms of testosterone with less potential to cause side effects are under development, along with new methods of delivery. We'll discuss these in Chapter 8. Until these new methods become widely available, however, TRT still will continue to have some drawbacks.

Testosterone Replacement Therapy for Women

A common practice in the past was to treat young, otherwise healthy women who had low sex drive with testosterone *supplements*. It is important to understand the difference between testosterone *replacement* therapy and *supplements*. Replacement means that testosterone levels are below normal and treatment with testosterone is intended to return testosterone levels to normal. As we've seen in male studies, there is probably little potential for harm, but also no benefit because the end result imitates the natural state, and does not exceed it. Supplements, by contrast, are given when testosterone levels are normal, to increase them to above normal for some specific effect. These higher than normal levels can cause many of the same complications as we've seen in men. In women, while testosterone supplements boost their sex drive, they also change their body, relocating body fat and muscle, giving them broader chests, narrow hips, small breasts, an enlarged clitoris, a mustache and a beard—too much hair in the wrong places, too little in the right ones.

Additionally, prescribing testosterone causes concern among some that women will not just develop secondary sex characteristics such as low voice and body hair, but also trigger some unpleasant psychological aspects of male sexuality: aggressiveness, surliness, and preoccupation with genital sex. Even certain criminal behavior such as sexual abuse and various other aggressive crimes committed by women could be attributed to improper testosterone treatment. Some studies have shown relatively high testosterone levels in violent female outpatients compared to nonviolent patient controls. There is also concern that treating women with testosterone could increase relationship difficulties by creating too much assertiveness and irritability and, consequently, conflict.

There is more research activity on the use of testosterone supplements for women in other countries (especially in Europe) than in the U.S. Momentum is building here, though, as the data filter in and work is being performed worldwide. Once brain-carrier molecule systems become available that transport testosterone directly to control centers without saturating the rest of the tissues, the negative masculine cosmetic side effects discussed should be eliminated. At that point, testosterone therapy for sexual dysfunction in women may get another look.

Even now, in spite of its potential side effects, testosterone is a possible treatment for sexual dysfunction in women with normal testosterone levels in carefully selected cases. Given the understandable tendency to avoid hormonal treatment other than to replace deficiencies, eventually nonhormonal drugs such as bupropion, deprenyl, oxytocin, and serotonin-reducing drugs may become more readily accepted as treatments for female (and male) sexual dysfunction, with or without testosterone. DHEA may also soon play a therapeutic role.

At any rate, this discussion of the value of testosterone in women's lives and its potential as a supplement in certain circumstances is not intended as the last word. There is truly not enough research to draw conclusive opinions with any degree of confidence. However, testosterone is becoming widely used in women in hormone *replacement* therapy, and I will discuss its role in more detail in the next chapter.

LHRH

Originally, there was great enthusiasm for LHRH—hopes that it would become the first successful reversible hormonal birth control pill *for men*. It didn't work out, because although it successfully arrested sperm production, it usually reduced drive and canceled erections along with it. Giving men the requisite constant and relatively high doses of LHRH (unlike the body's natural pulsatile doses) turns off the body's own system —of both sperm and, unfortunately, also testosterone (and thus desire and penile functioning).

However, LHRH treatment (in less extreme pulsatile doses) may actually have some nice effects on a man's libido, although it hasn't been particularly helpful in treating impotence. Increases in sexual desire may occur, but these changes are relatively small. Indeed, LHRH does improve sex in men with hormone deficiencies but hasn't shown anything but slight benefits for erection when tested against placebo.

In one study using LHRH replacement, testosterone levels peaked between weeks two and four and then gradually declined over the following four weeks. Firmness of erection peaked at week four, and sexual desire was increased through week six. In spite of these improvements, the frequency of sexual relations did not improve much—in fact, remained low—throughout the study. So, in spite of the obvious advantage of TRT in many testosterone-depressed men, in some cases, improvement of ability occurs without commensurate increase in sexual behavior. These observations suggest the need to explore other factors. It may be that the lack of sexual desire and response occurs because dopamine (responsible for anticipation, motivation, and movement) and certain peptides are absent or decreased. If dopamine enhancers were combined with testosterone therapy, the end result might be increased desire *and* increased activity.

LHRH has not yet been tested for sexual problems in women in spite of abundant findings in female animals of LHRH-induced increases in sexual activity.

Pumps: Special pumps can be worn to deliver LHRH therapy. This pumping action increases testosterone, testicle size, spontaneous erections, nocturnal emissions ("wet dreams"), and acne.

LHRH Sprays: LHRH has been used as a nose spray by Europeans for many years in hopes of increasing sex drive and improving erections. Although LHRH usually triggers testosterone, new potent LHRH analogues such as leuprolide (Lupron) and goserelin (Zoladex) inhibit the production of testosterone by overwhelming the feedback loop with constant stimulation.

As we've discussed, if too much LHRH is given, LHRH receptors stop responding altogether, preventing LHRH and FSH release from the pituitary and decreasing natural testosterone production. Essentially, the system is overwhelmed and shuts down. Consequently, men are actually "chemically castrated" with these LHRH sprays, losing both their libido and their erections.

The Male Pill: Continuous LHRH has been tested as a male contraceptive, but because men with rare exception lose their desire (along with their sperm and their erections), it works for the wrong reasons.

Other Alternatives

Reduce Stress: In a study conducted at Maharishi University on men, serotonin and prolactin were found to be decreased during actual Transcendental Meditation. TM raises DHEA. All three effects support the aggressive sex drive and favor sexual responsiveness.

Serotonin Depletion:
- Diet Pills: Most diet pills reduce a woman's serotonin levels.
- Dieting: Successful dieting also lowers serotonin levels.
- Exercise: Physical activity increases DHEA and indirectly decreases serotonin.

Dopamine Enhancement: In both men and women, the most mystifying aspect of testosterone has been its positive influence on sexual desire often absent a commensurate effect on sexual behavior and frequency. I am going to speculate that dopamine is the missing link; that where TRT alone fails, adding a dopamine supplement can make all the difference—for both men and women. In other words, when testosterone replacement revives desire and function, but results in no behavioral or frequency change, add a little dopamine to motivate and mobilize.

There is no research to prove my point, only to suggest it. The only way to find out for sure is to perform double-blind placebo-controlled studies on low sex drive men and women using testosterone with and without dopamine.

Drugs like bupropion/Wellbutrin and deprenyl/Eldepryl can raise dopamine levels. Testosterone also elevates it.

Bupropion/Wellbutrin is an antidepressant produced by Burroughs Wellcome that is widely used to treat various forms of depression—especially lethargic and sedentary depressives. Its dopamine and stimulant properties get you up and moving.

Deprenyl/Eldepryl is used together with L-dopa to treat Parkinson's disease. L-dopa gives these patients more dopamine. Eldepryl slows the progression of the disease by somehow protecting the dopamine receptors from further deterioration, making whatever dopamine is present (i.e., from L-dopa) more effective. Eldepryl is also being used in Europe and by physicians in the United States to treat cognitive symptoms of aging, although it has not been FDA approved for that purpose.

Other Factors: As you may know, there are many other substances that affect sex drive in specific ways. Here, however, we are focusing our discussion on the aggressive aspects only. As previously discussed, increasing DHEA, oxytocin, and vasopressin all contribute to the sex drive in various ways.

ME TARZAN, YOU JANE

According to your sex hormones, men are still swinging from vines and women are still answering the call of the wild, whether we approve or not. Fortunately, that isn't all there is to the story. Through DHEA and testosterone, women have developed the aggressive dimension of their sex drive. In addition, as we take an in-depth look at our receptive sex drive, it isn't as passive or as receptive as it first appears. Women have evolved considerably beyond our furry relatives, and we are not destined to repeat senselessly the rituals of these ancestors. Nonetheless, it would be wise to honor our heritage and recognize the forces that draw us toward primitive patterns and influence the nature of relationships we form today.

Clearly, testosterone is not the only molecule that generates lust. There is, in addition, a complex interaction of chemicals, including peptides, various neurotransmitters, and other hormones that play a leading and/or supporting role. I think, in time, we will discover that certain vitamins, minerals, and age-old substances first introduced by witch doctors, herbalists, and alchemists also play a valuable part.

This chapter, together with the next, concentrates on the mysterious forces that govern our aggressive and receptive sex drives, explaining the dramatic differences between how men and women process lust. This chapter has helped you to understand, perhaps for the first time, the dynamic relationship between sex and anger, tenderness and aggression, as well as the conflicting drives between sex with a partner and masturbation. It has also introduced you, with a backstage pass, to the aggressive sex drive most characteristic of the male, but more familiar to the female than anyone until now has been willing to acknowledge.

Next we are going to explore the mysteries of estrogen and the receptive sex drive, which is the predominant, but by no means the only, sexual pattern for the woman. Within her, testosterone and estrogen weave as the yin and yang, opposite but complementary. You will come to appreciate this relationship in ways you may not have been exposed to before. You are also going to learn surprising new things about your old friend estrogen—even though most people think they have heard it all by now.

6.

$\mathcal{Y}$OUR RECEPTIVE SEX DRIVE

She poured herself into a black spandex velvet dress. It fit like a second skin—a personal pelt. Her hair was loose and wild, her sleek muscular legs in sheer black hose. In another age, she would have worn long satin gloves. She was feeling almost predatory, but not quite. A persistent animal instinct had her under its control. She wanted something and was intent on getting it.

The man about to pick her up was a colleague from work who had asked her to accompany him to a social event. They were just acquaintances, nothing intimate, no chemistry cooking. He was actually a little dull. Not bad looking, but overly serious and somewhat shy. Clark Kent-ish.

So what came over her? She could have pretty much her pick of men, and ordinarily he wouldn't be her choice. Besides, she knew he would be shocked. He had only seen her in business dress, never like this. In a way, she was toying with him, like cat and mouse. She thought it would be amusing to watch his reaction, see if he could cope. Then again, she was "in the mood" and he happened to be handy.

When she answered the door, he stumbled backward. His eyes darted behind her, as if looking for the woman he had come to pick up.

Stammering, he blurted, "Kate? You look so—different. I hardly recognized you."

She took a step closer and asked, "Is this all right? It's not too tight is it?"

"Oh no, you look beautiful."

She turned, arching her back and lifting her hair. She said, "Would you zip me then? I couldn't reach. These dresses aren't made to be worn alone."

His hand shook. She noticed, with a big smile he couldn't see spreading across her face. It never reached her eyes.

His tongue was tied, and he became uncharacteristically clumsy. God, he hoped he wouldn't stumble down the stairs as they left her condo. Once they were in the car, she asked, "Has the cat got your tongue?" He laughed, sounding like the idiot he felt, and said, "No. I just can't get over how changed you are. I've never seen you like this—so—female. No, no . . . I apologize. That's not how I meant it to come out. I hardly know what to say."

She put her hand on his thigh as he drove, with a little pressure and a rub, somewhere between a comforting pat and a caress, saying, "You're cute. I like you."

She left her hand there, absently kneading him from time to time. To his mortification, his erection joined them, eager to be rubbed too. Hoping she hadn't noticed yet, he said, "You better stop that or we may end up in a ditch."

"Mmmn, that would be nice," she purred, moving her hand a little farther up and in, using her nails softly. His hardness was obvious to both of them by now. The side of her hand was feeling its heat, but it was being studiously ignored.

By this time, her body was pressed against him, her hair tickling his neck. Somehow, she had gotten under his arm so he only had one hand left on the wheel. The other one was dangling strategically in the air over her right breast, not knowing what to do, feeling as if it belonged to someone else.

The only way he could free himself was to fall out the door. All he could think about was how she was going to get out of that dress without him.

Compare her with this other feminine feline:

She rubs against your leg, stroking herself. She wanders across your body as though it's her birthright while you try to read, punctuating her affection with sharp staccato nails. She sweeps her tail under your nose, getting her hair in your face. You try to ignore her, then push her away. She mews, crumpling your newspaper as she curls up on top of the column you were just reading, purring like a semitruck. You can't possibly concentrate with all that racket. Besides, she is now kneading your lap with weapons drawn. Okay! You drop everything. She's got your full attention. You can't help being impressed with her strategy and persistence, but you are more than a little annoyed. She starts to walk away, indignant, but in the process manages to slink under your hand getting a full body stroke. She stops with your hand on her rump, waiting. You can't help yourself. You would like to resist, but you are under her power. You hate yourself for it, but you give her a little pat anyway. She arches her back, growls softly, pushes against your hand and flips her tail to the side, backing up all the while.

This is the proceptive sex drive. She is not just available (receptive), she is seductive (proceptive). In animals it is called, simply, lordosis. Humans don't have a similar term because we don't yet admit to this reflex although we experience it regularly. High heels help. They throw women into a lordosis posture—breasts pushed out front, back arched, butt on display. *Seductive* comes closest to describing it. *Feline* is more captivating. *Lordosis* is accurate. It won't happen without the interaction between estrogen and oxytocin, though. (You can even get male animals to achieve this stance if you inject them with enough estrogen.)

ESTROGEN: READY AND WILLING

Remember the stickleback fish in the last chapter? Somehow, the randy male has to be able to detect a willing female and distinguish her from an adversary. The same is true of a male monkey, a ram, or a human suitor prowling a bar. Lordosis enables this interaction. The female of the species sends out signals that are sometimes blatantly obvious, other times subtle, that say: "I'm available" or "I *might* be." "At least I won't bite."

Dinner Date or Courtship Feeding?

An article by Helen E. Fisher in *Psychology Today* draws a parallel between Western dinner dates and what is called "courtship feeding." She explains that courtship feeding probably predates dinosaurs because it has such an important reproductive function. It is the quintessential ploy by which males of the species hope to get sexual favors from females. By providing food to prospective mates, males show their talents as hunters, providers, and worthy partners.

In insect life, the black-tipped hang fly, for example, sets out a juicy morsel hoping to attract a passing female. While she pauses to enjoy her newfound meal, he mounts her and copulates before she is done. This way, I suppose, he always gets a return on his investment.

Think of the dinner scene in *Tom Jones*—a sensuous feast, an old-fashioned ritual? Fast forward to the movie *Nine ¹/₂ Weeks*. Separated only by time, not nature. In both cases, doesn't it seem as though they were enjoying sex right before your very eyes? They were courtship feeding.

When a man is courting, he pays the dinner tab and a woman instinctively knows she is being pursued. Once the human male makes a move, he prefers some feedback to know if his overture was accepted and if further advances would be welcome. When a modern woman reaches for the check, she doesn't realize that she is interfering with an ancient biological imperative, transmitting signals she may not intend. If you want to establish equality between the sexes, this is not necessarily the best way to do it. However, if you want to sabotage the relationship and cut it short, pull out your American Express card.

Perhaps I sound old-fashioned, but the fact is that despite the now-famous guidelines issued by Antioch College, which require verbal consent before each new "level of intimacy" can be experienced, most go-ahead signals throughout time have not involved spoken, written, or taped informed consent.

Regardless of the actual steps we now must take, the basic dance is usually the same: she attracts, he initiates, or he solicits, she responds, he escalates. Testosterone drives the aggressive pursuit of sex. What drives the proceptive, responsive, receptive mechanism is estrogen, with the collusion of a few of her allies.

What Is Estrogen?

Estrogen is the Marilyn Monroe in all of us. It is the hormone in the diaphanous dress, the chemical with the cleavage, the molecule with the glowing skin and ruby red lips. It gives women their curves, their softness, their rounded breasts and moist, inviting vaginas. When estrogen is in power, the females of every species signal, "Take me. I'm yours." The term estrogen originates from "generates" and "estrus," and what it generates is sexual excitement, receptivity, and heat—at least in all other mammals.

Estrogen makes a woman more attractive to a man—they say it makes a woman glow. We see this in the animal world. In primates and lower mammals, females with reduced or absent estrogen fail to attract males, lacking the "sexual scents" that are dependent upon estrogen. By contrast, when female primates were given daily injections of estrogen, male apes thought they were beautiful. They pursued them enthusiastically with sex obviously in mind. The females, in response, became aggressively soliciting as well as more competitive, outclassing other females. Does this strike a familiar chord? Check out any bar.

Estrogen plays a covert role in sex drive. This female hormone scripts a woman into being willing and available, but does not necessarily inspire her to initiate the act. That's "Rambo's" job—he's overdosed with testosterone.

Estrogen is part of the reason why some describe romantic love as a form of temporary insanity, a dramatic emotional detour. You moon over each other and act like idiots. Whatever your partner does is marvelous, cute, or clever. Isn't he adorable when he snores? His filthy habits just prove that he needs you there to pick up after him and keep him organized. Sickening, isn't it? Then why do women persist? Because they're full of estrogen, stoned on dopamine, and high on oxytocin.

Estrogen is the background music in women's lives. Sometimes it is louder than others, but it is ever present. It fluctuates, peaks, and cycles throughout a woman's life beginning with conception and lingering through-out old age. It is produced in response to the X chromosome and women have two of them, which is why we are saturated with estrogen. During pregnancy, it shapes all embryos into its own female image, suggesting that God is a woman after all. If testosterone is introduced somehow, either naturally through the Y chromosome, or artificially through medication or disease, the fetus transforms from female to male.

At birth, estrogen withdrawal (as the newborn is no longer awash in her mother's estrogen too) sometimes causes a little vaginal blood flow—a microperiod—demonstrating that baby girls are born reproductively intact, just waiting for estrogen and other hormones to make their move. Estrogen, however, lays low until puberty, when it announces its presence through budding breasts, cramps, blood, and mood swings—quite a repertoire. With this dramatic debut, a young lady then begins to cycle, menstruate, and ovulate, which usually continues throughout her adult life unless surgery, stress, disease, excessive weight loss, physical overtraining, or birth control interrupt her normal rhythms.

Men have relatively little estrogen. It is converted from DHEA and testosterone in the prostate and produced in small amounts by fat cells. If men have more than they should for any reason—cirrhosis of the liver, or estrogen treatment for prostate cancer—they grow pudgy breasts and get hippy. Their skin becomes smooth and hair pattern changes. Unfortunately, estrogen does not appear to stir the libido in men the way it does in women. It pushes the wrong button and causes impotence instead.

As menopause approaches, estrogen begins to fade, until eventually ovulatory cycles stop. Estrogens continue to recede with all the attendant consequences that you will learn more about in the next few chapters. But first, let's look at what estrogen does to you when it is present in full strength.

Women's ovaries are basically estrogen factories, generating fluctuating amounts in response to the pituitary-regulating hormone, LHRH, which stimulates LH and FSH, which trigger the ovaries to produce estrogen. When there is enough estrogen in the bloodstream, LHRH is inhibited—another very efficient hormone-feedback loop. Estrogen, and the menstrual cycle in general, is regulated by these hypothalamary/pituitary hormones, stress, and the phase of the moon.

Estrogen is processed and decomposed by the liver, which is why extensive liver damage can raise estrogen levels. It just continues to circulate and accumulate because it can't be destroyed and eliminated. Should the ovaries be removed or damaged from chemotherapy or as a consequence of menopause, our fat cells still make some, and our brain probably does as well. Fat not only fills out your wrinkles, giving you a more youthful look, more fat means more estrogen, improving the tone and glow and scent of skin as well. Perhaps that's why some men prefer Rubenesque women.

NOT JUST LUST

A woman's lust is ambivalent and multifaceted. There are four main flavors (the typical man has just one):

- active (aggressive)
- receptive (passive)
- proceptive (seductive)
- aversive (reverse)

ESTROGEN PROFILE

MOST PEOPLE DON'T KNOW THAT ESTROGEN:

- improves cognition
- improves and stabilizes mood
- increases performance, reaction, and vigilance
- protects against schizophrenia
- protects against Alzheimer's disease

ESTROGEN:

- is produced by the ovaries; also produced and stored by fat cells and the brain
- is an MAO inhibitor (mild antidepressant)
- potentiates oxytocin
- maintains skin tone, collagen
- prevents osteoporosis
- prevents heart disease

AS TO SEXUAL ROLES, ESTROGEN:

- maintains receptive sexual drive and promotes frequency
- generates attractive body odor and texture
- sustains texture and health of vagina and vulva
- promotes vaginal lubrication
- improves vaginal, urethral, and genital tissue
- prevents senile vaginitis
- promotes lordosis

AS TO BEHAVIOR, ESTROGEN:

- prevents depression
- prevents or reduces stress
- improves sense of taste and smell
- decreases appetite

ESTROGEN HAS BEEN USED:

- to treat symptoms of menopause
- to treat prostate cancer
- as a contraceptive

HOW WE CAN INFLUENCE ESTROGEN:

Increases estrogen:

- soy products
- chronic alcoholism
- intercourse
- chronic liver disease
- obesity
- oxytocin

Decreases estrogen:

- oophorectomy
- menopause
- sexual abstinence
- anorexia nervosa

In addition, desire can be general or genital. It can be orgasm driven, penetration driven, masturbation driven, or touchingly tender. The skin hunger can be for hot, steamy sex or gentle body comfort. The emotional need at one time is for intimacy, a connection, the next time for uncomplicated, don't-even-want-to-know-your-name sex.

Among women, then, there is apparently a wide spectrum within two extremes—those who prefer genital sex to those who only desire nonsexual contact and touch. The nervous system has evolved so that these behaviors—from hottest sex to gentlest snuggling—are all part of the healthy continuum. Sexually well-adjusted women experience the full range at different times, depending on their hormones, their circumstances, where they are in their cycle, and how their mate treats them. Many Americans remember the voluminous emotional reaction when Ann Landers asked whether women preferred hugging and cuddling "to the act itself." Twenty thousand letters were received in response. She struck quite a nerve!

Fear and apprehension, for example, inhibit desire as well as performance. Comfort enhances it. Newness inspires it. Tenderness promotes it. Danger and a sense of the forbidden intensify it. Then love enters the picture and confuses everything. When all these forces come together and have their way with you, is it any wonder you feel a bit off-balance? To get things into perspective again, let's take a closer look at the four main sex-drive manipulators.

A woman's sex drive is far more subtle and complex than a man's, laced with emotions and not as obvious to the naked eye.

Some Like It Hot—Active/Aggressive Sex Drive

Testosterone, as you learned in the last chapter, is responsible for the active pursuit of sex and the drive toward orgasm. She wants one, and might get angry if you don't help. Testosterone motivates her to seduce and pursue, but also promotes masturbation. Women with higher testosterone levels masturbate more frequently, even when a sexual partner is readily available. In this respect, her sex drive is much like a man's, but that's where the similarity usually ends, and why men find women so confusing.

Take Me, I'm Yours!—Receptive/Passive

Yes, I know, lots of you men would just love to "be taken," but when it comes right down to it, you get uneasy unless you are the one initiating (in control). That is because estrogen governs receptive sex: the desire to be penetrated and taken, causing longings for "having a baby." Consequently, desire for intercourse and other heterosexual activity is stimulated by estrogen. Normal levels of estrogen generally produce a sense of well-being, the need to be intimate and to be held. But there is more. Estrogen also puts out the welcome mat, quietly, unobtrusively, and gently. Being awash with estrogen, a woman welcomes and enjoys sex. Orgasm, to the astonishment of most men, and some women, becomes incidental.

Estrogen in primates and other mammals dictates receptivity, limiting their interest to estrus. It also enhances their sexual attractiveness to males. Human females are not governed so ruthlessly by estrogen responses, although it is still a major sensual force within their hormonal cycling, as we shall see. They can seem downright fickle at certain times.

Sex Kittens—Proceptive/Seductive Sex Drive

Estrogen also rules a woman's proceptive sex drive with the help of oxytocin, progesterone, and a little LHRH as you saw illustrated above. With estrogen and oxytocin working together, the lordosis response is at its most intense. The cat is loose and on the prowl.

This seductive, persistent, but gentle "in your face" sexuality is manifest by the "sex kitten" in women, named after the cat who so richly deserves the honor.

There is nothing passive about this behavior, but it is quite different from the blunt and direct testosterone approach. She wants a man, not a vibrator. She chooses rather than waits to be chosen. But she persists, rather than insists.

It may come as a surprise, but 60 percent of men in one study say they have had sex with women when they didn't really want to. This is the closest that most men come to the receptive sex drive. They weren't forced into it, but most had succumbed to women who touched them with sex being the obvious intent, or who started taking off their own clothes or undressing them. One man explained that a man can't say no when a woman makes an advance or she will think there is something wrong with him.

Her lordosis maneuvers have a very high rate of success. Aside: Since men are stronger and never consider themselves overpowered or in danger, is that why most can't comprehend the concept of date rape?

Ready Rats, Neutered Cats, and Smelly Goats: When a man pats a woman on the behind, it's a sexual overture left over from our evolutionary past. Whether she arches her back—a human demonstration of lordosis—or shoves him away depends on which peptide is pulsing—oxytocin or vasopressin (which we'll get to in a moment).

The presence of oxytocin in sensory nerves makes the skin of the whole body complement that special intensity within the genitals and the breasts. From there, oxytocin plays a leading role in taking lovers from affectionate contentment to the peak of orgasm and back again.

The lordosis response in rats is a dramatic example of the female reaction to touch. Oxytocin can be released in rats just by gently stroking their backs.

It is a reflex guided by the touch of the male on her rump, literally. But the full reflex will not proceed unless it is sensitized by sufficient estrogen, which defines periods of sexual receptivity or estrus.

Goats have been studied, too, for their hormonal profiles. Increases in oxytocin occur during estrus in female goats just before sex begins, triggered when the male is nuzzling the female. This increased level lasts for fifteen minutes after the male has been removed. Courtship nuzzling or even just the smell of a male goat can release oxytocin in the female. Actually, the presence of the male goat is the strongest cue, followed by his smell. But once the smell of the male releases oxytocin in the female goat, she will sometimes show lordosis even though no male is present, just in case.

LHRH also gets involved. In ovariectomized marmosets spiked with low-dose estrogen, LHRH enhanced proceptivity within two hours. Move over, alcohol. Men will be slipping LHRH in women's drinks instead.

You Think Too Much?:   In female sexuality, curiously enough, vasopressin inhibits receptive sex (lordosis) while improving our ability to think clearly. That's nice! Studies have shown that vasopressin injections into the brains of female rats prevent lordosis, while surgical removal of their cortex (the thinking center where inhibition is located) in female rats *increases* lordosis. In sum lordosis improves when thought areas of the brain such as the cortex and septum are removed. Vasopressin stops it cold. Do you realize what I am saying? Less thought, more sex! Does this mean the less a woman thinks, the more receptive she will be? Stupid is as stupid does?

Aversive/Reverse Drive

Women, much more so than men, have a fourth gear: *reverse*. Their gearshift stops at high (testosterone), low (estrogen-oxytocin-proceptive), neutral, (estrogen-receptive) and reverse (progesterone, prolactin, vasopressin, and serotonin). It may be hard to accept this other side of woman, but looking clearly through vasopressin eyes, this phenomenon won't disappear by pretending it does not exist. Sometimes our hormones just won't accommodate. Rather than being sabotaged by this aspect of our nature, it might be wise to incorporate and harness its powers.

PROGESTERONE: SEXUAL DISORIENTATION

Let's learn more about this resistant sex drive. When her hormones put on the brakes a woman goes into reverse. While estrogen motivates a woman toward sex, progesterone and, to a lesser extent, prolactin have the opposite effect, tripping a woman into her fourth gear. Progesterone just says no to sex. Like many of its relatives, however, progesterone sends its share of mixed messages. By contrast to DHEA, which sensitizes special sexual spots, progesterone numbs them. If a woman has sex while she is high on progesterone, she isn't likely to feel much. Yet it does sometimes increase a woman's receptivity, and proceptivity (primed with estrogen) making her feel nurturing and somewhat sedate.

What Is Progesterone?

Progesterone is produced by the ovaries, the adrenal glands, and, during pregnancy, by the corpus luteum. It is a grandchild of cholesterol, which breaks down into pregnenolone and transforms into progesterone.

Progesterone is a sexually disoriented hormone with a schizoid personality. It isn't sure whether it is male or female and gets quite cranky about it. I suspect it is bisexual. Progesterone is primarily an androgen, namely a male hormone (like testosterone), treacherously involved in the female menstrual cycle, pregnancy, sex drive, and mood. We know it best as the hormone that goes up during the last two weeks of the menstrual cycle and drops abruptly, shortly before menstruation, causing women to bleed. Males have very little of it—and prefer to have none of it. No wonder; it is used to chemically castrate perpetrators of violent sexual crimes by cancelling testosterone and consequently depressing their sex drive.

The sexual effects of progesterone are mostly the same for both sexes until after puberty, when it increases in girls to adult levels but remains the same in males. Just as testosterone is making boys preoccupied with sex, progesterone is cooling down adolescent girls, muting whatever testosterone effect they might otherwise exhibit.

PROGESTERONE PROFILE

MOST PEOPLE DON'T KNOW THAT PROGESTERONE:
- can reduce sex drive
- can cause depression, irritability, and weight gain
- can aggravate PMS
- has anticonvulsant properties
- reduces sexual attractiveness of females (scents)

PROGESTERONE:
- is primarily inhibiting
- inhibits LHRH/LH
- decreases testosterone
- increases MAO activity (mild depressant)
- is a sedative at low doses, an anaesthetic at high doses
- is more prevalent in women than in men
- reduces dopamine
- increases brain opioid levels
- increases fluid retention/weight gain
- decreases sensitivity to oxytocin
- increases body temperature at ovulation
- releases LHRH

AS TO SEXUAL ROLES, PROGESTERONE:
- decreases active sex drive
- sometimes increases passive receptivity
- decreases genital sensation and perception of touch
- decreases transmission and perception of pheromonal cues
- decreases uterine contractibility

AS TO BEHAVIOR, PROGESTERONE:
- promotes nurturing behavior and defensive aggression
- dulls perception
- can cause depression, irritability
- supports breast-feeding

PROGESTERONE HAS BEEN USED:

- to treat acne
- as a contraceptive
- in postmenopausal hormone replacement therapy
- to treat PMS
- to treat certain menstrual disorders
- to inhibit prostate growth
- to treat prostate cancer
- to chemically castrate sex offenders
- to treat certain problems with pregnancy

HOW WE CAN INFLUENCE PROGESTERONE:

Increases Progesterone:

- pregnancy
- nursing
- second half of menstrual cycle
- birth control pills
- LH

Decreases Progesterone:

- menopause

Here are some other ways progesterone affects our sex drive (in both sexes):

- It is a sedative in moderate doses and an anaesthetic at high doses, dimming your drive while making you more receptive.
- Because it depresses nerve arousal, it dampens your sexual fires by blunting sensations in general and in the genitals in particular.
- It has anticonvulsant properties (the opposite of estrogen, which promotes seizure activity). This sounds beneficial. After all, who wants convulsions? On the other hand, while orgasm isn't a seizure, the same drugs that stop seizures can stop orgasms, too.
- It also short-circuits uterine contractions, blunting orgasmic response —the opposite of oxytocin.
- Progesterone is a depressant. It increases MAO activity, which tends to cause depression (again the opposite of estrogen, which is an

177

MAO inhibitor and a mild antidepressant). If you are depressed, ordinarily so is your libido.

- It decreases LHRH reactivity, which also dampens drive. Paraphrased, it means that the sex-drive hormone produced in the pituitary is muted.

- It is responsible for the postovulation temperature rise that we all look for when trying to get pregnant. We don't know exactly what's cooking other than the egg.

- It makes you gain weight and feel fat by causing water retention in particular and physical discomfort in general.

- It can trigger irregular vaginal bleeding, which is a physical and aesthetic inconvenience for sex.

- It reduces dopamine in the rat brain reinforcement center. If progesterone reduces dopamine in the human, as some speculate, it would further reduce sex drive and pleasure.

- It increases brain opioid levels, sedating and diminishing perception of touch.

- It can cause fatigue and decrease information processing and verbal memory.

Primate and rabbit studies suggest that progesterone may reduce pleasant sexual scents and increase unattractive ones, even though sex drive is unchanged. So a rat doesn't smell as alluring even though it would still like to seduce another rat. This dimension of progesterone has not yet been studied closely. If the same holds true in women, their progesterone, while not reducing their sex drive, *may turn off their partner*. A woman trolling for men probably radiates estrogen, just like a scented lure. Progesterone cancels this effect, so women might be better off flirting during the first half of their cycle, before progesterone levels rise. (We'll discuss cycling later in the chapter.)

In general, what a sexually subversive hormone! It's also, as you can tell from reviewing its profile, contradictory and confusing. When progesterone occurs naturally, it appears to complement sexual desire and interest in its own ambivalent, untrustworthy way. Certainly most women do not seem to lose their desire or response entirely during the second half of the menstrual cycle (luteal phase) when progesterone levels are the highest.

The dual nature of progesterone can be seen in the behavior of rats. At first, it boosts sex drive in female rats, then it reduces it. When progesterone is added to estrogen, it makes female rats with ovaries go into heat, enhancing both passive and active sexual drives. Female soliciting toward males increases. Consequently, one is tempted to conclude that at natural levels, progesterone acts as a complement to estrogen stimulation in women as well.

When used as an artificial supplement for birth control or menopause (both situations which we'll discuss in detail), progesterone's effects are the opposite. At the higher dosages originally used, progesterones seem to have a marked dampening effect on sex drive and mood.

There are two basic types of progesterone that can be used in these treatments: natural and synthetic. Natural progesterone has to be administered by needle or suppository while synthetic progesterone (called progestogens or progestins) can be taken orally. Synthetic progesterone is chemically engineered either from relatives of natural progesterone or the nefarious male hormone testosterone. Those that are derived from the testosterone branch of the family (e.g., medroxyprogesterone) have the most powerful antitestosterone effects of all. These progestogens are superb sex offenders themselves. As noted, they chemically castrate men. The progesterone supplements derived from natural progesterone seem less toxic to sex than those derived from testosterone (progestogens), and some of the newer synthetics are kinder to your sex life.

In fact, presently used lower dose progesterone appears to have fewer negative sexual effects and perhaps a few positive ones.

The sexual side effects of progesterone have not been fully investigated. Clearly, the dosage women receive—whether for contraception (high) or as therapy for menopause (low)—varies, and consequently so does the impact on sex drive. There is much still to be discovered. The progesterone patch called Estracombi (Ciba-Geigy) only available in Europe may be free of most of these unpleasant consequences.

Progesterone, Prolactin, and Pregnancy

While this two-faced hormone can cut men off at the knees (and elsewhere), it is essential to protect and preserve pregnancy. So it supports reproduction while sabotaging the sex that makes babies possible in the first place. With a little bit of information and a fair dose of some common sense, however, a couple can weather a normal pregnancy with minimal sexual distress.

Working in concert with progesterone during pregnancy is prolactin. When a woman bears a child, this hormone helps her nurture (through nursing) along with progesterone. Unfortunately, it cancels sexual receptiveness when it rises, making these two hormones a deadly duo.

Prolactin fluctuates daily (about a forty-five-minute cycle in both men and women), typically dropping after meals, but during pregnancy it increases overall from trimester to trimester. By the time a woman is ready to give birth, her prolactin level is at approximately ten times its normal amount. Progesterone decreases uterine sensitivity to oxytocin, diminishing her sex drive and perhaps making it more difficult for a woman to be orgasmic because she won't be experiencing uterine contractions too often. It follows, then, that progesterone protects the pregnancy from premature labor.

LH peaks in the first few weeks of pregnancy, and estrogen increases rapidly during the first trimester (you recognize it through breast enlargement and tenderness), continuing to rise throughout the duration of the pregnancy. So, sexually speaking, if you are feeling well enough, you are likely to be *receptive* to sex (due to the high levels of estrogen), but not *proceptive,* due to the high levels of prolactin and progesterone, your natural sex offender drugs.

Sex During Pregnancy: Some women sail through the first trimester sexually robust and physically exuberant, but this is not the general rule. Yet by the second trimester, even women who felt awful during the first trimester tend to do pretty well. The nausea and vomiting have probably disappeared. Their abdomen is not yet enormous. The healthy glow of pregnancy can show. They may not feel great enthusiasm for sex but are quite capable of participating and enjoying intercourse as long as the man is gentle with his thrusting.

During the third trimester, as her girth increases, face-to-face sex becomes increasingly awkward. Here is where common sense comes in. If she is still physically comfortable with penetration, proceed accordingly but find a position where her abdomen doesn't get in the way. If she is physically or emotionally uncomfortable, find nonintercourse alternatives to enjoy until after delivery and healing.

If the pregnancy is not going normally, she should consult her doctor for individual advice about sex. Some general considerations can serve as good safety precautions. If there is bleeding or any sign of threatened abortion, avoid both penetration and female orgasm (including through masturbation). As I've mentioned, orgasm causes pronounced, expulsive uterine contractions that are similar to labor and could conceivably precipitate premature labor in a vulnerable pregnancy.

To prove the point, if you are pregnant and *have had no problem* called to your attention, enjoy a few orgasms by any method you prefer, and watch your uterus contract and move. You can see it knot up. It is not subtle. It is most obvious after the fifth month. Feel the force with your hand. Usually, the contractions squeeze the baby awake, and you get a few kicks for your pleasure.

As your due date approaches and if your cervix is ready to cooperate, an orgasm or two might induce labor. Some doctors encourage overdue women to have orgasms in hopes of accomplishing just that. Again, if you are at any risk of delivering prematurely, any unnecessary irritation of the uterus, especially orgasm, represents additional risk.

During a perfectly normal pregnancy, your husband or partner may avoid sex for several reasons. Most commonly, he is proceeding under the misguided but well-intended notion that he might hurt you. By abstaining, he thinks he is protecting you. Some men find it psychologically difficult to have sex with a woman once he sees her in the exalted position of mother (madonna) and assumes, mistakenly, that sex is no longer desirable. On occasion a man may find pregnancy physically off-putting. More often than not, men feel quite the opposite.

Progesterone and the Pill: Progesterone's talents aren't limited to carrying babies and castrating men. It can prevent babies altogether. It is widely used as an oral contraceptive in women, with or without estrogen. Hundreds of thousands of young women get zapped with progesterone without realizing it in birth control pills like Ortho-Novum, Lo/Ovral, and Demulen, to name just a few. In this role, it is better known as the minipill. More lately, it is developing a worldwide reputation as the notorious Norplant implant.
Remember Andrea and Will? At first he strayed after their marriage, but not for long once his vasopressin and oxytocin levels committed him. In fact, their sex life

couldn't have been better. Over time, nothing changed. Whenever they were together, which seemed to be less and less, they shed their clothes and made love. Then Andrea changed her form of birth control from a diaphragm to the Norplant implant. She was first inconvenienced by the irregular bleeding and intermittent spotting it caused. Along with it, she got irritable, depressed, and grumpy, which was disturbing to Will. He approached her less often because he found the frequent bleeding bothersome and her crankiness unappealing. Andrea's sex drive also evaporated in the process.

The Norplant implant seemed to be working great, but not in the manner originally intended. Basically Will and Andrea stopped having sex, she because of the bleeding and her reduced sex drive, he because of the bleeding and her irritability.

Norplant is a pure progesterone. And, as you now know, progesterone diminishes sex drive. So the very thing that was supposed to ease and enhance the experience of making love was ruining it. No one had discussed the sexual consequences with either Will or Andrea and it tossed their relationship into turmoil. After a few months, Andrea decided to get the Norplant implant removed, which, incidentally, wasn't as easy as she had been led to believe. But once it was done, their relationship returned to normal as the last of the chemicals cleared Andrea's system.

What is the lesson here? Many birth control methods contain high doses of progesterone in one form or another. Without essential knowledge of the effect these substances have on sexuality and mood, you are not in a position to make an educated choice.

A gynecologist told me recently that during the 1960s, when he was a resident in Canada, droves of coeds were coming into the health clinic to get birth control pills. They got a three-month supply, as was the custom then, but never returned to get their prescription refilled. This was the subject of lively discussion among the ob-gyn residents and staff, who concluded that their initial enthusiasm for sex had obviously disappeared. They joked about what kind of lovers these girls had chosen to bail out so fast.

In retrospect, it is quite clear what happened. The progesterone in the birth control pills abolished their sex drive along with their fertility. Most likely their relationships probably died of neglect and they had no further need for contraception.

Indeed, studies before 1970 on the sexual mood effects of oral contraceptives found loss of libido and an increase in depression in women taking contraceptives with higher progestogen content. However, with the lower estrogen and progestogen content of modern-day oral contraceptives, sexual and mood effects seem to be less frequent and pronounced. There are also studies showing no adverse effect and even sexual improvement using lighter dosages. Sometimes, however, these pills cause nausea and breast tenderness, which by themselves can discourage sex.

Bear in mind that when ovulation is suppressed and the normal cycle of sex soup altered, a woman's usual sexual peaks will change accordingly.

Indeed, consider this: Women who are breast-feeding, although less fertile than others, still require additional protection. To minimize the danger to the newborn, they are given progesterone-only birth control. Prolactin has already taken away their drive. Introduce progesterone and what effect can you expect? If prolactin didn't do the whole job, progesterone finishes it, diminishing enjoyment and responsiveness. Women still nursing after three months had higher levels of oxytocin, however, than nonbreast-feeders.

There have been few studies on the effect of progesterone on the nursing child, and while they have shown no adverse *physical* side effects, it will be years before we know for sure. Remember DES babies? Yes, that was given to pregnant mothers, not newborns, but the concept is the same in that until these babies become adults and there are enough of them to study, we really won't know for sure what the long-term effects are.

When progesterone is given abruptly by pill or injection and withdrawn, it can cause postcoital contraception—"afterthought" birth control. It seems to work but isn't well tolerated due to various side effects—bleeding, cycle disturbances, et cetera. Oddly enough, when progesterone is blocked altogether by another chemical, we can achieve the same effects—a paradoxical response. Mifepristone is a progesterone *antagonist,* meaning that it does just about the opposite of progesterone in every respect. The sexual side effects of RU-486 are not yet fully appreciated either, but could be promising. For termination of pregnancy, treatment doesn't last very long. However, RU-486 may have other medical values that could require longer courses of therapy. Keeping an eye on this highly controversial drug will provide more insights into progesterone's sex life. If its sexual effects oppose those of progesterone, the probabilities are that it will increase a woman's aggressive and proceptive sex drive, intensify her genital sensations, promote her drive for orgasm, facilitate orgasm, and intensify it—all speculative, of course.

Progesterone and Menopause: Older women know this hormone better as Provera, Depo-Provera, or Norlutin. Thousands of menopausal women take progesterone at the end of the month without really knowing why. Some are taking lower doses daily, so they don't have to complicate their schedules or their libidos. We'll discuss these treatments in detail in the chapters that follow.

AGGRESSION OR REGRESSION?

Under ordinary circumstances, just as men are the warriors, women are the peacemakers. Women don't have as much testosterone as men, and so they are somewhat spared the belligerent, self-destructive forces that plague males. At the same time, they may be deprived of a certain competitive edge. But since women have more progesterone than men, they become more irritable, moody, and sexually removed. Mellowing them are their relatively high levels of serotonin compared to the male, oxytocin in abundant supply, and estrogen, a gentle, ordinarily soothing antidepressant hormone.

These are the main forces that determine their aggressive and passive patterns. Since the chapter on testosterone covered male aggression, let's now take a look from the woman's side of the fence at her experience of aggression, fantasies, rape, and violence.

Women can be violent and aggressive. Of that there is no doubt. Just look at Betty Broderick, Lorena Bobbitt, and other battered women who turn on their tormentors. However, due to their hormonal makeup, socialization, and body size, women are proportionately less prone to anger and physical aggression than men. They may throw things, but have lousy aim. They don't generally get into fistfights. In fact, when someone attacks them, they tend to freeze—just like the female stickleback. It is perhaps an inborn evolutionary reflex that began with the earliest sea creatures and still persists today. Being still protects them against attack and doesn't provoke violent behavior in return.

Consider the criticism women have been subjected to when they are raped. "Why didn't you fight back? I didn't see any bruises. You must have liked it." Or how about men who think, "Well, you secretly want it. Besides, we know that when you say no it is just a game. You dress like that, walk like that, move like that, and then expect us to just stop? Who are you trying to kid?"

What has gone wrong here? How can there be so much misunderstanding between men and women over this issue?

It's easy. We are not alike. We are at opposite poles on the matter. If we look at these same issues through the filter of our hormonal differences, here's what you would see:

Rape Fantasies

Fantasies in large part, from what we know, are testosterone dependent, which seems to explain why men have more of them and why many women have such difficulty formulating any.

Women's fantasies evolve from childhood to include sexual content, but usually include a relationship theme. The question on the table is if women's rape fantasies represent a conscious or unconscious desire to be raped.

Here is one of my patient's favorite fantasies:

"I was coming home from work and stopped off at the store to pick up a few groceries. This guy I recognized as a neighbor said hello and gave me a look. I thought he was handsome in a Wuthering Heights *sort of way, but brooding and a little dangerous. He got behind me in the checkout stand and offered to carry my groceries to the car.*

"Later that night my phone rang. It was him. He said he had been watching me every night for months. That his window was across from mine and he would sit in the dark watching me get ready for bed. I was careless with my shades sometimes, and wondered if he had ever seen me masturbate, but didn't dare ask. The thought excited me. We hung up with no more words. I walked to my room, feeling aroused. I left my shade up intentionally this time. I took it very slow and rubbed lotion all over my body, giving special attention to my breasts and nipples, wondering if he was out there. I brought the lotion to the bedside, and turned the lamp down to a low glow, just enough so that he could still see. Then I slipped the penis-shaped vibrator out of the drawer.

"Sometime later, there was a knock on the door. I realized it must be him and that I had taken this game too far. I put on some clothes, intending to apologize, explaining that this was as far as it went. When I opened the door, before I had a chance, he pushed his way in, pinning me against the hall closet door, slamming the front one behind him. I tried to scream and push him away, but his mouth was covering mine. He was in a frenzy of passion, and couldn't be stopped. He threw me down and easily overcame my struggles, being alternatingly rough and tender, doing whatever he wanted, taking his time, savoring every move, saying, 'I had to have you. I have wanted you since the first moment I saw you. You are driving me out of my mind.' "

Who is in control here? She is getting raped, but according to her script, *he is the helpless one.* This man who could have any woman he wanted has to have her. Most rape fantasies of women are not violent fantasies at all. Instead, they are actually desirability fantasies. She is so irresistible that he loses all reason and ravishes her. Control? She has it. He lost it. He is her conquest. This is a power-over-men fantasy. I beckon, you fall.

This particular patient came to see me because she wanted help to get rid of these sexual scenes. She felt like a hopeless hypocrite. Here she was, a sincere feminist. She even worked with rape victims and yet she got aroused by these disgusting fantasies. What was wrong with her? Nothing was. She didn't want to be raped or hurt in real life. This was not wishful thinking. Just an erotic fantasy. More completely perceived as seduction fantasies, women who can have them without pangs of guilt are more sexual, more arousable, and sexually well adjusted.

Men typically don't have much in the way of rape fantasies. Assuming that they don't fall into the S&M end of the sexual spectrum, they don't find fantasizing about raping someone or being raped by males or females erotically appealing. Their fantasies tend to have minimal plots with greater focus on genital sex, oral sex, and other specific genital acts. They often use pictures or read colorless (from a woman's point of view) explicit porn.

When you combine the reputation women have for having rape fantasies with their markedly different patterns of aggression and resistance (from those of men), it's no wonder that some men hear "yes" when she says "no" —especially those men who have trouble taking a woman seriously unless she gets angry. This is not to imply that women have any fault whatsoever in this situation, but there is enormous territory for misunderstanding between the

YOUR RECEPTIVE SEX DRIVE

sexes on this score—especially when you add to the equation abused women, or the large share of women who have been socialized to be passive and to accept sexual aggression as inevitable in the male.

Forms of Rape

Meg is definitely interested in Alan, or she wouldn't go out with him. She wants him to like her in hopes that it will lead to a nice relationship. She dresses to kill, showing off her best features. At the same time, she needs time. She wants the relationship to build into something important. If all goes well, would she have sex with him eventually? Of course. She is attracted to him, after all. She just doesn't know him very well yet and wants to get to know him better, hoping to keep him interested in her long enough to do so. Therefore she puts her most sexual and alluring foot forward, but holds physical contact in reserve.

Alan is dazzled by the sight of her. "She wouldn't dress like that if she didn't want me. I am going to get lucky tonight." The last thing on his testosterone-laden mind is love, bonding, and attachment. He hasn't even touched her yet. That situation, however, he plans to correct as soon as he can.

They have a great evening. She thinks he's really exciting. He feels things are going just fine. She laughs at his jokes and lets him hold her hand. He put his arm around her in the car and she didn't pull away while they were dancing, even though she must have been aware of his erection against her thigh.

Estrogen was welcoming him, but no more. She was not yet proceptive, but he didn't know the difference. Also, vasopressin was in charge so she was using her head. She had no intention of spreading her body indiscriminately, and even though her estrogen might have been a little more wanton, she had no problem holding it in check.

Later, he got more physical. She held him off, saying she was attracted to him, but it was too much too soon. He kept touching and cajoling. She kept some distance, but thought he had gotten the message, so she relaxed into some more touching, with no intention of going on to intercourse. Her oxytocin levels were being stirred up though and she was starting to think a little less clearly. He was getting hotter and hotter. She wanted more touching, but within the line she had previously drawn. He kept trespassing and testing her limits. With all that oxytocin pumping, her gears shift. Meg is now not only fighting to keep him at arm's length, but with estrogen and oxytocin ganging up on her it is all she can do to resist a lordosis response and all that that entails. In fact, she is starting to slip—her brain and her body are responding to other cues. She can't seem to think straight. He is becoming more forceful as his testosterone shifts also from aggressive pursuit to orgasmic urges. Besides, she isn't fighting him. Her lack of resistance turns him on even more. She is doing her level best to hold back and not encourage him, but by now she either wants him too or feels emotionally paralyzed by his attack and indifference to her nonverbal and verbal signals.

Boy, do they have their signals crossed. When she gives up, he reads her as giving in. Because she cooperates, even though she does not participate, he is convinced she was by then as enthusiastic and willing—in fact as in need of orgasm as he. How could she have touched this long without feeling just as he did? Easy!

It is over. She is upset. He is oblivious. If he were to notice, he would be annoyed, not sympathetic. He was there. He knows what happened. She enjoyed it just as much as he did.

The evening ends on a sour note. She confides in her friends. He never calls again. She was just a conquest. The more she broods on how he treated her, the angrier she gets. Her friends spell out for her just how exploitive he was.

A few weeks later, when Alan is accused of date rape, he is stunned and angry. As far as he is concerned, he did nothing wrong. Nothing different than with most of the other women he had known. This one just had a bad attitude and was out to get him for some reason.

Violent date rape is an extension of this scene by a man full of too much testosterone, alcohol, and victory, who has much the same attitude—that "yes" means "no," and deep down she wants it. Then he gets too drunk or high to care to stop himself.

Violent rape is quite different in that the rapist is not deluding himself into thinking he is just a nice guy being led astray by a seductive woman. He is aggressive and full of intent. He usually knows just what he is doing, if not why. The woman under assault typically responds much like the stickleback female, and doesn't resist in hopes of saving her life. Her passive behavior may arouse him more, but usually not to violence. Should she fight him, though, it will usually provoke a physical reaction.

Women during date rape, and sometimes during forceful, brutal rape by a stranger, find their bodies responding against their will. In the same vein, men often ejaculate moments before they are hanged or shot. Yet these are not erotic moments. Anything but, in fact. Stress of that intensity can trigger orgasm. A woman is in fear for her life during a rape, even when no visible or verbal threat has been made. She knows she is overpowered, and if he chooses to, he can kill her. She is not much different from the man standing before the firing squad, blindfolded, wondering when the gun will go off—except that he is also invading her body and touching or hurting the very body parts usually used in sexual stimulation of a gentler sort.

It is not out of the ordinary for a woman's body to react sexually out of fear and/or habit. She is not at fault or to blame, but often feels profoundly guilty and can tell no one. Where she is in her cycle will influence how her body behaves. If it happens to be the peak time of sexual interest in her cycle, she may orgasm. If oxytocin and testosterone are relatively high, the same is true. If estrogen is flowing, she may lubricate copiously against her will, but if progesterone is at its peak, she is unlikely to respond physiologically at all, and the rapist may find himself with a tiger by the tail.

As you can see, the daily and monthly peaks of a woman's hormonal cycle have a profound influence over her sexuality.

WHAT DAY IS IT? SEXUAL PEAKS AND CYCLES

Anytime soon, men will stop asking women, "What is your sign?" and try to find out what number their day is instead. "Day twenty? No thanks! I'll call you in a couple of weeks."

Roseanne, in one of her comedy skits, plays several of the twenty-eight women who live in all of us, taking a tour of our monthly moods. No wonder men are confused about women and say they just can't figure them out. Each and every day this huge group of hormones changes—often just a little, sometimes quite a lot.

Now that you know the powerful characteristics of each of these chemicals, it is obvious that emotional, physical, and sexual forces are shifting daily. Women are predictably unpredictable. Those with regular cycles do follow a pattern. Those with irregular cycles can make pretty good guesses about where their hormones are, based on how they are feeling and behaving. Here are typical profiles of women when each hormone peaks:

Testosterone-High Woman:

Dinner Talk: *I'm taking* you *out tonight.*

Evening Dress: *Nude*

Sexual Overture: *The Jacuzzi is hot and so am I. Let me tear off your clothes.*

Estrogen-High Woman:

Dinner Talk: *I fixed some appetizers for you. There is champagne in the ice bucket. Would you set up some candles and music for us?*

Evening Dress: *Black lace lingerie*

Sexual Overture: *Would you like a back rub? I'll heat up some oil.*

Ovulating Woman:

Dinner Talk: *Want me to cook you something, hon? What would you like?*

Evening Dress: *Silk pajamas*

Sexual Overture: *Sure.*

Progesterone-High Woman:

Dinner Talk: *I don't want to move, I'm so tired and depressed. Besides, I feel so fat and bloated, the last thing I want to do is go out. I can't fit into anything anyway. Look at me. I'd rather just read a book or watch TV. Anything good on?*

Evening Dress: *Flannel night gown*

Sexual Overture: *Not tonight dear. Just hold me.*

PMS Woman:

Dinner Talk: *Get your own. I don't feel like cooking tonight. As a matter of fact, why don't you get mine for a change? You never do anything around here but sit around watching your stupid sports. You always expect me to wait on you, and never offer to help. I don't know why I put up with you. When was the last time you did anything nice? I haven't seen any flowers since the last funeral.*

Evening Dress: *Your old torn T-shirt*

Sexual Overture: *Where's my vibrator? Did you borrow my batteries again? I told you never to do that. Now what am I going to do?*

Does any of this mean that women are unstable and untrustworthy? Of course not! Compared to men, we are peace-loving, easy sorts, not irritable warmongers cycling several times an hour. That argument is no good anymore. We don't need to vote on whose mood swing should be elected. Suffice it to say that we both have them and they are quite different.

As we discussed in Chapter 1, during the course of a woman's monthly cycle there are also several sexual peaks. Here's a review of various different studies reporting sexual peaks during the menstrual cycle:

- First half: Eighteen studies showed follicular peaks (days six through seven).
- Ovulation: Eight studies showed peaks at ovulation (days thirteen through fifteen).
- Second half: None of the studies showed peaks between ovulation and PMS.
- Premenstrual: Seventeen studies showed peaks premenstrually (days twenty-five through twenty-eight).
- Four identified peaks during menstruation.

It is also important to note that many women notice no sexual cycles whatsoever. That does not mean that none occur, just that they are not sufficiently pronounced to be noticeable. It is also important to appreciate that it is almost impossible for a woman to identify the presence or absence of a receptive sex drive unless there is someone involved who initiates.

Soup du Jour

The First Two Weeks (Estrogen Dominates):   Estrogen · pumping. She is flirting, soliciting, and arching, flexing her sexual muscle. She is feeling great, confident and social, on the prowl.

Ovulation (Estrogen is high; Both Testosterone and LH Peak; So Does PEA):   LH triggers ovulation and boosts libido. PEA, our natural amphetaminelike stimulant, may potentiate the sexual urge and/or increase the desire for a mate. FSH (follicle stimulating hormone) is high and estrogen is already at its peak, but has gotten there gradually. Testosterone and LH (leutinizing hormone) probably dominate sometimes, because the rate of change here is probably more important than the absolute amount of each hormone at any given time.

In women, blood testosterone levels increase progressively from the time menstruation ends, peaking at ovulation and gradually decreasing as menstruation approaches again. This mild rise and fall of total testosterone occurs for free testosterone as well. Most people, including scientists, think that women manifest their most intense aggressive sex drive at this time (to promote conception and ultimately survival of the species). Yet while some women do identify a sexual peak at ovulation, the most common peak is premenstrual. The key difference between these two sexual peaks is that the ovulatory drive is for penetration, and is predominantly receptive (which is obviously good for conception). Even though testosterone peaks and progesterone is low, at this time apparently estrogen dominates because there is just so much of it. Consequently, feelings of well-being and compliance increase along with receptiveness to sexual advance. The premenstrual drive, on the other hand, is for masturbation and is predominantly active.

Assertive sexual advances may be present at ovulation, but they are subtle by comparison to the fact that at this time of the month, a woman's sexual receptiveness is at its peak—that alone is enough to promote sex. Oddly enough, one study that evaluated *physiological* sexual arousal during different phases of the cycle showed that physiological arousal capacity was *lower* at ovulation than at any other time. There are many contradictions and much to be learned.

As always, we can take some clues from the primates. Researchers have found that female monkeys become more sexually attractive to males when estrogen peaks at midcycle, prompting sex when females are in their most receptive state.

Second Half: After Ovulation, but Before PMS (Progesterone Dominates; Serotonin and Estrogen Are Low; Testosterone fades):   When progesterone levels are high (and estrogens are low) during the second half of her cycle, a woman may feel nurturing and prefer touch and cuddling rather than "genital" sex. She is neither interested nor particularly receptive, and if she does allow sex, orgasms might be elusive, not even of interest.

We made love last night, not on my part because from a compulsive need to do it, but because we sort of drifted into it. In fact, to begin with I wasn't even sure I wanted to. But once we'd begun, I was very glad because I was in one of those marvelously physical states where all my sensations were velvet. Anywhere I was touched and any touch I put out to him felt floating and exquisite. Each bit of my flesh was full of tiny air bubbles, all receiving stroking delight. I didn't orgasm in the end, because it would have taken too long. I could have stayed stroked and touched all night. It was marvelous. Discovered in the morning why I'd felt so sensual. My period began, two days early. If someone could market whatever it is that floats through my body the night before a period, they'd definitely be the world's greatest millionaire. (Anne Hooper, The Body Electric: A Unique Account of Sex Therapy for Women*)*

Did you notice that she was too tired for an orgasm, but could have touched all night? A natural and marvelous state common to women during certain times of their cycle.

Paradoxically, instead of making her feel calmer, progesterone can also have the opposite effect, producing irritability, belligerence, and snappishness. We have no good explanation for this contradiction.

During PMS: The premenstrual sexual desire peak that most women experience has several explanations. Whichever one is accurate, the fact is that a complex combination of molecules is terrorizing the bloodstream at this time, with each hormone offering its special brand of torment.

When progesterone drops premenstrually, there is a rather sudden reversal in the testosterone-progesterone ratio that can cause a rebound testosterone-driven, aggressive sexual surge. Although testosterone does not increase, progesterone stops blocking testosterone receptors, making them responsive again. During this time, genital sensations revive, and orgasm is easier to achieve. Sexual sensitivity simmers. Endorphin and estrogen withdrawal contribute to this sexual crescendo by increasing the physical hypersensitivity that occurs from the simultaneous progesterone withdrawal.

Interestingly, estrogen withdrawal has been associated with menstrual migraine and seizure. Migraines and seizure have been associated with sexual hypersensitivity and orgasm. So it's not too surprising that some women describe orgasm, sexual pleasure, and arousal as greatest during this time.

Often, the impulse to masturbate intensifies, predictably, premenstrually, since the separatist hormone testosterone favors self-directed and self-controlled behavior rather than other-directed, cooperative behavior. Even though testosterone is low, estrogen is also at its lowest level and progesterone has all but disappeared: again, it is the *relative* relationship of hormones at play here, and the *rate* of change. Remarkably enough, this urge for sex can still occur in spite of severe premenstrual tension, perhaps because of it. This increased sexual desire is different from the peak that occurs around ovulation for some women, when the sex drive is predominantly receptive.

Since serotonin decreases premenstrually it may account not only for some increased sex drive, but the accompanying irritability and downright meanness. Along with all this, a woman also becomes hostile and irritable due to the absence of the tranquilizing effects of progesterone—some combination! Masturbation is safer for everyone concerned. And, in fact, can be a much-needed release of mental and muscular tension.

Menstruation: Just prior to menstruation, progesterone drops even more, triggering menstruation. In Catholic school, I was taught that the disappointed uterine lining, not getting a fertilized egg to raise, weeps blood.

Bleeding, due to the mess and the distinctive odor, discourages sex among some men and women. Yet this is one of the peak times of interest for more than a few females. Perhaps it is that the uterus is blood engorged, swollen, irritable, and more sensitive than at other times of the month. Perhaps it is because a few orgasmic contractions, if they could be inspired, would help the natural process of menstrual flow along, or perhaps it is just the refreshing aspect of the absence of progesterone, and the relative absence of estrogen allowing the little bit of testosterone to exercise more power.

PREMENSTRUAL SYNDROME (PMS)

Premenstrual syndrome can be quite a curse for those who have it—and for those who have to live with these women. When a woman becomes pre-menstrual, her sense of humor disappears fast. A healthy, even-tempered, pleasant, delightful person can become a moody, irrational, irritable witch overnight. The problems that bothered her a little bit yesterday bother her a lot today. She cries for no reason. The future looks gloomy. Her mate of the moment is the cause of it all. Of course.

Too many women don't even realize that their biochemistry is involved and take their unhappiness very seriously. One professional woman described her PMS symptoms this way:

"Normally, I am fairly even tempered, easy to get along with, soft and warm, pretty decisive. When my hormones go haywire, I transform into an emotional terrorist, a miserable misanthrope. I quibble, quarrel, and grump. I cramp, bleed, and gush. Brushing my teeth becomes an insurmountable obstacle. Washing is not an option. Moving? Unthinkable. I have no future. I can't do anything right. Money is slipping through my fingertips. Bills and creditors that aren't yet overdue take on the dimensions of Stephen King.

"I swell. I bloat. I ache. Teenage acne breaks out. Where did everybody go? I cry at the drop of a hat and scrap for no reason: Ma, they're being mean to me!

"The bank made a five-dollar mistake with the government. I know it will be a zillion-dollar fine and cost a fortune in accounting fees. Geologists are predicting another earthquake. My computers are freezing up. My car is overheating. Life is an obstacle course!"

Over one hundred fifty symptoms have been associated with PMS. Some are more common than others. PMS messes with your mind. You have difficulty concentrating, remembering things, get easily confused, can't make decisions, vacillate, withdraw, use poor judgment, feel silly, unreal, as if in a dream.

One patient expressed her transformation eloquently:

"Just before my period, I seem to lose interest in just about everything. Nothing matters. I can't make decisions. This morning I was trying to order breakfast at the cafeteria and couldn't make up my mind. When I finally made a choice, I couldn't stop thinking of all the other things I would have liked better,

and was unhappy with my selection but it was too late to change my mind. Besides, I couldn't remember which waitress took my order. When she finally returned, I added sweet rolls, hash browns, and pancakes on top of the sausage and eggs—more than I could ever eat, but I get such cravings."

You get intense cravings and indulge yourself. Your self-control takes a vacation. Your appetite seems bottomless. (Serotonin has dropped and you can't seem to get enough carbohydrates to appease you.) You will binge— sweets, salty foods, peanut butter lust, carbohydrates galore. Alcohol, drugs, and smoking intensify. Energy? You don't have any. Fatigue, lethargy, malaise, and sleeplessness alternate with the urge to sleep all the time and take daytime naps.

Emotions go haywire too, with mood swings, crying spells, impulsiveness, irritability, anxiety, depression, insecurity, sadness, and guilt. Much of the emotional upheaval is due to the relative estrogen/progesterone ratio and rapidly changing levels. It fluctuates as the lines cross: estrogen high/progesterone low promotes anxiety and irritability. Progesterone high/estrogen low promotes depression, crying, and confusion. You can't cope, feel empty, and may even have suicidal thoughts. Women actually become violent, committing crimes, child abuse, and even murder more often during PMS than at any other time of the month. PMS has even been used as a defense in murder trials. Extremely violent behavior, sexually provocative behavior, and psychosis have also been associated with PMS. Interestingly, those women who abuse their children only during this time, when treated with progesterone, find that these aggressive symptoms ebb and often disappear.

Women also complain that they become clumsy and accident prone. Their reflexes desert them and they run into things. Let's not even think about its effect on driving. On top of that, they may get dizzy or faint, sweaty, nauseous (prolactin may be up). Add ringing in the ears, tingling of the skin, itching, trembling, palpitations, headaches/migraines (thanks to rapidly dropping estrogen levels), bloating (aldosterone levels are up), weight gain, abdominal cramps (prostaglandins), backache, diarrhea, constipation, breast tenderness and swelling, acne, joint/muscle pain, recurrences of yeast infections and herpes. On top of that, your gums bleed, your contact lenses don't feel right, and your ankles swell up. Who wouldn't be a bit annoyed?

Just what is going on? There are many theories, but no one answer. Possibilities that have been proposed include hormonal imbalance (which is clearly a natural state), prolactin excess, endorphin metabolism, neurotransmitter disorders, and prostaglandin imbalance. (Prostaglandins are the fatty acids whose secretion stimulates uterine cramping, triggering labor and perhaps orgasm). There could be some medical conditions that aggravate symptoms, such as chronic yeast infections, low-grade uterine or chronic pelvic inflammatory disease, acid/base imbalances, and nutritional deficiencies. What all these theories mean in practice is that no one fully understands the whole picture, and there is no single dependable treatment even though combinations of some remedies may work beautifully for some women.

In spite of all these physical and emotional symptoms reported by many —not all—women, much of the medical world persists in perceiving PMS as a psychological disorder. A man with half these symptoms would be put into intensive care, but women are called neurotic. Indestructible is more like it.

Make Your Own Mood Chart

Medical books disagree about what percentage of the female population suffers from PMS. Some say less than 5 percent, others suggest 80 to 90 percent. I suspect that almost all women experience it from time to time to a greater or lesser degree, but most of them don't even make the connection. Too many times over the last twenty years, I have had women patients come to me with relationship conflict or mood disorders who insisted they did not have nor had they ever experienced PMS. I would ask them to keep a chart of arguments and troubles for three months, then go back and compare it with their cycle. There was almost always a clear correlation retrospectively.

Those of you who do not think that you have PMS, double-check your perceptions this simple way. Each day on your calendar make a note in the morning and before bedtime of your mood and state of mind on a +10/−10 scale. If you experience any of the symptoms, physical or emotional, that I have described earlier as associated with PMS, note them. If PMS is sabotaging your life, you can't correct it if you don't even realize it.

Getting Rid of the Monthly Menace

There have been many natural remedies suggested, all of which may have some relief to offer. For those who wish to utilize all their natural resources,

June Konopka, M.S., R.D., a nutritionist in San Diego, California, recommends the following:

"Consuming meat, dairy products, poultry, and eggs may contribute to PMS symptoms. This is because of specific fatty acids contained in them which produce proinflammatory prostaglandins. Evening primrose oil and flaxseed oil supplementation, when coupled with an avoidance of the above-mentioned foods, may reduce the production of these prostaglandins and reduce symptoms. [Vitamin] B6 and magnesium are also important nutrients for the PMS sufferer to take note of in the diet. They can be increased by eating whole-grain products instead of the refined grains, and by increasing the consumption of green vegetables, respectively. A third factor to consider is the ability of the liver to clear excess estrogen efficiently since many PMS symptoms are thought to be caused by an imbalance of female hormones. For example, if a woman has been consuming alcohol frequently and/or eating a fatty diet, the liver may be overtaxed, sluggish, or even damaged, making it unable to remove the excess estrogen from the body easily. Finally, every effort to normalize weight is crucial since it is known that fat cells can produce extra estrogen which could contribute further to hormone imbalances."

However, these methods often don't work for the most severe cases. Moderate to severe symptoms should receive a medical workup to rule out any other disorders.

Some of the most potent remedies boil down to good common sense:

- Practice good nutrition.
- Reschedule stressful events and critical decisions at home and at work whenever possible.
- Educate your spouse and family. Wear a PMS button when necessary, saying Don't push me. Look for PMS support groups.
- Don't binge. Resist compulsive eating and drinking, especially sweets, excessive carbohydrates, and peanut butter (because it's salty and promotes prostaglandins, also because it contains high levels of arachidonic acid, a precursor of prostaglandins, which can cause cramping). Increase intake of complex carbohydrates and fiber to fill you up without fattening you up.
- Decrease salt, coffee, tea, cola, chocolate. Salt makes you bloat. Caffeine products increase irritability and perception of pain. Alco-

hol depresses you and many of your hormones. Its most dangerous effect at this time, however, is to disinhibit you. You are already angry. It may remove your last vestige of self-control so you lash out at someone. Cigarettes will depress, irritate, and stimulate you. Smoking also *increases* menstrual pain, making "killer cramps" in fact.

- Get moderate aerobic exercise three to four times a week.

- I recommend in addition that you pamper yourself as much as possible. Rest. Take naps, bubble baths, cut your schedule to half time, get a manicure, a massage. Whatever soothes you at other times will please you all the more now.

- Vitamins that may be of special value are vitamin B6, vitamin B complex, vitamin A or beta carotenes, and vitamin E.

Other treatments that have been used for symptom relief include Midol for pain relief; diuretics to combat fluid retention and cerebral edema (swelling of the brain); nonsteroidal antiinflammatories (to combat the effect of prostaglandins); with birth control pills—progesterone combination (that lower prostaglandins); short-term antidepressants like Anafranil, Wellbutrin and Prozac—especially those that raise serotonin; calcium and magnesium to diminish muscle irritability and cramping. Bright-light therapy has also been used to advantage to treat the associated anxiety and depression. Because progesterone withdrawal or insufficiency is thought to cause PMS in some women, one of the most popular treatments is to give a woman extra progesterone, although the effectiveness of this approach has been questioned.

Treatment for extreme cases is to eliminate cycling altogether for three to six months with LHRH. When PMS is aggravated by a crisis such as a divorce or death, this method stabilizes hormonal balance long enough to allow the woman to identify the stressors and develop coping mechanisms for survival.

The most important thing is your attitude. Believe in yourself. Don't cave in to the neurosis/psychobabble theories of PMS. Don't judge yourself as neurotic or inadequate. Just make yourself as comfortable as possible. And don't use this time as an excuse to attack the people you love. Exercise self-restraint there, but enlist their aid by letting them know why it would be wise to treat you with kid gloves for a few days.

In particular, avoid scheduling elective surgery premenstrually or menstrually. Why have it when you are feeling your worst? Also, there have been some studies that indicate longer recoveries and poorer outcomes at this time. Also, some medications will affect you differently premenstrually, especially sedatives and antidepressants—even alcohol. It will take less to achieve the same effect.

Spicy Blood: There is one more thing you can do for yourself. Masturbate! Here's why: The three-day sexual peak during the premenstrual period can be partially explained by progesterone withdrawal, which saturates blood with sensory peptides such as oxytocin, thus triggering prostaglandin secretion. You now have "spicy" blood. This blood may cause pain and cramping (as with labor and orgasm) but also imitates the feelings of sexual arousal that occur when blood rushes to the vagina and uterus during the crescendo of sexual excitement.

In reality, as menstruation approaches, more blood flows to the uterus, making it boggy and congested (just as it does during sexual arousal). The uterus isn't especially smart and the signals to the brain can get confused. Consequently, the brain interprets these signals from premenstrual tension as sexual arousal, triggering the urge for orgasms. The urge to masturbate conveniently relieves physical symptoms of a boggy uterus due to its explosive contractions during orgasm, sometimes dislodging the cervical plug and triggering menstruation. The problem is that masturbation does not relieve the cramping uterus for very long, so a premenstrual woman has to do it often. Awkward, perhaps, but more fun than the alternatives.

THE CURTAIN CALL

You have now met all the players, the whole cast of characters. PEA, the instigator. Estrogen, the warm and seductive one. Progesterone, the resident schizoid—both nurturer and psycho. Testosterone—Rambo with a high dose of the Terminator mixed in. Luteinizing hormone releasing hormone (LHRH), the director, not above exploiting the casting couch. Prolactin, your duenna or matron, turning off your sex drive, making you feel motherly and inclined to breast-feed. Oxytocin, the sensual touching undertow, sucking you in against your will. Vasopressin, the family man, who keeps you from going astray and protects you from extremes.

Not to mention DHEA, which appears to contribute to the sex drive of both men and women in numerous ways, increasing your libido by sensitizing your erogenous zones to touch and promoting positive sexual scents. Dopamine and serotonin also play a role—sometimes delightfully and sometimes deviously, depending on relative degrees. Dopamine increases drive, while serotonin puts on the brakes. Other molecular forces influence sex drive beyond today's comprehension. And as we uncover more about the mysteries of oxytocin and vasopressin, our insights into the surrounding emotions will continue to expand.

You must see by now that the prevailing idea that sex drive is just about testosterone and estrogen is quite obsolete. The sexual cocktails that we mix in our blood using our natural hormones each have very distinct roles, but together can produce so many different combinations it boggles the mind— intellectually and sexually, literally and figuratively.

Another assumption reinforced by numerous experts *that we should discard right now* is that men and women's sex drive is basically alike and we should find someone whose libido is compatible with ours.

Men's and women's sex drives are more different than alike. This is normal. Imagine a man searching for a woman who wants sex on his frequency. What day of her cycle should he sample and what will it tell him?

Normal sex drives are different—from day to day, from person to person, and especially from man to woman. However, when communication and cooperation are working well for a couple it isn't hard to have wonderful sex frequently. It's not the ratio of our hormones as much as how we treat each other and how much we touch that determines both the quantity and quality of our sexual experiences. These hormones are so sensitive to our physical and emotional environment that we can influence them enormously for better or for worse by what we say and do.

Now that you know more intimately the hormones that move you sexually, emotionally, and physically, you are in a better position to decide which ones you would like to do without and which ones should get their contracts renewed. No doubt there are features of each hormone that most women would love to skip—the breast tenderness of estrogen, the moodiness of progesterone, the irritability of testosterone. Men might willingly trade in their testosterone temper and their lonely lives, but not the self-confidence, assertiveness, and the lust that they enjoy from it. But each of the hormones that dwindle as men and women get older take with them qualities that have been a familiar and important part of our identity since adolescence.

The greatest hormonal changes, next to puberty of course, occur at midlife—particularly for women. Yet thanks to modern medical technology, the "change of life" has become optional. Choice is involved. Today, menopause is avoidable—unless you prefer to age without hormone replacement treatment.

First, however, we will take a careful look at what happens when menopause and viropause (male menopause) proceed "naturally." Then, we will discuss how to prevent menopause and viropause by taking advantage of the newest and best of modern medical techniques.

7.

MENOPAUSE AND VIROPAUSE

Pat had never been vain. Natural good looks, but nothing extraordinary, had been life's gift and eased the way. But ever since passing forty, the mirror had been talking back. "Look at your eyes. Bags! Crows feet? More like the Eagles Have Landed. I don't even have to close my eyes anymore. My lids do it for me. My waist? What waste? You mean where all that toxic stuff I eat decided to make its landfill?

"How did age creep up on me without warning? Overnight I went from lean, taut, and tight to fat, soft, and flabby. I've done carbs, meat, and beer all my life —never did me any harm until now. The doctor says my arteries are clogged and I should take up golf.

"I thought about seeing a plastic surgeon. I hear they can suck fat out faster than you can inhale calories. No problem. A little facial work, cover that gray, and I'll pass for thirty again.

"But it's the sex thing that really gets me down. Maybe I could have some of that fat transferred farther south—enlarge my penis. Amazing what they can do these days. Thing is, I don't think it will make it stiff again.

"As if that weren't enough, these other changes really annoy me: I drip and itch all the time. I get these infections I never had before. The doctor calls it "senile penisitis." Says it's inevitable. Even if I could get hard, all I want to do is wrap my penis in bandages. Who would want to get near it anyway? Especially if they found out about its little surprises. For one thing, I lose urine every time I laugh. Who can enjoy a joke anymore? The joke is on me, so to speak.

"Then I have this pink bulge coming out the tip of my penis. It scared me to death. Doc said my bladder was dropping. Common at my age. It won't hurt me, though. He told me to learn to live with it. Easy for him to say.

"On top of that, my penis is shrinking and my testicles—can't even find them anymore. There is not enough left for a good scratch. Where did all life's pleasures go?

"My penis isn't the only thing that weeps. . . . I get these crying spells—right in the middle of a board meeting, I start to sob for no reason. Older guys should come equipped with handkerchiefs as part of the uniform. They could do double duty for hot flashes. One good one and my collar wilts, my glasses fog, and my suit sticks to me.

"With all this bad news on the outside, I wonder what's going on on the inside. If my heart doesn't give out, my brittle bones will break. Maybe my mind will go first. Come to think of it, I can't remember much anymore, anyway."

By now, you must think I have lost my mind. What doctor would allow a man to get into this condition? Sounds to me like he needs immediate admission, but I would have a hard time choosing between mental and medical.

Yet more than 43 million women in the U.S. experience variations of this theme, commonly known as menopause. Whether women are going through "natural" or surgical menopause, they are being told to grin and bear it. Their doctors say it is normal, and their gurus coach, "Just live through it." Even those who have built solid reputations championing causes for women have advocated the "natural approach," including Germaine Greer:

> *The object of facing up squarely to the fact of the climacteric is to acquire serenity and power. If women on the youthful side of climacteric could glimpse what this state of peaceful potency might be, the difficulties of making the transition would be less. Calm and poise do not simply happen to the post-menopausal woman; she has to fight for them. When the fight is over, her altered state might look to a younger woman rather like exhaustion, when in reality it is anything but. The difference between [the younger woman's] clamorous feelings and the feelings of the silent, apparently withdrawn older woman is the difference between the perception of the sea of someone tossing upon the surface, and of one who has plunged so deep that she has felt death in her throat. . . . I wouldn't have missed it for the world. (Germaine Greer,* The Change*)*

This "natural" approach to menopause is like an expedition with Outward Bound. A woman is thrown into an unfamiliar wilderness without any of the tools she used to depend upon, and expected to forge her way across dangerous territory—without the hormones, body, and mind she was once able to count on.

Men experience a "lite" version of menopause—physically, that is—called viropause. Their hormones and neuropeptides diminish, albeit less abruptly. Their bodies sag and change shape. Characteristic medical conditions like an enlarged prostate develop. Sexual functioning is often compromised by hormonal imbalance, disease, medications, mind, or mood. Their stamina and temperament alter as well. Emotionally, like their female counterparts, men can have repercussions from viropause of catastrophic magnitude—including severe depression and suicide. Yet often they are less well equipped to deal with these extremes than women.

The tragedy isn't that this happens, but that it doesn't have to happen. Modern medicine can prevent most of these problems.

Menopause and viropause are endocrine disorders, not figments of a neurotic's imagination. As with diabetes and thyroid deficiencies, if the missing hormones are not replaced, the patient will become acutely ill, chronically ill, and will die prematurely. Although some changes are correctable, irreversible general and genital changes will occur without treatment. However, by contrast to other endocrine diseases, menopause and viropause are *predictable*, and therefore *preventable*.

Chapters 7 and 8 will take you through menopause and viropause, their symptoms, the controversies surrounding them, their treatments, and the solutions. While this chapter will define and explain both syndromes, Chapter 8 will propose the possibility of preventing menopause and viropause altogether. For now, let's take a closer look at the physical and psychological conditions of menopause and viropause.

MENOPAUSE: FEMALE

Destiny began experiencing hot flashes and irregular periods along with night sweats in her late forties. By the time she reached fifty she was well into menopause. Her doctors had discussed hormone replacement therapy, but she declined. Having read every article on menopause in every upscale women's magazine—along with every popular book on the subject—Destiny was apprehensive of estrogens. A lifelong vegetarian who tended toward more natural remedies instead of pills, she was loath to put any foreign substances in her body. This reluctance was reinforced by the homeopathic brochures and booklets she found at her organic market. Besides, she had a family history of breast cancer.

Destiny prepared herself to age naturally, approaching the process with optimism and determination. Her family, however, wasn't as sanguine. Already they had found her in tears over a single dirty dish left in the sink. One morning, she asked her oldest daughter to pick her up and take her to the office because she had locked her keys in the car, only to discover later in the day that they were sitting right in her purse.

Tim found her preoccupied and sex was almost out of the question. She never really turned him down, but no longer welcomed his advances.

Destiny still enjoyed the idea of sex, but didn't think about it as often and found that the experience was vaguely uncomfortable. Although she had been told that lubrication would diminish, she didn't like the idea of using unnatural gloppy jellies. Because of recent infections, alternating between vaginitis and cystitis, she became additionally self-conscious about intimacy. By the time she was fifty-five, she was experiencing such pain with intercourse, she couldn't have it at all.

All the while, Tim was a model of understanding. He ran defense for Destiny when her sudden moodiness was aimed at their kids or friends. However, even he began to tire and her depression started to drag him down. He saw no end in sight and for the first time, he wondered if he could spend the rest of his life like this.

One day, a long-time client, who had recently run into the wrong side of Destiny, sat her down for a talk, saying, "I've noticed some changes in you and I'm worried. I know you're into natural foods and all that stuff, and forgive me for intruding, but I've been through this recently myself. It took me a long time to find a doctor who finally got my hormones back in balance. I want to give you her name and I hope you'll open your mind long enough to listen to what she has to say."

Stunned that a client of hers would move into more personal territory, and realizing for the first time the implications to her business, Destiny reluctantly made the call. After meeting with the doctor and learning about the different methods of hormone replacement therapy, she decided to find out for herself whether or not it would make a difference.

It took a year to get her completely back into balance, but she began feeling better within months. Her physician told her that it might take a little longer before *her irregular spotting was completely under control. It was only in retrospect that she realized how poorly she had been feeling.*

This was brought home by comments made by her family and friends, who finally stopped walking on eggshells around her. Tim was more relieved than anyone else to be able to recover the woman he had married.

Natural Menopause

Not quite two decades ago, premenstrual tension as it was then called and menopausal symptoms were still considered figments of a woman's imagination, or the "red flag" exposing an emotionally unbalanced female. Not so. Both are real medical conditions that affect your body and your brain in dramatically adverse ways. The hormonal upheaval most women experience directly impacts central areas of the brain, affecting emotional stability and creating physical disturbances.

Not so many years ago, most women were given the distinct impression that menopause began as a vague and subtle process; that it was gradual and gentle, but disturbing to those women who were probably somewhat unbalanced beforehand. Nobody mentioned lightning bolts out of the blue or a wake-up call in the middle of the night—the unsettling feeling of coming out of a deep sleep drenched and flushed—announcing, "Hello, you've just started menopause." I have heard many women describe an abrupt and sometimes frightening onslaught of symptoms while others do experience a quieter beginning.

Menopause, as most commonly defined, is cessation of the menses (bleeding, ovulation) for one year. This definition will soon be obsolete, because more women are opting for earlier treatment every day. Perimenopause and postmenopause refer to the periods immediately before and following. Altogether, the whole process can last from five to fifteen years. For practical purposes I am going to use the term menopause to cover it all.

The most noticeable and immediate effects of estrogen withdrawal (along with other hormones) are hot flushes and sweating. The less estrogen, the more perspiration. The body thermostat goes haywire. Hot flushes (skin reddening) are preceded by hot flashes, which are the subjective feelings experienced just prior to the physiological flush. It's a lot like a blush. You feel it coming on before you see it.

The flush is caused by a sudden increase in blood flow to the skin and lasts an average of one to five minutes. Due to the high concentration of blood vessels in the face and upper neck, flashing and flushing are most often felt there. Skin temperature increases by about seven to eight degrees. One woman told me she felt as if steam was rising from her when it was happening. Flushes are usually accompanied by night sweats, abnormal perspiration, and palpitations that often disturb sleep and cause daytime irritability. Sleep disruption typically occurs every ninety minutes or so in response to estrogen cycles. These bothersome temperature changes affect about 75 percent of women during and right after menopause. Weight gain is another unwanted side effect.

Various studies confirm that estrogen withdrawal can cause depression and that replacement can correct it. Some studies find the opposite. Consequently, there is still widespread skepticism among physicians regarding the relationship between menopause and depression, even though the weight of the evidence supports a connection. Women are often unhappy, tired, dejected, tense, nervous, and irritable. Many report poor memory, an inability to concentrate, lack of initiative, nightmares, and crying spells. Recent data suggests women suffering from estrogen withdrawal are more prone to Alzheimer's. Some women feel profoundly shaken and "uprooted" without fully knowing why. Others have suicidal thoughts. In the absence of treatment, understanding, and support, some women follow their impulses.

Unfortunately, only 10 to 15 percent of the women in the world are taking hormone replacement therapy. The rest haven't been offered the opportunity by their physicians or are so apprehensive and confused by the myths and opinions, they refuse treatment. That attitude helps researchers glean information in studies—for obvious reasons, we get most of our knowledge about what estrogen normally does in young women by studying what happens to these older women who don't have it—but doesn't address the powerful help ERT could be for the 85 to 90 percent of women suffering through menopause unassisted.

By now you've read about the impact estrogen has on a woman's life. Without this crucial hormone, a woman becomes progressively less female, less robust, less healthy, and less happy. I'm speaking of disease and deterioration, not just cosmetic effects, as many people think.

Reflect on the pros and cons of estrogen as you read along, those of you who have enough of it still washing around in your brains, and see what you think. And, if you find this story interesting, just wait until you hear about testosterone.

Gail Sheehy, author of *The Silent Passage,* describes in her eloquent style how menopause descended upon her as a most unwelcome intruder:

> *No more incongruous time or place could be imagined, the night I was hit by the first bombshell of the battle with menopause. It was a Sunday evening. Snug inside a remarriage not yet a year old, I was sitting utterly still, reading, in a velvet-covered armchair. . . .*
>
> *Then the little grenade went off in my brain. A flash, a shock, a sudden surge of electrical current that whizzed through my head and left me feeling shaken, nervous, off-balance.*
>
> *. . . [S]ome powerful switch had been thrown. I tried to go back to reading. It was difficult to concentrate. When I looked down at the pages I had just finished, I realized the imprint of their content on my brain had washed out. I felt hot, then clammy. I tried lying down, but sleep could not soak up the agitation. My heart was racing, but from what? Complete repose? I felt, for perhaps the first time in my life since the age of thirteen, profoundly ill at ease inside my body.*
>
> *. . . Usually optimistic, I began having little bouts of blues. Then little crashes of fatigue. Having always counted on abundant energy, it was profoundly upsetting to find myself sometimes crawling home from a day of writing and falling into bed for a "nap," from which I had to drag myself up just to have dinner.*
>
> *. . . For the first time since my early teens, when the sexual pilot light went on and I was warned not to want sex too much, I began to worry about not wanting it enough.*
>
> *. . . I went to see my conservative, male gynecologist, known as a superb clinician but short on communication skills. He measured my hormone levels. I was very low on estrogen.*
>
> *. . . "Could I be a candidate for hormone replacement therapy?" I asked.*
>
> *. . . "You're not in menopause, because you're still menstruating.*

You have to be menstruation-free for a year before I can give you estrogen replacement"

"But this, um, effect on my sexual response"—embarrassed, I fumbled for the words—"couldn't that be because I need more estrogen, like a vitamin supplement?"

"It's nothing I can help you with. Decrease in sexual response is just a natural part of aging."

The curt clinician washed his hands of me. I left his office feeling as though I'd just been handed a one-way ticket to the dumpster.

Today most physicians take the physical symptoms of menopause more seriously than they did a decade or two ago. The emotional symptoms, however, still don't get the respect they deserve. But now that women are speaking out and communicating information more freely with one another, progress is being made at a faster pace.

Men are considerably behind us.

VIROPAUSE (ALSO CALLED MALE MENOPAUSE, ANDROPAUSE, OR CLIMACTERIC)

Midlife crisis is no different from adolescence except that your face doesn't break out and you have more money. (Howell Raines, Fly Fishing Through the Midlife Crisis*)*

In the case of male menopause, we are still in the Dark Ages. Men have fewer guideposts to help them today than women had a generation ago. Only recently have we begun to understand the biochemistry of these events, tilting the scales toward a physiological explanation.

Will was only forty-five, but overnight, his life turned upside down. Having become one of the most successful, hard-driving businessmen in New York, he had accomplished everything in his career he had hoped for and more. But it meant nothing to him now. His recent sexual failure was so profound it clouded everything else. Despite his professional success, it was all he could think about. For twenty years he had lived and served the same company, eventually reaching top-

level management, expecting to retire someday with all the spoils of war. But his progress had been at significant cost. His marriage was shaky and his two children —well, he barely knew them. His career had always come first. He worked long hours, and traveled more than he was home.

His sex life had been all right up until now—or so he thought. His wife, Andrea, never complained. It was a bit monotonous, but what can you expect after twenty years of marriage? Will had been faithful to Andrea for a lot of years, but gradually he slipped back into his old habits. Sex on the road. It made monogamy bearable. Besides, it didn't mean anything.

Then Will discovered Andrea was having an affair and it devastated him. It wasn't at all the same when the tables were turned. Things weren't perfect, but he loved her. And look at how well he had provided for her and the kids. He did it for them, after all. What good was all his money and success without his family to enjoy it with? Ever since he found out about Andrea's affair, he couldn't function with her—or any other woman. He was completely and utterly impotent.

Will's sexual shock threw him into the depths of viropause, and he was in over his head.

As we've discussed, men have three testosterone syndromes. When they are younger, they have a more pernicious form of PMS: testopause, that fluctuates moment to moment, as described in Chapter 5. Annually, they are plagued by STS, the seasonal form. Be that as it may, men have one more obstacle to overcome. Their last testosterone hurdle, the one that cycles over them at midlife, is viropause.

Viropause involves a man's biological, physiological, and chemical changes in the context of his professional, social, and relationship values. While it usually occurs between the ages of forty and fifty, it can occur at almost any time.

Viropause can come on gradually or abruptly, triggered by the loss of a job, widowhood, divorce, illness, physical injury, financial setbacks, balding, decreased sex drive, or impotence—to name some of the most common psychological reasons for its occurrence. But there is a physiological process involved as well that is becoming increasingly distinct.

DHEA drops precipitously as men age (about 3 percent per year). By the time they are eighty, it is almost undetectable. When sixty healthy men twenty to eighty-four years old were compared with sixty healthy women in one study, DHEA was significantly *lower* in the men. Total and free testosterone also diminish each year, but more gradually (about 10 percent per year). Free testosterone fades faster than bound, perhaps because serum-binding proteins gradually increase by about 1 percent per year. Serum endorphin levels go up. Nocturnal penile tumescence goes down.

According to the results of the Massachusetts Male Aging Study, which studied aging males from forty to seventy, the combined prevalence of impotence (minimal, moderate, complete) was 52 percent. The presence of *complete* impotence tripled from 5 to 15 percent between ages forty and seventy. Much of the sexual dysfunction was correlated with disease, but impotence was inversely correlated with DHEA levels, meaning the lower your DHEA the more likely you are to become impotent. Cigarette smoking made complete impotence more probable in men with heart disease and hypertension.

It is difficult to distinguish between the sexual consequences of disease and those of "natural" aging, because so few studies exist (in scanning for references since 1985 on physiological aspects of male viropause under all its names, we were able to retrieve fewer than one hundred compared to nearly three thousand on female menopause. Nonetheless, it has become clear that many of the changes are independent of a specific disease, and merely correlated with advancing age—in my opinion, confirming the existence of a physiologically distinct male menopause.

When a man faces emotional, sexual, and physical changes without the comfort of guideposts and a game plan, he panics inside, not knowing where to turn, with no one to talk to. Doctors tend to discount his concerns, should he have the courage to bring them up. It is safe to say that most physicians today do not believe male midlife crisis has a physical basis, and treat it with psychotherapy and antidepressant medication—the same approach they used for female menopause decades ago.

Now that you have a better understanding of some of the causes of viropause, and that both menopause and viropause are physical conditions triggered hormonally, let's take a look at the sexual consequences of each.

SEXUAL CONSEQUENCES OF MENOPAUSE: WOMEN

A survey performed at a London menopause clinic zeroed in on some of the specific sexual effects on women. Out of a group of one hundred eighty-five women, the findings showed the following:

- Sexual problem first noted during perimenopause: 68 percent
- Loss of interest in sex: 39 percent
- Aversion to sex of any kind: 9 percent
- Vaginal dryness: 31 percent
- Painful intercourse (50 percent vaginismus): 37 percent
- Loss of clitoral sensation: 17 percent
- Loss of urine/fear of incontinence: 9 percent
- Altered touch (irritated by touch): 31 percent
- Decrease in orgasm: 23 percent

Sexual Scents: When trying to attract a new man, or the man you already have, pleasant body scent—or lack thereof—may play a powerful but subliminal role. As mentioned in an earlier chapter, primates and most other animals walk in a cloud of pheromones—sexual scents that provoke specific sexual responses, notably, attraction. Though the presence of these vaginal pheromones and their estrogenic nature has not been confirmed in the human female, it is probable that these secretions occur during sexual activity, throughout the day, and during sleep. Perhaps that's why youthful lovers like to wear each other's T-shirts.

With menopause, there would be a decrease in these pheromonal secretions, which we know depend upon estrogen in animals. Consequently, the ardor and intensity of attraction will lessen, with destructive physical consequences. Perhaps a woman will feel that she is less attractive, less appealing, by some fault of her own—not knowing the hormonal underpinning of this phenomenon.

A Touchy Problem: The drying and thinning of skin—which is the largest sexual organ in the body—comes with a steep romantic price tag since it reduces both sexual attractiveness and the pleasure of touching. As a result, touching diminishes, to everyone's loss. Sex could, and often does, disappear as well. Even if it doesn't, pleasure and orgasms might.

Philip Sorel has written about a perimenopausal and postmenopausal syndrome of "touch avoidance," which he attributes to decreased estrogen. He has eloquently described sexual deficits in touching and sensation and concludes that nerve transmission slows, resulting in a vague sense of numbness throughout the skin, including the erogenous zones. Not too surprisingly, this "numb love" is associated with other menopausal symptoms—loss of sexual desire, hot flashes, and insomnia.

Touch avoidance has several aspects. As estrogen decreases, so does oxytocin, which is dependent on it. Thus the desire to touch and be touched diminishes. When touching does occur, it doesn't feel as good because of the alteration in feeling and perception caused by the falling levels of estrogen. Irritability and itchiness of dry skin—also due to estrogen deprivation—contribute to touch avoidance.

Should touching continue and sex result, diminished sexual sensation in the genital area is sometimes replaced by "senile" vaginitis (itching, tearing, ripping, and bleeding of the skin), additionally discouraging contact and certainly taking away the pleasure. On top of all this, estrogen governs the receptive sex drive. If a woman doesn't have much, she won't be very inviting or available. If her testosterone has dwindled too, there goes her active (aggressive) sex drive. Just about all she has left is progesterone, which, as we know from the previous chapter, triggers her "reverse" gear. No wonder touching and sex go by the wayside.

Losing Your Grip: Once estrogens disappear, the vagina loses its grip. It becomes smooth, less elastic, and yielding.

The vaginal walls are composed of a wave of transverse folds called rugae. Without estrogen support, these rugae disappear and the vaginal walls becomes slick and smooth. A loss of vaginal tone makes sex more difficult and less enjoyable, not only for her, but for him, as well. Just when he begins to need physical stimulation the most, he gets less.

But that's not the worst of it. Without hormone replacement therapy, pelvic disease and other unpleasant conditions are more likely to occur.

"Senile Vaginas": This terminology represents a pretty disgusting array of menopausal perks that could turn just about anybody off sex forever, especially the women experiencing them.

Lower estrogens causes a woman's vagina to shrink, thin, dry up, and hurt. Cracks and lacerations of the vaginal walls can occur, inviting germs to grow. A rise in vaginal pH from acid to alkaline, along with a decrease in Döderlein's bacilli (a good kind), and thinning of vaginal and bladder linings increase the risk of vaginal revolt. The rise of pH with dropping estrogen levels cause drying, and along with decreased lubrication and moisture, aggravates itching and allows infections to take hold. While yeast infections (*Candida albicans*) are inhibited after menopause, lots of other infections come to visit. Trichomonal vaginitis thrives in the absence of estrogens.

The vulva changes, too—more subtly and more gradually, but also more relentlessly. There is a flattening effect, with loss of fat and moisture from the tissues. Small blood vessels below the surface react to minor tissue irritation, even in response to hot baths. Bleeding may be caused just from the friction of towel drying or from gentle penile thrusting. Itching causes scratching. Scratching causes bleeding. Small scars may appear around the lips, gluing the folds together. So much for your "sex skin"!

Sex after a long intermission can be traumatic to these tender tissues, causing rips and tears. A postmenopausal woman who hasn't had sex in years should see her doctor first to get rid of the shrinkage, "cobwebs," and "rust."

Somebody labeled these chronic discharges, itches, and related miseries that come along with menopause *senile vaginitis* before there were enough women in medicine to protest. The name hasn't disappeared. Neither have the symptoms. I suggest we find an alternative for both cases.

The Fallen Woman: Because of weakening tissues in general, the uterus sometimes "prolapses"—falls out through the vagina! It may even have company. Just like the uterus, the bladder can start drooping through the urethral opening—an unwelcome sight and quite a fright. Estrogens can keep these tissues toned and young so they stay put.

To make things worse, women can get miserable bladder infections, and problems including increased frequency of urination and stress incontinence with menopause. These symptoms can come on out of the blue, or sex can bring them on. Although they aren't inevitable, they are quite common in estrogen-deprived women. Ten to 15 percent of women over sixty have recurrent urinary-tract infections. Over half of women over seventy-five have some form of menopausally related gynecological problem.

Some women have a much less severe version of incontinence and lose a few drops of urine with a sneeze. The urethra and bladder trigone (a triangle at the bladder base, like a funnel) have high concentrations of estrogen receptors, and hence, are very sensitive to decreased estrogen. In fact, estrogen withdrawal aggravates chronic interstitial cystitis (trigonitis). Women in this condition may have to get up several times during the night and wear pads in the daytime to protect themselves when they cough, laugh, or dance. Loss of urine at orgasm is an additional treat. Some women avoid sex for this reason alone.

Too Hot to Handle: The indirect effects estrogen withdrawal has on sex are global. A woman with dizziness, palpitations, irritability, anxiety, insomnia, depression, and headaches does not make a very willing or responsive sexual partner. When you add to that the direct sexual complications that result from tissue atrophy that can lead to leaky bladders and painful intercourse, it is easy to understand how relationships suffer.

Not only do couples pay an unaffordable price, female menopause that is medically mismanaged can, in addition to the divorce, temporarily cost a man his sanity by pitching him into male menopause, if he isn't already there on his own.

Alan saw his wife transforming before his very eyes from a lovely peace-keeping lady who knew how to calm him down when he was in a snit, to a woman who knew how to throw a few herself. Of course, she had a good teacher, and boy, was he getting it back in spades, lips curled and all. Her mood and attitude aged her more than the years, and he saw himself growing old in her reflection. Her depression and anger were getting him down. She always seemed to have something wrong with her female parts—or maybe that was just an excuse. Whatever, the answer was usually no. Clearly she had no use for him. For a man who had always prided himself on his drive and performance, imagine his surprise when he discovered that he couldn't get an erection anyway.

Alan's job responsibilities were winding down as he edged toward retirement. Nevertheless, he would probably have had a lot more stable years in him if he hadn't had to contend with all these problems he had no way of understanding. This added burden tripped him over the edge of his capacity to cope, and he was unable to support his wife or anybody else. The emotional overload threw him into viropause.

SEXUAL CONSEQUENCES OF VIROPAUSE: MALE

From both a physiological and psychological perspective, we now know sexual response is much more complex than formerly believed. You've met a whole host of chemicals that contribute, yet there is one more that has great force at any age, but especially during viropause: adrenaline. The adrenaline reflex, which I first described in my book, *Bedside Manners*, in 1984, is the cause of *psychological impotence*. Also referred to as performance anxiety, it goes like this:

Alan enters the bedroom, now a forbidding place. He has had a couple of beers to get his courage up. He is anxious, to say the least (adrenaline surge), but saunters in with bravado, thinking that she is going to chop his head off. To his great surprise, she doesn't refuse him. Nonetheless, he approaches in somewhat the same frame of mind as the stickleback. His fear/fight/flight response is in high gear. Adrenaline is patrolling his system, battening down the fort and alerting the sentries. His pupils dilate so he can see danger—better peripheral vision. His heart beats faster to pump his blood to important parts—away from the brain and the skin to deep muscles and lungs for running or fighting.

The reason you get cold and clammy when you are apprehensive is that the capillaries that warm your skin have shut down. So can the blood supply to your penis. Direct orders from command central. Not necessary for survival.

Alan notices that it takes considerable stimulation from his wife before he even gets an erection—hands, breasts, mouth, tongue. It is obvious to him that she is having to work hard on him and her frustration is showing. She makes an exasperated comment and he wilts. As soon as Alan notices his erection has disappeared, he pumps another surge of adrenaline into his system due to abject terror, and ejaculates through a flaccid penis before he has even had a chance to enter her. She storms out of the room, completely furious with him.

Remember the guys who were about to be hanged? The adrenaline reflex also causes ejaculation if it is intense enough. That is how I can usually tell if a man is a premature ejaculator just by shaking his hand: cold, sweaty palms.

The point is that as a man enters viropause, he is much more susceptible to psychological impotence than he has ever been in his life. Let's add narrowed penile arteries from arteriosclerosis, the "sex offender" blood pressure medication he might be taking, a few drinks and the cigarettes that, interestingly enough, constrict the arteries in his penis and elsewhere, and this guy is a sitting duck for sexual dysfunction.

In Normal Working Order but On Strike: But let's suppose that he has no diseases, no sexually toxic medications, and a trusted, willing, exciting sexual partner by his side. He can still make full use of the adrenaline reflex and ruin himself sexually just by how he processes the normal sexual changes that come along with aging:

◆   Erection takes longer to occur.

◆   He usually requires direct mechanical stimulation to get an erection; a sexy sight or fantastic fantasy won't perk it up as it did before.

◆   The full erection doesn't get quite as firm as it used to. His "ten" in his twenties is an "eight" in his sixties.

◆   His urge to ejaculate is not as insistent as before. Sometimes he doesn't feel the need to have an orgasm at all.

◆   The force of ejaculation isn't as strong as it once was. The amount of his ejaculate is less, and he may have fewer sperm.

◆   The desire for and frequency of masturbation drops.

◆   The testicles shrink some, and the scrotal sack droops, so they may even seem bigger; The sack doesn't travel north and bunch up as much during arousal.

These are the normal changes associated with aging in the absence of disease. Add heart disease or diabetes, however, and sex is in serious jeopardy. Most experts still claim that desire feels as good as it ever did, so does intercourse, so does orgasm. And for some men who keep themselves unusually sexually and emotionally fit, this can be true. The sexual experience—

both the physical sensations and the emotional ones—can feel better than they ever did.

Usually, however, this is not so, and evidence is accumulating that builds a case for the existence of a predictable, measurable viropause in males. Ignorance and anxiety make it worse. Here is what often happens:

Alan sees a pretty girl; no erection. Hmmm. He wonders what is going on. (Adrenaline microsurge.) In bed with his wife that night, decides he'd better check it out. Turns toward her hardly aware of his concern, but pumps just enough adrenaline to overcome his incipient erection. When he gets no response from his penis—it's hibernating—he gets really anxious, his worst fears confirmed. More adrenaline squirts throughout his system, eliminating any remaining chance to function.

Unless he changes his attitude.

Fortunately, adrenaline leaves the system almost as quickly as it arrives, but not if he stays upset, which most men do under these circumstances. In fact, next time he gets into bed, it is unlikely that his equipment will work at all. He will be so convinced that he is impotent that he will be. Whether a man gets worried about the diminishing firmness of erections, his lack of desire to ejaculate, or no physical response to girl watching, it doesn't matter . . . enter the adrenaline reflex, and impotence becomes self-perpetuating.

Once a man in viropause develops erectile insecurity for any reason, certain other reflexes get set into motion that shift him into another sexual stage. He regresses to the libido-boosting methods of his youth: novelty. Remember the Coolidge effect? He starts thinking of having affairs even if he doesn't act on it, and many do. He develops a new or renewed interest in pornography. Hoping to inspire an enthusiastic reaction in his wife and thinking more stimulation will help his erection, he brings home a bag of tricks—sexual toys that vibrate and gyrate for them to play with together, and perhaps a movie or two. Some women are appalled. Most are less than thrilled by such "impersonal" mechanical and unusually noisy tactics.

Nonetheless, he searches for more and more intense visual and or mechanical stimulation which include, but are not necessarily limited to, magazines, prostitutes, sexual experimentation, threesomes, and younger and younger partners.

Tossed into turmoil due to his impotence, Alan had an affair with a woman younger than his daughters, bringing upon himself the disapproval of her family, his peers, and absolute strangers. In his own mind, he thought he could do things over again, recapture his mismanaged youth through her, and create the home and family he had always wanted.

Sound familiar?

Alan's behavior imitates those of his fondly remembered virile days: a renewed desire to procreate, fast cars, disregard of social conventions and age-appropriate attire, too-hip clothes, perhaps, or facial hair and a little dye to cover the gray so he still looks young from behind. He may even drink and do drugs as he used to. He goes to the gym, finds a woman half his age to match his fantasies, and tries to start over in almost every respect.

This sexual stage is typical of many men in their forties, fifties, and sixties who are going through viropause. But while he is less resilient physiologically, he is quite capable of responding normally sexually. The problem (or advantage, depending on your perspective), is that his erection is much more responsive to his feelings than his orders.

In trying to recapture that teenage testosterone high, there is also regression to adolescent emotional behavior: impatience, anger, sulking, and temper tantrums. Consequently, the very testosterone that empowers men sometimes defeats them.

Some men may even start developing an interest in genital-altering procedures like penile implants, enlargement procedures, testosterone shots, and penile injections. They are apprehensive about keeping up with the life they have created, and start having that old adrenaline reflex again. . . .

Tim and Destiny were thriving after she mastered her menopause. Her law practice was going so well, she went into business for herself. The romance revived, and everything seemed to fall into place. Sex had never been better—until Tim was unexpectedly demoted at work. He became anxious about keeping his job. (Adrenaline flowing.) Although he knew that they were financially secure on his wife's income alone, it didn't matter. Suddenly he felt useless. He was frightened and depressed, but couldn't express himself to anyone. He wanted to be strong and so he acted the way he thought he should, but underneath, in this competitive world we live in, he felt he had lost the battle. (Testosterone plunge.)

He started finding fault with the woman he loved and had been happily married to for thirty years, expressing resentments about everything, especially her work. In fact, he was so disturbed by his perceived reversal of their pecking order, that he quit his job and found work in another city. (Perhaps if he is alone his testosterone will rebound.) His urgent need to get out of town was a desire to reestablish a new territory. (Searching for testosterone.)

His leaving made Destiny sick. Oxytocin withdrawal and confusion overwhelmed her. With distance, the relationship became strained. He became defensive, uncommunicative, and progressively more angry with her lack of understanding (and his relative lack of testosterone). His whole personality changed. Instead of his usual energetic, positive, kind self, he turned into a virtual Mr. Hyde. This previously mellow man, who saw himself as a "banished male," started lashing out (testosterone surge) at the closest target, his wife, his love. He had to do it often, because the hormone high was short-lived. She didn't know whether he needed a therapist or an exorcist.

The last thing Destiny wanted with this creature was sex. Whenever he approached her, which wasn't often, she was cold and unresponsive. In response, so was he. When his penis refused to participate in this hostile atmosphere, he blamed it all on her. He was well into viropause.

Tim is suffering from a relative testosterone deficiency, and doing just about anything, scratching any itch, to rescue himself from the torture of his feelings. He is a strong and capable man in others' eyes, but sees himself as a defeated male with little hope for the future. What is worse, he not only lost to other males in the workplace, he thinks he has lost out to a more dominant female. Such is the curse of the modern world.

Self-Defeating/Self-Destruction: Another way that defeated (testosterone-deficient) males behave is to turn their aggression on themselves. Although it sounds contradictory, aggression is largely a social event. To be *successfully* aggressive, the animal or person must be mean to someone else. Increased testosterone precipitates this aggression toward others. Being mean to yourself doesn't count. Consequently, an animal biting off its own foot would not be considered aggressive or dominating.

Animals who lose fights do sometimes turn on themselves in a pathological manner. The most dramatic human equivalent of course, is suicide. The testosterone of men who jumped out of windows during the crash of 1929 was probably as low as the stock market. Self-punishment among humans after a major loss might involve substance abuse, "accidents," or sabotaging a valued job or a wonderful marriage. But watch what happens when a brooding man has even a small victory. His confidence revives, his future seems momentarily bright, and his libido recovers. All due to a testosterone surge.

So what do you suppose is happening to Tim's sex life? Well, with the depression, the defeat, and the subsequent testosterone deficit, naturally he lost all interest in his wife, except for the occasional times his verbal abuse and small victories gave him a temporary high. He needs a testosterone fix so badly, he is a sitting duck for just about any woman who comes his way, shows him a little sympathy, and agrees with his life's plan, flaky as it may be. Variety, novelty? He instinctively knows how to get his molecules in motion. But he has that vasopressin flowing too. If the prairie vole response holds true for us, then like the male vole, he too made a commitment, and even though he would like to erase it, with vasopressin circulating in his bloodstream, it has become a part of him now. The longer he stays away from Destiny, however, the fainter it will become. His oxytocin response will fade as well, and in not too long a time. By separating himself from her (another way to get his testosterone back), he will eventually succeed in leaving her for good, whether it is good for him or not. If he can't make the final break, sooner or later, she will do it for him.

Destiny has developed a successful career. Having recently confronted female menopause only to find that as her career soared, her relationship disappeared, she is bound to do some reflecting.

Albeit on a smaller scale, some women, usually hard-driving, competitive ones, also go through their own personal version of male menopause (the psychological part only, of course).

They reach a point of battle fatigue in which they reevaluate their priorities. Looking back at the emotional wastelands of health, family, and life, they conclude that the performance wasn't worth the price of admission. Lily Tomlin perhaps expressed it best in her Broadway show, *In Search of Intelligent Life in the Universe.* Portraying a woman in her thirties with babies, a thriving career—and a stress-related alcohol habit—she said, "I wanted it all and I got it. If I had known what it was going to be like I would have settled for less."

In essence, just like a huge number of her male colleagues, Destiny worked so hard to achieve success that she almost destroyed herself in the process—drinking too much, under chronic stress, children to raise, a prestigious job with tremendous responsibilities. Lots of money and no time to enjoy it. As Gloria Steinem said on reaching midlife, "I have become one of the men I used to hate." There comes a time when you are living like this (whether you are a man or a woman) when you stop and take a hard look at your quality of life. This is the psychological side of viropause, usually painful, but quite an opportunity for making major improvements.

SURGICAL MENOPAUSE

There are two distinct types of female menopause: natural menopause and surgical menopause. The general thinking is that this surgery is merely a premature natural event. In actuality, it is neither premature nor natural. Surgical menopause is *complete*. Natural menopause is incomplete and mild by comparison.

One third of women in this country have had their uterus removed by the time they reach sixty. In fact, most hysterectomies are performed on women during their early forties *prior* to natural menopause. I don't see a third of the men in the U.S. wandering around without precious parts of their genitals. Yet many physicians are considerably more reluctant to remove the testicles of a man for malignant disease than to remove the uterus and ovaries of a woman for benign conditions. I know from personal experience that doctors agonize a lot about whether or not to remove a man's testicle—just one, even for cancer!—no matter that the other one is all he needs. The ovaries, on the other hand, often get popped out by well-meaning surgeons, "just in case," while they're in there anyway amputating the uterus.

In a recent publication, Gail Sheehy described a Seattle divorcee who had just had a hysterectomy. Quoting her: "My doctor was a yanker instead of a saver." Says Sheehy, "It may sound like a nice midlife housecleaning, but in fact, removal of both ovaries is castration."

Most women about to undergo hysterectomy do not realize the full nature and implications of these surgical procedures. Here are some useful terms every woman should know. They are medical tongue twisters, but don't let that throw you. It is important to learn the distinctions in case some variation of this surgery is ever recommended for you.

When facing one of these procedures, women may know that the uterus and ovaries are coming out, but they don't understand what that will do to

---

SHORT CUTS

Partial (subtotal) hysterectomy—removal of upper two-thirds of uterus, but not cervix (rarely performed anymore)

Total hysterectomy—removal of uterus and cervix

Oophorectomy—removal of ovaries

Salpingectomy—removal of fallopian tubes

Adrenalectomy—removal of adrenals

Total hysterectomy with bilateral salpingo-oophorectomy—removal of uterus, cervix, tubes, and ovaries; only vagina remains

---

them physically and mentally. Much of this information has been discovered by studying surgically menopausal women. Here is what can happen:

Symptoms Following Total Hysterectomy

Uterus Only Removed: Removal of the uterus only does not cause menopause immediately, but accelerates it. This surgery is most like a premature natural menopause. The ovaries, even if they are left in place, will retire sooner than they would otherwise. In fact, some gynecologists routinely perform oophorectomy along with the hysterectomy in perimenopausal women because they assume the ovaries will soon stop working anyway.

One study of 164 women revealed the following incidence of symptoms:

- irritability—44.5 percent;
- nervousness—43.9 percent;
- sleeplessness—22.0 percent;
- aching joints, bones, or muscles—38.7 percent;
- headaches—33.5 percent;
- palpitations—30.5 percent;
- depression—27.4 percent;
- anxiousness—24.4 percent;

- vertigo—23.8 percent;
- discomfort during intercourse—17.7 percent.

Ovaries Only—Oophorectomy: Watch out for doctors who will remove your ovaries just because they are there. In the female, testosterone is produced by the ovaries and the adrenal glands. During natural menopause, testosterone decreases somewhat. During surgical menopause, women are deprived of the normal testosterone levels they would otherwise experience for most of their lifetime. The abrupt loss of testosterone and estrogens from surgical removal of the ovaries seems to cause much more depression than natural withdrawal from estrogens. Sex drive diminishes in some women, but not others, because the adrenal glands still pump some testosterone and lots of DHEA into their system.

If testosterone has a valuable role to play in women as a replacement or supplement after menopause, the most obvious void will show up in surgically menopausal women whose ovaries have been abruptly removed, and along with her two ovaries, most of her ability to produce testosterone. Given the frequency of this procedure, there has been an appallingly small amount of controlled clinical work to examine the sexual and psychological effects, especially the consequences of testosterone depletion.

Ordinarily, estrogen is the only hormone offered to surgically menopausal women (if any at all), but testosterone should not be overlooked in view of its potential sexual and psychological impact.

Adrenalectomy: Unless there is disease in the adrenal glands themselves, they are generally not removed, except in cases of certain cancers like breast and ovarian. The object is to get rid of some of the hormones they produce that might contribute to growth of the cancer. When these glands go, sex drive usually disappears completely because with them goes the rest of your testosterone and most of your DHEA.

Almost the Whole Works—Uterus, Tubes, Ovaries: Removal of the uterus, cervix, fallopian tubes, and ovaries (total hysterectomy with bilateral salpingo-oophorectomy) is an opportunity to research the most extreme sexual changes due to menopause.

With this procedure, only the vagina remains. The vagina shortens if some of it is removed along with the cervix, which can sometimes result in painful intercourse. In surgical menopause with the removal of the uterus and ovaries, osteoporosis can occur within two years. After four years, over 60 percent of women demonstrate osteoporosis because of estrogen deprivation.

Chemical Menopause: Chemotherapy commonly causes a menopause most like a total hysterectomy. Chemotherapy causes an abrupt form of menopause because these chemicals can destroy ovarian function and perhaps some adrenal functioning. This facet of chemotherapy has been little studied, and we don't yet know which "chemo cocktails" are better or worse as far as menopausal symptoms go. It is not unusual for women facing chemotherapy to enter treatment unaware of the probability of precipitating or aggravating menopause. It hits them instead as an unpleasant surprise for which they have had no opportunity to prepare.

Surgery and Sex: We have discussed the effects of hormone withdrawal on sex, but the sexual effects due to the mechanics of the surgery can also be serious. Accidental nerve damage can decrease genital tissue sensitivity. For some women, the cervix—rich in sensory nerves—is enjoyably sensitive to deep thrusting and this reaction is lost when it is removed. Also, most women have uterine contractions with orgasm, but few feel them. Those who do register these contractions savor them and experience them as part of their orgasm. For them, the loss can be profound.

And what about the G-spot? Is there such a thing? Some women swear by it. Most doctors don't believe in it. During hysterectomy and certain other procedures like resuspension of a prolapsed (dropped) uterus and bladder, the surgeon may carelessly remove it. If you are attached to your G-spot, protect it!

A doctor's wife I know had her uterus removed just because she was bored with bleeding. This may sound astounding, but is not uncommon. Depending on what actually gets removed, women who get a "hysterectomy" experience sudden unwelcome consequences of menopause they never expected or anticipated. Unknowingly, some women are trading a healthy or marginally diseased uterus for sexual problems, ill health, and reduced life span, along with all the rest.

Considering that the hormonal consequences of surgical menopause are usually more severe than natural menopause, even when the ovaries are not removed, surgical remedies should be approached conservatively. Certainly, if the uterus is diseased, treatment of the disease should be weighed against the medical problems caused by the cure. In addition, it would be thoughtful to provide a woman beforehand with comprehensive information about the postsurgical consequences, instead of limiting informed consent to what "structures" will be taken out. In California, there is a state law specifically requiring that physicians provide informed consent prior to this procedure, and that they present alternative therapies.

Sometimes for no explicable reason, except perhaps heredity, premature natural menopause occurs in the twenties or thirties. A thorough endocrine evaluation must take place to rule out serious pathology. Otherwise premature menopause can be managed much like ordinary menopause.

PROTECTION OF THE PROSTATE

I know it sounds like women are a medically mismanaged minority, subject to all sorts of surgical procedures, but they are not alone. The corollary for the man is his prostate. Protect it.

With age, it enlarges. And that's not good. Prostate enlargement increases urinary frequency day and night, interfering with men's sleep and their ego. A busy bladder is second only to baldness as a key marker for "over-the-hill" thoughts and is a competitive setback. No matter how young men look, or how virile they seem, excusing themselves at a dinner before another man or, God forbid, a woman is a testosterone-down situation.

Enlarged prostates eventually demand medical attention. The medicines used to prevent their progression typically are testosterone-defeating drugs, not good for mood or sex. The surgical procedures are not fun to contemplate either, but work well when done properly.

When urinary hesitancy, nighttime urination, or other symptoms become severe enough to disrupt day-to-day living, surgery becomes an option. It is also recommended when testing reveals urinary retention *or* impending obstruction. The worst-sounding procedure, but best if you can get by with it, is the transurethral prostatectomy, fondly known as the Roto-Rooter procedure.

I'll spare you the gory details, but in short, an instrument is inserted up through the penis, and chips out the prostate like coring an apple, opening the passage for better urine flow. The new and improved version uses a tiny roller to cut and cauterize, which reduces bleeding and healing time considerably. Usually, there is vast improvement without complications. Sometimes the shut-off valve at the bladder is damaged and a man backfires when he ejaculates. This sounds worse than it is. During an otherwise normal and enjoyable orgasm, the fluid shoots into the bladder rather than out the top of the penis. Some women have told me they love this dividend. They don't drip all day after sex, and with oral sex there is no ejaculate to swallow. The next time the man urinates, it is cloudy with sperm. To some men, this effect is very disturbing. For most, once they understand what is going on, it is incidental, unless they still want to have children. With this kind of procedure sex won't otherwise be affected. Men say orgasm feels just as good as before.

When prostatectomies are performed through the lower abdomen, or when the approach is between the rectum and testicles (the perineum) there is more chance of nerve damage.

Prostate cancer is the other, more serious, potential problem. It will be discussed, along with related treatments, in the next chapter.

THE ROLE OF VARIOUS HORMONES IN MENOPAUSE AND VIROPAUSE

The visible changes of estrogen withdrawal are clear harbingers that tissue, bone mass, and muscle mass suffer when normal estrogen levels no longer surge through a woman's body. To the same extent, a man's journey through viropause is heavily based on testopause.

But estrogen withdrawal and testosterone depletion are not the whole picture. Progesterone and DHEA need to be considered as well, not to mention oxytocin, growth hormone, and various others.

Growth hormone is produced by the anterior pituitary and is ordinarily secreted in six to eight pulses per day. The strongest surge occurs shortly after a person goes to sleep. The frequency of the pulses does not seem to differ between the sexes or with age. However, the duration of each pulse, and the amount of growth hormone that is released diminishes substantially with age. These changes can be found in men as young as thirty. After age fifty, growth hormone production stops altogether in as much as 50 percent of the population. Both men and women experience the following changes as this hormone declines: bone mass and density drop, muscle mass drops, fatty tissue increases up to 40 percent. Kidneys shrink along with the stomach, small intestine, liver, and spleen. Immunologic resilience diminishes.

DHEA, which decreases in both sexes from the thirties on, may also play a pronounced role in the midlife changes of both men and women. In women, its midlife effects may be masked by the greater trauma of estrogen withdrawal. At the same time, DHEA may actually buffer the severity of some of the symptoms a woman experiences initially. DHEA is more closely related to testosterone, however, and, as it diminishes, most probably accentuates the adverse effects of a man's testosterone withdrawal. In both sexes, however, decreasing DHEA increases weight, depression, and decreases sex drive, thereby contributing in part to this complex process.

What happens to oxytocin and vasopressin with age? It is difficult to say because so little research has been done with humans and it is not possible to study menopause in animals. They just don't seem to go through what we do. We also know that there are alterations in LH and FSH (follicle stimulating hormone) which both sexes have in differing amounts.

With untreated menopause, the hormone profile of men and women becomes considerably more alike. DHEA is pretty much the same in both sexes. Testosterone and estrogen are still quite different, but his testosterone has dropped while his estrogens have probably gone up. At the same time, her relative testosterone levels are higher since her estrogen has dwindled. Because they are approaching a much more compatible stage hormonally, from this point forward their relationship prospects improve considerably.

Indeed, the age-old conflict between the sexes seems to disappear as the balance of power shifts somewhat from the man to the woman of the same age. This new man rather enjoys leaving some of the driving to her, learns to relax and smell the roses. And she has assumed her new role with grace, not force, contributing her strength to the relationship. Together, perhaps for the first time, they walk side by side.

How can you take advantage of modern medicine and the most advanced technology to counter diminishing levels of critical hormones and preserve quality of life, sex, and relationships as the years go by? The next chapter will address these issues.

It is crucial that women and men become sufficiently informed about these molecules to become good, intelligent consumers of their own medical care. To treat vaginal and penile complaints only to the extent that local symptoms are taken care of means treating only part of the problem. Worse havoc occurs where it cannot be seen.

8.

PREVENTING MENOPAUSE
AND VIROPAUSE

Zothara was a biohistorian in charge of Molecular Time Travel. She researched, recorded, and reconstructed early human biological conditions the way ancient man resurrected dinosaur skeletons. She also ran the Experiential Subjective Education Unit for the Federation. No human had experienced natural menopause or viropause for centuries. Zothara's Federation had mastered menopause—eliminated the nuisance along with the associated health hazards. With the ability to take precision measurements of all the delicate fluctuating hormones involved and the technology to replace them with the same rhythmic patterns as those of a man or woman in their prime, perpetual youth, good health, and exquisite sex became their reality.

Zothara, although beyond two hundred years (she would admit to no more) had the biochemistry and hormone composition of a thirty-year-old—without the fertility, of course. Should she wish to bear children again, it could be arranged. During her monthly macromolecular analysis she would simply request a fertile cycle, unless she preferred the convenience of contributing an egg for extrauterine cultivation, and hope that her application was accepted.

Gender-specific molecular manipulation could create any hormonal state at will, and almost everyone preferred to keep their hormonostats set somewhere between twenty and thirty years. For instance, if a man wanted a little more free testosterone in preparation for a contest or a con-genital visit to one of his home ports, he could merely adjust the concentrations in his bloodstream accordingly. This was accomplished with an EFM (electron force microscope). Just by dragging the device's tip lightly over the skin, precise atomic and molecular readings could be taken of any body fluid or tissue to determine its biological age and state of health. After each lunar diagnosis—for it was now known that the phase of the

231

moon influenced hormonal fluctuations—the deficiencies were identified and precisely replaced through a variety of sophisticated drug delivery systems, the most usual being Transdermal Electrical Molecular Transport. This method could pulse hormones electrically into the system at exact intervals, imitating nature's own rhythms. The old subcutaneous pumps had been out of use for generations.

Not only had scientific technology succeeded in reproducing and sustaining ideal biochemical states indefinitely, it had improved on a few features. Men and women could actually change places—biochemically—temporarily or permanently. If they found their current gender sex or gender orientation unsatisfactory or inconvenient, or if they merely wished to vary their experience, they could request a special transformation, which could be performed in one of three ways: brain only, body only, or both. You could have almost any combination imaginable: a man's mind in a woman's body—complete transformation with cloned body parts to match—or any setting on the rheostat in between. A couple could even decide to adjust their central nervous systems (brains) to match at any position along the spectrum they preferred, while keeping their bodies gender specific—male and female.

Even so, uromates still argued over their respective settings:

"Honey, would you please turn up your oxytocin level, or I am just going to have to raise my progesterone? If you won't touch me, I'll just become untouchable! And don't forget your promise to boost your vasopressin next time you go into hyperspace. I want you all to myself."

"Okay, but only if you raise your serotonin level when I get back home. I think you should lower your aggression quotient."

What most people did, at least once in a lifetime, was to change their "brain's" sex to the opposite one for a month or two without changing their physical features, just so their uromate would stop saying: "You'll never understand what it is like to . . ."

Zothara had done so often, and was considering adopting the male molecular mode again to better endure her fourth uromate, Galen's, absence. The man's endocrine state was so much better at long separations. All Zothara had left of Galen, now that he was stationed indefinitely on the planet Darth, were his holograms, scents, and packed cells of strategic tissues to keep her cycles regular, her sensuality fresh, and her mood high. It had been illegal to clone ever since the Zeth Wars (except for replacement body parts). Even with the assistance of virtual reality to recreate the molecular memories of their erotic adventures and her *portable orgasmatron, she was discontent. Old-fashioned as it was, she wanted his warm arms to enfold her.*

Sound fantastic? Of course! But not for long, according to some experts. We already have the precursor to my fictitious EFMs: Operational AFMs (atomic force microscopes) can scan molecular surfaces, split molecules, pick them up, and relocate them with such extreme control and precision it is astounding. Physicist Eric Drexler is convinced that humans will be able to control matter at the atomic and molecular level, through the development of nanotechnology, biological systems, and synthetic chemistry, in the not-too-distant future.

We also have drug-delivery systems that drive molecules through the skin with battery-operated electric currents, patches, subcutaneous pumps, pellet implants, inhalants, nasal sprays, eye lozenges, medication transport molecules that release their passengers at predetermined destinations, linguets, and more. But where does that leave those of us who won't be around to benefit when the breakthrough comes?

PREVENTING MENOPAUSE

Today we can already reproduce and sustain the chemical composition of our hormones as they are at our prime of life. We cannot do it with the sophistication described above, but sufficiently well to successfully bypass the clinical symptoms of menopause—male and female—altogether. I have done it and helped guide patients through the process. A handful of doctors across the country and around the world are doing it every day. Menopause doesn't scare me or many others anymore.

Welcome to the world of hormone replacement therapy (HRT), modern medicine's Twilight Zone—your ticket to better sex, fewer flashes, and a happier disposition. Unfortunately, it's not quite that easy. For most women, the decision to undergo hormone therapy as they approach menopause is complicated by concerns about cancer, and conflicting medical opinion.

What woman hasn't anticipated menopause with dread—hasn't been afraid her sexuality would dwindle? What man hasn't had to cope with a woman coping with these issues, as well as his own worries? Hasn't he ever wondered about what sexual changes are in store for him, through her, as time goes on? Not to mention the ones he has to face himself. Instead of waiting to find out, I am in favor of bypassing them altogether whenever medically sound to do so.

I met a woman the other day who told me that as she approached forty her doctor put her on low-dose birth control pills to protect her against menopause and ease the transition. She never had a hot flash. No mood swings or any other of the common complaints you have just heard about. Compare her experience to that of Gail Sheehy and the other women who beg for help only to be told that they must wait until they have gone a year without menstruating.

As for men, there are the problems in having their viropause symptoms taken seriously. Those doctors who look for solutions are challenging traditional approaches and risk the derision of their colleagues. There is a daring, adventuresome, and increasingly notorious physician in London named Dr. Malcolm Carruthers who is treating viropause with various experimental hormone combinations in an effort to break the code and restore the missing ingredients. He is held in disrepute by much of traditional medicine, but not by some of the men who swear by his care.

Even though many men and women who need it are not getting hormone replacement therapy, *what nobody is telling you is that you don't have to go through menopause or viropause anymore.* We can prevent them altogether, much like Zothara. Listen to what this man wrote in to Dear Abby:

Dear Abby: My wife and I are in our fifties and have been married for thirty-six years. We have had a solid and happy marriage and a satisfying sexual relationship until a few years ago, when I had difficulty performing. We have both kept ourselves physically fit and can both get into our wedding clothes and I could see no reason for this problem. The guilt I felt was devastating.

Finally, out of desperation, I sought out a urologist who specializes in sexual dysfunction. I felt embarrassed and defeated, and after many tests and some lab work were completed, the verdict was that I had gone through a male midlife change. (I thought that happened only to women. I was wrong. It happens to men too.)

For me, the solution was an injection every three weeks and a small pill three times a day.

Now I feel like I'm twenty-five again—and so does my wife.
I'm writing to encourage other men to seek help for this problem.
Many therapies are available. What is right for one man may not be right
for another, but help is available if you're man enough to seek it out.

If appropriate treatment is delivered *in a timely fashion* our generation can simply skip most of the midlife miseries our parents and grandparents had to endure. Shocking, isn't it? If this is so, why aren't people yelling the good news from their rooftops? Perhaps because no well-respected authority has stated this fact in print as explicitly and bluntly before. In fact, the concept of preventing female menopause is considered radical and probably labeled treason by women (and by the majority of the medical world) who oppose hormone replacement therapy in favor of natural methods.

In the preceding chapter, I suggested to you that menopause and viropause are not natural conditions but *endocrine disorders*, like diabetes. Dr. Paul Brenner, professor and vice chairman of the Department of Obstetrics and Gynecology at the University of Southern California Medical School, also considers menopause an endocrine-deficiency disease and thinks that it should be treated just like any other illness. In addition, he states emphatically that, "American women have an average life span of almost eighty years. One third of their life is spent in the postmenopausal years. Hormone replacement therapy (HRT) increases their life span and improves their quality of life."

In fact, estrogen replacement therapy can prolong a woman's life considerably. After seven and a half years of treatment, women had a 20 percent overall reduction in mortality for all causes. Mortality decreased with increased duration of the treatment. Current users with more than fifteen years of estrogen use had a 40 percent decrease in their overall death rate!

Dr. M. E. Ted Quigley, a clinical associate professor in the Department of Reproductive Medicine at the University of California, San Diego, Board certified in both ob-gyn and reproductive endocrinology, is one of the few physicians in the country who routinely anticipates menopause in his patients and prevents symptoms. He initiates treatment sometimes as early as the thirties and maintains a woman's hormonal profile throughout her ensuing years.

Dr. Nancy S. Cetel, also of San Diego, takes this timeless approach. "There are many different treatments and preventative therapies for women to choose from," she says. "No single approach works for every woman. Physicians should spend enough time with their patients to cover the issues involved. One quick visit does not do justice to the situation."

Some of the more aggressive physicians are including testosterone in this hormone regimen, sometimes also using thyroid supplements. DHEA and LHRH are being used in Europe. Growth hormone, oxytocin, vasopressin, and a few others may receive consideration down the road. It is therefore important that you remain current as new information becomes available—not so that you can jump at every fleeting promise—but to enable you to modify your program if warranted. For example, utilizing the recent trend of adding progesterone to estrogen therapy to protect against uterine cancer.

Create the Demand

At the turn of the century, we didn't have to worry about menopause because most people died long before its symptoms had time to surface. Now we spend one third to half of our adult life in this stage—postmenopause. It is time to deal with the issue more aggressively, to press for more research, and insist that physicians pay attention. Don't wait for someone else to take the lead.

When Carolyn told her ob-gyn that she was having disturbing mood swings along with her hot flashes and other symptoms, he said that her condition was not due to a problem with hormones. When she countered that these emotional disturbances were at times disabling, he responded that she should see a psychiatrist. His opinion was that it would be premature to treat her physical symptoms.

She refuted his suggestion by insisting, "Doctor, I am not crazy now, nor have I ever been in the past. Something abnormal is happening to my body that must have an underlying medical cause that happens to be affecting my mind. If you won't take my symptoms seriously, I will find a doctor who will."

Tough talk, but wise and courageous. In order to stand up for yourself in the face of authority, you need to be well informed, have a good grasp of the medical alternatives available to you regarding menopause and viropause, and sound information on the potential risk and benefits of each approach. It is terribly important that a woman not make significant medical decisions based on old data. Only in this way can you make intelligent decisions for yourself.

It's hard to imagine a physician who wouldn't rush to provide treatment for a man, if withholding such treatment caused his sex organs to shrink, his testicles to flatten, his muscles to melt away, and his brain to mildew. In fact, at the first sign of impotence, lots of doctors prescribe testosterone shots that the man doesn't even need—just in case.

Sadly, just as men are loath to face emotional issues, women are often too ready to accept them as being the culprits, while being equally resistant to hormonal solutions within easy reach. But no man would ever go through what women have endured for millennia without doing something about it. You must assert yourself just as Carolyn did and not be so ready to accept that your changes are "all in your head."

It's a different battle for men. You don't need to tell a man to be assertive with his physician as long as he knows what the issue is. That's the problem. The symptoms of viropause are vague and ill defined. Men need to press for more studies and better research on viropause, as women have recently begun to do on their own behalf. Currently, men are too quick to turn to mechanical solutions—vacuum pumps, penile injections, implants—bypassing the identification and consequent solution of the underlying hormonal imbalances.

Remember when birth control pills came on the market? Women drove the issue. Physicians were reluctant to prescribe them, but women wanted them and soon got them. We needed the cooperation of our physicians, but the consumer created the demand. Physicians and women embraced estrogens and progesterone *supplements* for birth control in a dosage many times higher than nature allowed, in order to prevent pregnancy. But, ironically, many of these same doctors and women are afraid to take or prescribe a *much lower dose replacement therapy* to keep a woman healthy, happy, and alive. Men and women must drive this issue forward just as women have in the past, and make sure that the medical community is responsive to their needs.

I want this chapter to serve as ammunition in the battle for good hormonal health—to help those who already have it to preserve it, and for those without it to reclaim it. Armed with the information that follows you will be better prepared to evaluate and deal with your doctor. You will ask the right questions, get the best help available, and settle for nothing less.

Do not expect absolute answers. Because we cannot study menopause in animals, the main way we learn about hormone replacement therapy is by using it or studying the effects on women deprived of it. Each new approach requires a complete life cycle to fully evaluate. Thirty or forty years from now, you will probably not be in a position to benefit from the results of research that began today. That said, there is a lot that we do know and can put to use now. Not surprisingly, parts of this chapter are heavily weighted toward women simply because we currently know more about women.

PREVENTING VIROPAUSE

As I've mentioned, our knowledge of viropause is lagging about twenty years behind our research on female menopause. Nevertheless, there is still lots to offer and even more to think about.

Considerations to Take into Account

There are some general considerations for men entering viropause to heed, the foremost being overall health: If you drink and smoke and are overweight, your sex life is in jeopardy and so is your heart. In that case, forget trying to prevent viropause; let's work on preventing sudden death.

Cardiovascular Functions: For men, cardiovascular health is as important to the penis as the heart. In fact, cardiovascular disease, its complications, and treatment are the single greatest cause of the sexual dysfunction, as well as diminished quality of life of viropause. Adding insult to injury are most of the medications used to treat cardiovascular problems. Ask your physician if your prescriptions can cause any sexual problems and if there is an alternative available that is not sexually toxic.

A good thing to note is that the baby aspirin you may be taking to protect your heart and your brain will also help you keep your penile blood flow pumping. There are no studies on the penis to prove this, but good support for the rest of the circulatory system. Why would the genitals be an exception?

Depression: Mood is another consideration to keep in mind. As with women, the psychological and actual changes that occur at this time of life have a strong impact on a man's outlook. As we've seen, job loss, demotion, or the approach of retirement are among the most powerful triggers of viropause.

Men often don't even recognize that they are depressed. More often it masks itself as chronic anger, irritability, and hostility. Anxiety, depression, and anger are common features of viropause, however, and need to be addressed preferably without the use of traditional antidepressant medication, most of which cause some form of sexual dysfunction. Don't let yourself get to that point. Monitor your moods, keep open lines of communication with your spouse, seek therapy if necessary.

For more serious mood disorders, antidepressants may be necessary. The two least likely to disrupt your sex life are Wellbutrin (bupropion) and Deseryl (trazodone). Both have been reported to have some positive sexual influence, but as with all medications, there can be adverse side effects.

Sexuality and Viropause: As discussed, due to viropause, various diseases, and medication, men often struggle with a decline in psychological and physical sexual response during their forties and fifties. Further, the woman's perimenopause and menopause creates a time when sexual dysfunction commonly occurs within a couple, becoming a permanent pattern.

This mutual deterioration of sexual functioning can cause psychological depression, anxiety, anger, aversion, frustration, and reticence within a couple. Associated sleep problems and irritability promote further tension. Bitter about their own sexual decline, men may blame their difficulties on their wives' "growing old and unattractive." Unbeknownst to many men, while they may look elsewhere, some women go in search of younger partners, too.

Consequently, it's not so difficult to understand why some men go to such extreme lengths to pursue anything that promises sexual stimulation— spending fortunes, eating poisons, injecting their penis with exotic substances. As his partner, a woman might be inclined to think her man, in such a state, is motivated by ego, idiocy, or superficial macho things. Some men are. But there is a far more powerful motive: intercourse. To participate, a man must have the cooperation of his penis.

All a woman has to do is to make herself available. She can lie there calculating her financial statement during sex, although she may need a little artificial lubrication to do the trick. She can even fake orgasm at will if her body is not responding naturally, as the deli scene in the movie *When Harry Met Sally* demonstrated to the world.

Such devious opportunities do not exist for the male. In order to have intercourse, he must have an erection. To have an erection, he must pay attention. If he is thinking about mowing the lawn or writing legal briefs, his erection gets bored and naps. This compromises a man's ability to sustain a relationship and experience intimacy. He needs dependable erections. Without them, he does not feel adequate, which naturally diminishes his self-esteem. In fact, he often feels desperate.

But assuming that a man is in good general health, none of the sexual signs and symptoms of menopause described in the last chapter should be sexually disabling. He will not become impotent, nor enjoy sex less—unless he is blindsided for lack of timely information on these changes and their meaning. If they alarm and distress him, he will tumble right into the adrenaline reflex and become impotent.

To prevent this development, be sure you are educated *well in advance* (if possible, in your thirties at the latest) about all the specific sexual changes to anticipate with aging, and that you fully understand the implications. Be prepared to ask for more physical stimulation to the penis as your fantasy and visual responsiveness diminishes. Discuss these potential sexual changes with your partner beforehand so you will both feel more at ease. Learn to distinguish between these sexual changes and those associated with any chronic illness you may have.

It is also only common sense to make sure your partner is not headed for problems of her own. Accompany your wife or partner to the gynecologist to discuss her sexual health and any existing or anticipated sexual problems, should she be approaching menopause. Sexual problems are catching. Difficulties in one partner will cause them in the other.

Educate yourself on how to rebound from the adrenaline reflex should it plague you from time to time so that it doesn't stick with you. The best way to practice doing this is to focus on the moment. When anxious thoughts crowd in, triggering the adrenaline reflex, try thought substitution. Conjure up something sexy and erotic to you that tunes you in and replaces your worry about your erection, or of ejaculating too fast. You will feel those toxic thoughts in your hands when they recur, because they will get somewhat cold and clammy. Tell your partner that you are anxious about getting or keeping an erection. Sometimes just putting the feeling into words makes it disappear. Another way to get rid of the adrenaline reflex is to concentrate all your attention on your fingertips touching some wonderful part of her body. If you tune in successfully, you will tune out those distracting thoughts and lose yourself in the moment.

During periods when you cannot have sex for more than a week or two for any reason, avoid the "use it or lose it" consequence by masturbating once or twice a week if it is allowable with your value system. You will diminish the risk of developing impotence this way.

Testosterone Replacement Therapy for Men

Various studies on men with low testosterone levels have confirmed that testosterone replacement restores sex drive, erection, orgasm, ejaculation, and nocturnal erections. The biggest effect is on sexual desire, as expressed by sexual thoughts and fantasies. Interestingly, studies also report a general improvement in mood. This suggests that testosterone is not only a natural aphrodisiac but an antidepressant as well. And why not? Depression often sinks libido right along with spirits. However, as discussed in Chapter 6, giving testosterone to men with normal levels will do nothing for them sexually, and will only increase testosterone's negative effects. Also, testosterone won't counteract the adrenaline reflex. Unfortunately, measuring the total amount of testosterone does not correlate well with the hormone's impact on sexuality. The free or available fraction, however, seems to be critical. There also seems to be a minimal threshold for free T, below which men will not function. Since the difference between free and bound testosterone became clear only in the last few years, this distinction is not reflected in the main body of research. However, current research usually distinguishes between free and bound, so our understanding of testosterone's influence will be much more precise in the future.

The normal range of total testosterone (free and bound) blood levels in the man is 250 to 1,200 ng/dl. The normal value for free testosterone is: 1.0 to 5.0 ng/dl for the male (or a high of 41.0 pg/ml in the low 20's to a low of 9.0 in the 80's) and 0.1 to 0.5 ng/dl for the female.

Testosterone replacement therapy works for men with total levels below

250–300, usually reviving sex drive and erections. If free testosterone approaches 1.5, I recommend treatment.

Currently testosterone can be taken in one of five ways:

- injection
- orally (swallowing)
- sublingually (under the tongue)
- implanting a pellet
- transdermal patch

Injection: The most common method used for TRT in men is injection into a willing muscle every two to four weeks. The typical injectable testosterone dosage is 250 mg every three weeks. Testosterone injections (testosterone enanthate) can result in a threefold increase in testosterone levels, and a significant decline in LH and FSH. This means that your body will stop producing testosterone naturally. This is one of the reasons testosterone therapy is not given to men who don't really need it. Effects will linger after treatment. The highest blood levels are achieved within the first week and then gradually diminish.

Injection is currently the most powerful route of administration, getting highest blood levels the most efficient way. The drawbacks involve the inconvenience of a doctor's visit and the discomfort of the shot. Add to that the roller-coaster ride of the spike and drop over the injection cycle, and you have a less than ideal system.

As described in Chapter 5, this artificially induced rise and fall of testosterone is quite different from a man's normal testosterone rhythm. While the aggressive sex drive generally rises and falls consistently with the blood levels of testosterone, mood effects have not been studied as extensively. But a valid concern to raise is the potential anger and irritability that might be provoked by a sustained period of abnormally high testosterone, and what effect peak levels might have on a relationship, especially if the man was already predisposed toward domestic violence.

Orally: Oral testosterone must be absorbed through the digestive tract. Consequently, high blood levels are not readily achievable. Absorption is unpredictable and dosage can be difficult to control. There is, nonetheless, a valuable role for oral testosterone, particularly for men with borderline testosterone deficiency who may dread injections.

Testosterone undecanoate (TU) is marketed in Europe by Organon as Restandol, Andriol and under various other names, but is not marketed in the U.S. The usual dosage is 120 to 160 mg a day (60 to 80 mg taken twice a day), avoiding the need for shots and frequent doctor's visits.

Other oral forms of testosterone include mesterolone (Proviron, which is not available in the U.S.), fluoxymesterone (Halotestin—5 to 20 mg a day) and methyltestosterone (Virilon, Testred, et cetera—10 to 50 mg a day). A dose of 75 mg a day of mesterolone has been suggested to treat male climacteric in men aged forty-five to sixty.

Side effects of taking testosterone orally include liver damage, increased cholesterol, salt retention, weight gain, enlarged breasts (due to testosterone's conversion into estrogens), edema, acne, high blood pressure, and prostate hypertrophy.

These side effects, while they can occur, do not occur to everyone. Just as aspirin and penicillin can cause side effects, including death, the adverse effects are not frequent enough to outweigh the benefits for the population at large. Nonetheless, it will be a significant improvement when these side effects can be eliminated by new methods of delivery or newer, cleaner drugs.

Sublingual: The third form of testosterone therapy, popular throughout the world, is sublingual administration. A testosterone linguet (i.e., Testoral) is placed under the tongue where it is rapidly absorbed directly into the bloodstream, bypassing the digestive process and the liver. Blood levels are more predictable than with oral preparations, but do not match the high levels achieved by injection.

Transdermal Testosterone Patches: During recent years, transdermal patch delivery systems for drugs have become quite popular. Treatment with patches for nicotine (to stop smoking), nitroglycerin (for angina), clonidine (for blood pressure), scopolamine (for seasickness) and estradiol (for menopause) are routine. A testosterone patch is under development and should become commercially available soon.

It will be a self-adhering hormone patch applied directly to the scrotal skin. The rise and fall of testosterone absorbed from the daily patch more closely resembles the normal circadian variation of natural testosterone secretion. Consequently, men will experience fewer of the highs and lows that are by-products of injections.

There will also be relatively little superfluous testosterone to be transformed into estrogen, so that men are less likely to develop enlarged breasts. For women for whom estrogen is medically inadvisable, use of this form of testosterone can be safer than others for the same reason. One potential drawback for men is that by delivering testosterone directly to the genital areas, the risk of benign and malignant prostate complications could be increased. The effects of high concentrations in the genitals are unknown.

Other Methods: New forms of testosterone with less potential to cause side effects are under development, along with new methods of delivery. Given the pulsatile nature of testosterone release that occurs from hour to hour, the ideal delivery system is probably a testosterone pump. This mechanical device is now in clinical use for some conditions like diabetes, delivering insulin in small divided doses at regular intervals. The pump is implanted under the skin and squirts medicine into one's system at preprogrammed doses and intervals. It is safe and painless once installed.

Testosterone can now be attached to special carrier molecules designed to transport it directly to the brain. In this fashion, it can act specifically on the brain without getting into the general bloodstream and causing body hair, balding, hypertension, and prostate cancer growth. Until these new methods become widely available, however, TRT still has some limitations.

Discuss the pros and cons of testosterone therapy with your doctor to find out his or her philosophy and approach in advance of needing this form of treatment. You want a physician who is open to it when appropriate. Ask, "Under what circumstances do you recommend testosterone treatment? What would be your approach if I were to develop erection problems as I got older?"

Also ask your physician to measure your *free and total* testosterone levels, thyroid hormone DHEA, FSH, and LH every few years and yearly after forty. This will show you the trend over time, and give you the opportunity to identify and treat changing hormone levels before unpleasant sexual symptoms develop. This is one of the ways to *prevent* viropause.

Testosterone's Role in Prostate Cancer

As we've discussed, testosterone can cause high blood pressure, liver damage, and prostatic hypertrophy (benign enlargement of the prostate). Further, while testosterone doesn't cause prostatic cancer, it aggravates and can trigger its growth. Routine autopsies of men over sixty who have died from other causes show that many of them have prostatic cancer that had not yet flourished. Had they lived to their eighties or nineties and not died of a heart attack, perhaps the prostate cancer would have killed them. Since so many men harbor quiet prostatic cancers, giving them testosterone if they do not have low levels to begin with is tantamount to fertilizing weeds.

Approximately 25 percent of all men will require treatment for benign prostatic hypertrophy if they live to be eighty. More than four hundred thousand prostatectomy operations are performed in the U.S. each year. Most (96 percent) are done transurethrally.

Consequently, it is essential that before considering testosterone replacement therapy or supplements, a man receives the following blood tests and procedures:

- Serum testosterone levels—free and total (to justify the need)

- PSA—prostatic specific antigen test (to screen for evidence of prostate cancer metastases)

- Prostatic ultrasound (to look for hidden tumors)

- Prostatic rectal examination (to check for prostate size and tumors)

About 60 to 70 percent of prostatic cancers greater than 5 mm will be detected by prostatic ultrasound. The blood test for PSA detects about 70 percent of prostatic cancers. When used together, they give excellent information, but nothing is a 100 percent guarantee. A physical exam that includes digital examination of the prostate (rectal exam) is a necessary evil. Used together, these tests are the best prostate cancer screening system that we have available today. The object is to detect prostate cancer early enough that it is readily curable.

Treating Prostate Cancer: Often, the treatment of prostate cancer causes impotence. There are, however, nerve-sparing surgeries, such as the Whipple procedure. This is a special operation for prostate cancer that avoids damage to sexual nerves while still removing the tumor. It is more painstaking and time consuming for the surgeon, but you might ask your doctor about it. I also wonder why men are not given the opportunity to receive reconstruction with a penile implant at the time of surgery, just like women for breast cancer. Imagine the boost to your mood if you came out of surgery without the cancer *and* with an inflatable, dependable erection (instead of impotence) that you can control at will. The inflatable implant is permanent, but the erection is only there when you want it. There are also semirigid implants that are permanently installed that provide a permanent erection. When not being used for sex, the penis hangs down one pant leg or the other, and does not show. Most of these procedures should be covered by insurance, since they are only necessary for the direst of medical reasons. This oversight is something I would encourage men to squawk about.

While, as we've seen, progesterones and their relatives are generally so unkind to the male sex that they can be used as a form of chemical castration, they can also be used in the treatment of prostate cancers—if the choice boils down to your life or your sex life.

As I've noted earlier, some prostate cancers are testosterone dependent, much like some breast cancers that thrive on estrogen, which means that the more testosterone you have, the faster your cancer is likely to grow. The progestogen megestrol (Megace) is widely used to treat prostate cancer. Progesterone stops this testosterone effect. Negative sexual effects occur, however, with marked loss of sex drive in up to 70 percent of men.

Newer antiandrogen drugs such as flutamide (Proscar) have recently been introduced in hopes of reducing these sexual symptoms. They block androgen receptors without depressing circulating testosterone. While there seem to be fewer sexual consequences with these newer drugs, what happens in the long run is still an open question.

PREVENTING FEMALE MENOPAUSE

Some women thrive on hormones, others can't take them even if they want to because the side effects are too unpleasant.

Here is what Lena Gorwood, of Lafayette, Tennessee, had to say about hormone therapy:

> *I am an eighty-year-old lady. I have been taking estrogen since some time in the 1960s. I swear by my estrogen. I fell down a flight of stairs (eight steps) from top to bottom, also fell off a golf cart twice and nothing has been broken yet. I am still active, have no arthritis, play golf and do most of my housework. I sincerely believe that taking estrogen has kept me going for eighty years. I go to the doctor once a year and have a good physical. Otherwise I only go if I have a cold or sinus problem, only something minor.*

Pat Stotts, of Manchester, Tennessee, on the other hand, couldn't tolerate estrogen:

> *I am thirty-eight years old. When I was thirty, I underwent an abdominal hysterectomy. . . . Approximately three years ago, upon a yearly exam, my gynecologist explained that my periods of depression were probably due to a shortage of estrogen. . . . Shortly after I started taking the hormones, I did feel better, but that was short-lived. Within two weeks I was experiencing severe migraine headaches. . . . So I stopped [taking estrogens].*
>
> *Recently I have experienced the same sort of symptoms: depression, loss of ability to think clearly . . . poor sleep habits. . . .*
>
> (from* The Tennessean, *Tuesday, November 15, 1994)*

For women, three hormones available and in use throughout the world (U.S. included) can be used to prevent the symptoms of menopause: estrogen, progesterone, and testosterone. Together, they can prevent or correct most of the sexual, emotional, and physical aspects of this disorder (unless irreversible changes already have occurred).

Since estrogen begins its decline in the early thirties and FSH and LH rise, prevention of menopause strategies ideally should begin before then. Life-style is again the key variable. Developing healthy lifelong habits is a nice dividend of preventing menopause.

There are two constants for women from age thirty forward:

- good nutrition, including gradually increasing calcium supplements, and
- weight-bearing exercises.

Both, in addition to contributing to general health, give your bones a head start to buffer the adverse effects of osteoporosis later on.

Another challenge to sustaining sexual health beyond menopause is to remain free from the multitude of vaginal and bladder infections that can occur when a woman is young. They can end up plaguing her throughout her lifetime, predisposing her to some of the more unpleasant problems associated with menopause, such as "senile vaginitis" and incontinence, which in turn will aggravate any chronic condition. The most effective way to avoid these earlier problems is to limit the number of sexual partners, which additionally protects against cervical cancers. Should an infection develop, get treatment promptly.

To prevent menopause, at the first signs of hormone change in her bloodstream (not after she already has the symptoms), a woman should begin receiving replacement estrogen. Testing for FSH should begin no later than the midthirties. This hormone is the first readily measurable indicator for change, although it is volatile and should be rechecked periodically as long as a woman is having regular periods. Estrogen and testosterone should be checked every two to three years until after the forties, when testing should occur every year.

The most user-friendly method of replacing estrogen is to place a woman in her mid to late thirties on low-dose birth control pills, containing both estrogen and progesterone, which accomplishes several things at once. It prevents pregnancy during this unpredictable period. It regulates her periods and her moods throughout what would otherwise be increasing perimenopausal irregularity. Birth control pills also help to prevent the severe anemia that so often accompanies heavy irregular periods during perimenopause.

This treatment can ensure that a woman will bypass most, often all, meno-pausal symptoms. She will not have hot flashes. She will continue to lubricate and to function as she always has. The transition is almost problem free for women who can tolerate the pill. It's that simple. If there are no predisposing health factors to discourage this approach, a woman can be put on birth control pills sometime in her late thirties or early forties.

The birth control pill also contains progesterone, so a woman is protect-ing her uterus (if she still has one) from cancer. We still don't know quite why this is so, but suspect that it is because estrogen alone can cause endometrial hypertrophy (a benign overgrowth of the uterine lining) which somewhat predisposes the development of uterine cancer. Progesterone keeps that lining thin and trim, having the opposite effect on cancer risk. Up until lately, women who have had hysterectomies have not been prescribed progesterone because they had no uterus to protect. That position is being reconsidered in light of the fact that progesterone has other beneficial metabolic effects to offer them.

To prevent or treat menopause, testosterone should also be considered, but the timing is variable. Blood measurements should be taken periodically and replacement offered when they start to drop subtly. There is disagree-ment about whether replacement alone is sufficient, or if some additional supplementation should be done. The desirable therapeutic range to achieve are blood levels between normal and 20 percent above normal, based on the few studies that are available.

It continues to be important that the perimenopausal woman take calcium supplements. Although there is debate about the cost effectiveness of bone density studies, I believe they are of value, both to establish an important base line and to monitor progress periodically. Mammograms and breast exams should be performed according to the current age recommendations. Self-exams should be done monthly.

Once it is clearly demonstrated that a woman is through with her fertile cycles (has stopped ovulating), it is then sensible to change her over from the birth control pill to one of the lower-dose menopausal replacement combina-tion therapies. Assuming there are no contraindications to the use of HRT, the formula for preventing menopause in women is straightforward. The complex part can involve finding the right dose and type of each hormone that is best tolerated by each woman. There can be a period of months during which time it is necessary to switch and mix, requiring close teamwork between physician and patient, patience and persistance. Regardless, that process is usually a breeze compared to the real thing!

Estrogen Replacement Therapy: Pros and Cons

There is a great deal of information in this book devoted to estrogen, estrogen replacement therapy, and alternatives for women who cannot take estrogen. There is a compelling reason for this: the majority of a woman's adult life is spent past menopause, and estrogen, or the lack of it, has an enormous impact on women's health, sexuality, quality of life, longevity—and, indirectly, on those of their spouses and families.

Perimenopause is really the culprit that causes all the trouble. Once you get past this transition into real menopause, you will feel pretty good for some time even without hormones. (Which is why I think it is pretty sadistic for physicians to wait until full menopause to treat women.)

Perimenopause can be a real trip. It is like running white-water rapids where two strong rivers collide: you get to enjoy PMS, unannounced periods, and the full smorgasbord of menopausal symptoms all at the same time. Just imagine this exciting raft ride. Sometimes you just have to hold on and hope you get through. And you usually do. But it can be a real nightmare—literally:

Here is how one woman I know described it:

"When I turned forty-eight, I stood unprepared in front of a runaway hormonal train. It hit me broadside and sent me reeling—both physically and mentally, it could not have been worse.

"As far as I can tell, menopause affected me in four distinct ways: I gained weight; I lost my memory; I lost control of my financial stewardship; and I was terrorized and practically immobilized by nightmares.

"First, weight: Within a year of feeling those 'hot flashes,' I gained some seventy pounds, almost all of it in my buttocks, hips, and thighs. Since I am fairly long-legged, it actually affected my balance by shifting my center of gravity. A staircase and high heels was asking for disaster! My weight gain was so sudden and grotesque that I could see the shock in people's faces who had known me all my life and would not recognize me in the supermarket. I tried to control it; I cut down on my food. I ate one third of what I used to eat, and still I would gain weight. I

used to be so slim a man could span my waist with his hands, and now it seems like I gain weight just fantasizing about food. I don't dare even think of it.

"Next, my financial inertia. I was used to living carefully and frugally, but I was not prepared to deal with what can only be described as fiscal irresponsibility. I would not pay my bills. I would refuse to think of ways I could pay my bills. I would take the phone off the hook. I would throw away unopened letters. My way of dealing with financial pressures was to go to bed and stay in bed for days. I am ashamed to say that I behaved like the proverbial ostrich, sticking my head in the sand and hoping that my troubles would dissolve out of their own volition. I never overspent, but I let responsibilities slide until they were unmanageable. It frightened me because it was uncharacteristic. I knew it wasn't me.

"Third, memory. I used to have a photographic memory. When menopause hit, I actually became convinced that I was getting Alzheimer's. There were times when I could not remember my phone number, which absolutely panicked me. Phone calls that were important professionally would 'disappear' on me, sometimes within ten minutes. It was as though somebody with a sponge was following my life, erasing every word I heard or spoke.

"And finally, my nightmares. They were the worst. I still don't have the means to tell how horrible they were. They never let up. I couldn't sleep without being attacked by demons. It was terror—base and raw and brutal beyond words. I was beaten, burned, sliced, crushed, dropped from horrid heights and buried beneath mountains. I have always had a vivid 'nightlife' through my dreams, but my menopausal dreams were not dreams—they were torture. I am a trained psychologist and I am convinced my nightmares were hallucinations of the psychotic kind—distortions in both sight and sound.

"When I was put on estrogen (eight years after these symptoms started) it was a huge relief. I could think again. I could sleep again. I could act responsibly again and take back my life from my creditors. I took part but not all of my memory back. Regrettably, though, my metabolism never reversed, and to this day, I struggle with weight. I have managed to lose over forty pounds, however, and plan to keep going."

ESTROGEN TREATMENT—SUMMARY OF EFFECTS

Advantageous Effects:

- Improves and stabilizes mood
- Gives cognitive benefits
- Improves performance, reaction, and vigilance
- Decreases irritability, anxiety
- Increases memory, well-being, optimism
- Acts as a mild anti-depressant (MAOI)
- Protects against schizophrenia
- Protects against Alzheimer's disease
- Decreases appetite
- Resolves hot flushes/flashes
- Improves sense of taste, smell
- Maintains skin tone, vulva, collagen
- Prevents osteoporosis
- Prevents or reduces stress
- Prevents incontinence, urethral, bladder, uterine prolapse
- Prevents heart disease

Sexual Effects:

- Maintains sexual receptive desire and promotes frequency
- Generates attractive body odor and texture
- Sustains vaginal rugae
- Promotes vaginal lubrication
- Prevents senile vaginitis
- Improves vaginal, urethral, genital tissue
- promotes touch and touchability via oxytocin

Possible Adverse Effects:

- Nausea
- Malaise
- Peculiar taste aversions
- Breast tenderness
- Can disturb tryptophan metabolism
- Vitamin B deficiency, causing depression, fatigue, irritability—correctable with vitamin B6 supplements
- May increase risk of breast and endometrial cancer
- Migraines

The Good News

With estrogen, there *is* life after menopause. Studies have found both sexual and emotional benefits with estrogen replacement in menopausal women, including improvements in well-being, mood, cognitive state, appetite, and the absence of physical complaints (such as headaches) and particularly genital complaints—like the urge to urinate (at the wrong times!). Translated: "Neurotics" get better because they weren't neurotic in the first place—just estrogen deprived.

Aggression is significantly reduced on estrogen. Without estrogen, testosterone spikes, making a woman more assertive, aggressive, self-confident, and "sex driven" in typical male patterns. One physician I know said that he had his wife on high doses of estrogen only and didn't want her on progesterone or testosterone because, "I like the Stepford Wife model and am not going to give her anything that would help her stand up to me or have moods." (After I recovered my capacity for speech, I gave him a dose of my testosterone with a dash of progesterone tossed in!)

Estrogen therapy affects mental shortcomings due to hormone deficiencies. These deficiencies can be subtle. In fact, certain tests that measure short-term memory, number recall, clerical speed and accuracy, and recall of text material—all low-grade symptoms that do not show obvious impairments in job performance or during everyday activities—have revealed shortcomings in these skills in women who were surgically menopausal. But if you can't remember people's names or retrieve verbal data easily, it complicates your life. Estrogen replacement therapy corrected the problems. With ERT, verbal ability is sharpened, dexterity improves, and reaction and vigilance perk up. While the full extent to which the mind slows down without estrogen is unknown, what we know so far is formidable.

Just a few years ago, respected researchers came out with pronouncements that estrogen deprivation did not cause cognitive damage. Today the position is reversed. Not only does estrogen withdrawal impair memory, reduce cognition, and cause depression, it promotes Alzheimer's disease and in some cases, it predisposes women to schizophrenia.

Estrogens also help fight off the doldrums, since these hormones are nature's antidepressants. There is a clear explanation of their antidepressant

253

effect. MAO inhibitors are commonly used to treat depression, and estrogen has MAO-inhibiting qualities. Estrogens may be enough to prevent mild mood disorders but wouldn't be likely to correct severe ones. In one animal study, helplessness and immobility of female mice was doubled after their ovaries were removed. They returned to their old mouse selves with estrogen replacement.

For mood disorders associated with menopause, a physician has a rather bewildering array of prescription alternatives from which to choose. Instead of estrogens, menopausal symptoms are often treated with multiple psychotropic drugs, including benzodiazepines for insomnia and anxiety, tricyclic antidepressants for depression, and serotonin uptake inhibitors like fluoxetine/Prozac for almost any other emotional condition ranging from phobia to obesity. These medications are often prescribed by a woman's various physicians and are not monitored as carefully as needed. Unless estrogens are medically inadvisable, wouldn't it be better to be given a trial of estrogens first? A fair share, perhaps *all* of a woman's emotional symptoms, may be relieved by estrogen replacement therapy alone.

Sex: Unfortunately, maturing women with sexual problems often don't associate them with low estrogen levels. Even if they did, they might be too reserved to mention the problem to their doctor.

Embarrassment of this sort—with physician and partner—leads to less and less sex, and eventually, perhaps, to none at all. Add painful intercourse, decreased libido, a decrease in touch sensitivity, depression, and a host of other griefs, and who is even considering orgasms?

Some husbands of postmenopausal women with sexual symptoms develop an understandable fear of hurting their wives, both physically and emotionally. Concern about making them bleed, insecurity, rejection, and anger are an unholy mix squelching the ordinary man. Even if a woman later receives estrogen replacement, too much damage may already be done. Hurt feelings and toxic patterns may be so well established they are difficult to reverse by then.

That's one of the many reasons it is so crucially important to keep the sex between a married couple good—especially around midlife. Sexual intimacy during the late forties and early fifties is sometimes the only bond that enables a couple to bridge the gap that children, work, and life circumstance have brought to bear on their relationship. When sexual continuity is broken due to lack of ERT in these vulnerable situations, it may destroy the marriage. Had ERT been offered before the divorce, the wear and tear on the relationship might have been avoided.

Since estrogen provides mood-mellowing benefits, it creates an emotional atmosphere more receptive to sex than depression or grumpiness, and no wonder. Who wants to make love to a grouch? When estrogen replacement is given to a woman, she experiences more sexual desire, enjoyment, and orgasmic frequency than with progesterone alone or a placebo.

While estrogen, when used alone, is often enough to maintain normal female well-being, including sexual desire and activity, testosterone often needs to be added to estrogen therapy to ensure robust, dependable sexual desire and response. Adding progesterone to estrogen decreases the estradiol benefits somewhat—perhaps taking the edge off sexual desire, making orgasm a little more difficult and making women irritable. None of these symptoms are dramatic or discouraging as long as low-dose progesterone is used.

Estrogen replacement shifts pH back toward normal, while increasing lubrication and blood flows so vaginal dryness improves. In fact, estrogen vaginal creams can reverse many local symptoms within a week or so. Within days, vaginal tissue begins to thicken and itching common to menopausal women often disappears. But bear in mind that vaginal estrogens are absorbed into the bloodstream as well, so if you have medical reasons preventing the use of estrogens, they extend to the vaginal creams. Sometimes testosterone cream is also necessary to resolve all the symptoms because the health and thickness of some of the tissues is dependent on testosterone, not estrogen, and if it is not replaced the tissues atrophy even if you get enough estrogen.

The improved health of vaginal tissue with estrogen replacement extends to the health of urethral tissue and other genital structures. For many postmenopausal women with stress incontinence, vaginal estrogen is sufficient to solve the problem. Treatment with oral estrogen alone may be adequate, but resolution of symptoms is more likely with vaginal cream because it gives higher dosages of estrogen density to the troubled area, instead of a weaker form that has been diluted by the bloodstream before it gets there. If that doesn't work, adding phenylpropanolamine (PPA), an over-the-counter drug, can help take care of it. PPA is a urinary tract muscle stimulant that adds to the effort your urinary tract is making to contract in the right place at the right time. Strengthening the muscles of your pelvic floor with Kegel exercises (if you had children, you may remember these from pregnancy) can also help.

Better yet, more estrogen makes sex more likely, and frequent intercourse can raise estrogen levels even more. Having sex more often is also correlated with fewer female reproductive system deficiencies such as surprise bleeding —the joy of menopause.

In the absence of estrogen, oxytocin does not work its wonders; for reasons we don't fully understand, they influence one another in powerful ways. What does that mean for menopausal women? Let's review what oxytocin does:

Sexually, oxytocin spurts at orgasm in both sexes. As we discussed in earlier chapters, it spikes with nipple stimulation, priming both men and women for orgasm. In females, it causes uterine contractions during orgasm. In males, it speeds erection and orgasm, intensifying response. It makes the skin of the whole body complement the special intensity within the genitals and breasts. It plays a central role in our urge to touch and be touched, to bond with others and, primed with sex hormones like testosterone and estrogen, oxytocin spurs us to seek and pursue a mate.

If lack of estrogen causes a depletion of oxytocin, then the reverse probably also holds true, More reason to opt for ERT, since to lose all these marvelous effects would be very destructive. Its absence could cause reduced drive, intensity, and ability to be orgasmic, less desire to touch, be close or intimate, along with a feeling of being less connected, less needed.

There are fair numbers of women who have never enjoyed sex who welcome menopause as a great relief. For them it signals a graceful sexual retirement. However, women who want to avoid sex should still weigh the pros and cons of estrogen replacement therapy carefully from the perspective of their general health, well-being, and longevity.

Morning Breath: Estrogens lubricate your teeth and prevent tooth decay and tooth loss. Without estrogens, you get dry mouth, bad taste, decreased saliva, decreased oral sensation, and bad breath. What do you think that does for kissing? With estrogens you won't have morning breath but will be better able to smell his.

Cosmetic Effects: The cosmetic aspects of HRT—estrogen replacement in particular—have been scoffed at and generally held in disrepute. Conservative physicians, with the notable exception of those in southern California, balk at giving a woman ERT just to help her avoid the cosmetic effects of aging. I want to suggest another point of view.

If a substance markedly improves the health and appearance of our exterior, I think it is fair to question what advantage it might present for our interior structures. Men have a reputation for aging better but dying younger. Women who live longer are now narrowing the gap—aging well as well.

Look around you today, at women in particular. Do you see dowager's hump anymore? Rarely. This deformity is a direct consequence of osteoporosis, now mostly preventable with ERT. Remember when you were a kid? Women in their forties and fifties looked *absolutely ancient.* Today, so many are vibrant, young, energetic, and, yes, even glamorous.

Estrogen maintains subcutaneous skin tone, collagen, and elasticity. This translates into younger skin and fewer wrinkles. Without estrogen, a woman's skin loses collagen and becomes thin, flaky, dry and bruises easily. Collagen reduction is greatest (up to 30 percent decrease) during the first five years after menopause. Thereafter, it levels off. Estrogen replacement prevents postmenopausal collagen loss and even restores collagen levels back toward normal in older postmenopausal women—much less costly than cosmetic collagen injections. Not surprisingly, estrogen gels can give a boost to skin components, particularly collagen. They are available at various department store cosmetic counters in low dosages, or at higher dosages through your physician.

Nothing's Perfect

Estrogen is not without drawbacks. Reported side effects include nausea, malaise, breast tenderness, peculiar taste aversions, estrogen intolerance, postmenopausal PMS and menstruation due to certain methods of prescribing estrogen/progesterone supplements. Estrogens can also cause depression by disturbing tryptophan metabolism causing a vitamin B6 deficiency, which by itself can cause depression, fatigue, and irritability. If recognized, however, vitamin B6 deficiency can be easily corrected with over-the-counter vitamins. Although some women consider some of these effects as positive, since the nausea decreases appetite and helps them lose weight, clinically they certainly must be seen as negatives.

There is no doubt that estrogens are a controversial hormone. During a woman's reproductive years she may decide to use estrogen for birth control, which is of itself controversial in some quarters. In medicine per se, there has been a continuing debate about the safety of long-term treatment with contraceptives and much concern for cancer risk with estrogen treatment for menopause. Today, the medical perception is shifting. Birth control pill use is considered protective rather than harmful regarding ovarian cancer and breast cancer. The doses used today are now so low compared to the past that the side effect profile has changed. For example, blood clots (thromboembolisms) that used to be a big worry are no longer such a great concern. New research could change our minds again, but that seems to be the case for now.

The medical debate over estrogens becomes a greater dilemma as women approach menopause. To be or not to be on ERT may be one of the most important medical decisions a woman makes in her lifetime, and it is important that she receive enough information and guidance to make an intelligent choice. But in an effort to make a wise decision about ERT, women may find that well-respected experts disagree on what is "good" for them.

Often women do not know whom to believe, or whose judgment to follow. Many physicians categorically refuse to prescribe estrogens or are so conservative in doing so that few of their patients receive them. Curiously enough, these patterns vary more according to geography than philosophy. It is easier to get ERT in southern California where people worship youth than in Alabama, a more conservative area. More hysterectomies are performed in the United States where physicians are paid by the procedure, than in Britain where physicians are paid a flat salary under their national health plan. Clearly, economics and fashion often influence these trends.

Sometimes the advice given to women *by* women can be part of the problem. There are certain women's groups who are militantly antiestrogen, preferring what they refer to as a "natural means" of coping with menopause. While such attitudes show admirable willingness to take "responsibility for wellness," and are well intended, these women often fail to recognize the degree to which estrogen withdrawal can undermine relationships and emotional well-being, while destroying physical health. Nonestrogen remedies can help but they can't come close to the real thing, any more than a man could compensate for low testosterone with exercise, vitamins, and vegetables.

One example of well-meaning women giving misleading advice originated in *The Ms. Guide to a Woman's Health* by Cynthia W. Cooke, M.D., and Susan Dworkin (1979). It stated, "Warning! Accept temporary estrogen replacement therapy only for:

a: Severe, prolonged hot flashes
b: In cream form for severe vaginal dryness and thinning after menopause
c: For premature menopause, resulting from surgical castration or ovarian failure before age forty (to be taken only until age forty-five to fifty)."

The authors recommend sedatives or mild tranquilizers (albeit for "short periods of time") and even progestogens which ". . . frequently control hot flashes to a large extent and are probably much safer for many women, especially those with complicating problems such as obesity, hypertension, cystic breasts, heart disease and diabetes." This fourteen-year-old advice has gained momentum and is heard even more loudly and frequently today.

For the record: progestogens are seldom adequate to eliminate hot flashes. They can *cause* obesity, hypertension, heart disease, sexual dysfunction, depression, and other problems, which are often avoided by ERT. Furthermore, depending on dosage, vaginal estrogen can reach circulating blood levels just as high as oral forms, sometimes higher. These recommendations are just not medically sound. Who can you believe?

Relative Risk: When one takes *all* health issues into consideration, the pros outweigh the cons and ERT wins out. For one thing, the incidence of osteoporosis and heart disease increases after menopause, and estrogen has been clearly shown to reduce the frequency of these illnesses. But let's look at the situation from a statistical perspective: the cumulative risk of death in white women aged fifty to ninety-four from breast cancer or hip fracture is 2.8 percent, 0.7 percent from endometrial cancer—*and 31 percent from heart disease!* Estrogen replacement protects women against heart disease and osteoporosis, both of which individually cause more death among women than breast cancer and endometrial cancer added together. Taking a hard look at these figures, heart disease kills almost ten times as many women as breast cancer and endometrial cancer combined. Don't take my word for it. Ask your doctor this very specific question: "What is my likelihood of dying from heart disease or the complications of osteoporosis without estrogens compared to my likelihood of dying from breast cancer or other complications if I take estrogens?" When, not if, he or she tells you how much higher your odds of death from heart disease are without estrogen, then ask, "If that is so, why are you advising me against estrogen treatment? In fact, why aren't you encouraging me to take it?"

The relative importance of these two risks is clear. Estrogen replacement therapy keeps the majority of women alive longer even if it makes them somewhat vulnerable to breast cancer.

That said, there will be some women who can't do HRT at all. There are some legitimate issues that a woman must consider for the sake of her overall health that will interrupt or prevent hormone therapy altogether.

ERT and Breast Cancer

An estimated 182,000 women will have been diagnosed with breast cancer in 1994, most of whom have been told they cannot take hormones. The women related to them are apprehensive about taking hormones because of their family history. This extended group of women will live a markedly different quality of life than those on HRT, and need two things: full disclosure with information about the adequacy of data behind these recommendations and the best advice available on how to negotiate natural menopause without hormone therapy.

It is important to put the breast cancer risks into perspective without minimizing them. Overall, cancer-related deaths are reduced from 20 to 34 percent in ERT users, indicating that ERT may be a health and longevity factor—whatever the reasons may be. This decreased risk of death from cancer may partially be due to the fact that women at less risk are put on ERT. Women on estrogen are also examined more frequently and monitored more closely, with the result that breast lesions would be detected considerably earlier—often while still in stage I, at which point breast cancer is usually completely curable.

The weight of the evidence regarding ERT and breast cancer suggests that conventional ERT with 0.625 mg of estrogen (Premarin or generic equivalents) with or without progesterone does not seem to add to breast cancer risk and may even reduce the overall cancer mortality rate.

In 1994 the *Journal of the American Medical Association* published an article on ERT and breast cancer, concluding that there is no persuasive evidence that estrogen replacement accelerates the growth of breast cancer, even in tumors that are estrogen dependent. However, other researchers disagree.

Also consider the fact that the most pronounced increase in breast cancer rate occurs between the late thirties and early fifties when natural estrogen levels are declining. An elevated risk continues throughout the postmeno-pause despite markedly lower estrogen levels. Furthermore, breast cancer has been increasing in recent years in *both* ERT users and nonusers. This increase suggests that factors *other* than prolonged ERT contribute to the problem.

However, common sense dictates caution, so for safety's sake, let's assume the worst until we know differently—that there is some unknown measure of risk. Then let's weigh the pros and cons of ERT in that light.

This discussion is dedicated not only to women with breast cancer, but to every woman concerned about getting it.

I speak from experience. My maternal grandmother died of breast cancer in her fifties. My mother was diagnosed with it in 1991, and I discovered I had it in 1993.

Women with breast cancer are almost always told to eliminate their hormone therapy the moment the diagnosis is made, tossing them into estrogen withdrawal at a time they most need their wits about them. They face critical decisions regarding surgical, chemical, and medical treatments that will challenge their strength and require enormous emotional resiliency, even with the advantage of a full battery of hormones in their system.

My doctor made the same recommendation to me. Since I had already been taking ERT for years and thought the consequences of stopping it abruptly would be worse than maintaining continuity for another few months, I chose to continue on it until the major decisions were made and surgery was behind me. I wanted the advantage of both the physical and emotional buffer of estrogen during this time.

After having bilateral mastectomies, I continued taking a combination of estrogen, Provera, and testosterone, because I opted for particularly aggressive surgery, and because my nodes showed no evidence of metastases. My physicians considered me cured, although there was a small margin for error. I felt that the adverse medical repercussions to my life and health that would result from estrogen deprivation would be far more serious, statistically, than the chance of my cancer recurring. *I would not have taken this risk, however, if I had had any evidence of spread.*

Most of my doctors disapproved. There were some, however, who felt I was pursuing a reasonable course. One had a fifteen-year follow-up study of breast cancer patients who continued estrogen treatment without a recurrence.

However, I still worried about the consequences. Intellectually, my approach was logical. But each morning when I swallowed the pill, I felt that I was poisoning myself. I had changed from the estradiol patch to Premarin hoping that it would minimize the impact on any residual breast tissue that remained. The problem I faced—just like so many other women in this situation—was having to make decisions on life-and-death matters in the absence of enough dependable data. (Even with complete mastectomies, doctors can't get every cell.) Is Premarin (estrone) safer for the breast than the patch (estradiol)? We don't know for sure. It is an educated guess. Will estrogen replacement cause a recurrence? We don't know that either.

After surgery, I learned that my tumor had tested positive for estrogen receptors, so I decided to go off of estrogens to see if I could get along without them. (The tumor tissue itself is tested to see if the tumor contains hormone receptors. Some do, some don't. The ones that do are thought to grow faster in the presence of estrogen, which serves as a fertilizer for it. This was only important in my case if, unbeknownst to us, my tumor had already seeded elsewhere [metastasized] even though the studies said no.)

Here is what happened: I noticed no adverse effects at first and was quite relieved—prematurely, as it turned out. About two weeks later the symptoms hit me one day with full force. Intolerable flash floods as I call them: hot flashes day and night, drenching me during business meetings, distracting me with patients. It was difficult to concentrate and hard to stay interested in just about anything. Being low on energy and easily fatigued, I gained weight without appearing to eat more. I did not have the stamina to work out. I couldn't sleep at night and was worn out during the day. I flattened out emotionally as well. When I woke up in the morning, I had to give myself pep talks to make it through the day. I was cranky and without my usual emotional resilience. For the first time in my life, I uttered the words, "I give up." And I said them more than once. I felt depressed, exhausted, as though I were running on empty most of the time. I saw no hope, no solutions, only dilemmas. My sense of well-being was gone. I did not recognize myself.

After about six weeks of what felt like a constant case of PMS in the deep tropics, I decided to go back on estrogen. I couldn't imagine going through the rest of my life as a wet, irritable zombie. The questions that came to mind were: How long would these symptoms last, and would they ever get better? Hot flashes eventually disappear, but not always, and it can take years. Just as some women never have them, some women never get rid of them. What about the moods? Would they improve, or would I just have to redefine my sense of normal?

Within five days of taking my first dose of estrogen, I woke up happy for no reason and remained basically unflappable throughout the day. Nothing else had changed. I still had the same stresses and pleasures as in the weeks before in the weeks ahead. I went to a movie that night—*Wolf* with Jack Nicholson—I couldn't help thinking throughout the film what a werewolf I had become without estrogens. In retrospect it was appalling. The better I felt, the more apparent it became how terribly I had been feeling.

I have chosen to continue my hormone replacement therapy for the sake of my quality of life (QOL), a term that has become popular jargon among doctors. I don't mean to imply that I am completely at ease with this decision. How could anyone be? But I have come to terms with it as best for me at this time, and I am aware of the risks involved. This is my gamble, however, and not advice I would press on anyone else: Quality of life is not very valuable in the face of perpetual fear. And there are many women who are profoundly apprehensive about this issue. For someone to feel physically good but emotionally terrorized is not an accomplishment.

Fortunately many women do not have such severe symptoms from estrogen withdrawal, and consequently need not face such extreme choices. Some women waltz into menopause without missing a step.

I asked my mother what it was like for her to go through menopause. She said, "What do you mean?" I said, "Menopause, what did it feel like to you?" She said, "I don't know." Frustrated, I asked her how old she was when it happened and again what symptoms she had. She responded that she couldn't remember when it happened exactly because she didn't feel much of anything. Never had a hot flash. Didn't know what one felt like. She discontinued her estrogens when her cancer was diagnosed and is doing fine without them today, except for her displeasure with her wrinkle count.

Are You Sure You Can't Take Estrogens?

Women with breast cancer themselves or in their family should still investigate HRT before deciding arbitrarily not to take it, even if their doctor advises against it. They should not simply settle for no as an answer until they thoroughly appreciate the full scope of their options and the associated risk/benefit ratio. The risk/benefit ratio is determined by balancing the emotional and physical consequences of taking HRT against the consequences of not taking it, and developing a strategy that gives you the most favorable balance between the advantages and the drawbacks.

It is important to understand that the *type* of ERT affects the degree of breast cancer risk. Differences in breast cancer risk may exist depending on the type of estrogen taken:

estrone (Premarin)	may have lower breast cancer risk	oral	conjugated estrogen made from pregnant horses' urine
estradiol (Estrace) (Epinil)	may have higher breast cancer risk	oral patch injection	synthetic

Estrone appears to be associated with lower breast cancer risk. Some studies have found higher breast cancer risks with estradiol, which is synthetic. Here is a partial explanation: estradiol (rather than estrone) is the chief estrogen found within the breast—meaning that perhaps estrone won't fertilize a breast cancer while estradiol could. Therefore, while estradiol replacement is perhaps more beneficial psychologically and sexually, it may result in greater breast cancer risk, because estradiol receptors are the main estrogen receptors found in the breast. Unfortunately, the current estrone form (pill) does not pass through to the brain, so you may miss out on the cognitive benefits. Pellet implant or transdermal skin patch have scarcely been investigated for breast cancer consequences because they haven't been available long enough for studies to be meaningful.

European studies have shown significantly increased breast cancer with ERT more frequently than those performed in the U.S., perhaps due to the prevalent use of estradiol rather than estrone. An important study reported in the *New England Journal of Medicine* found an overall increased breast cancer risk of 10 percent in ERT users. The risk was increased to 20 percent in estradiol users, and further increased with duration of exposure.

Some researchers have found a decreased breast cancer risk in estrogen-progesterone users. European studies, however, also found a higher risk of breast cancer in estrogen-progesterone users. Possibly, the difference between these two studies is again the greater use of estrone products in the United States and estradiol in Europe. This again suggests that estradiol is more dangerous than conjugated estrogens as far as breast cancer is concerned. However, estrone can be converted into estradiol once it enters our system, so there is no guarantee of better protection. The research isn't definitive.

Beyond breast cancer, there are other conditions commonly mentioned that most physicians say rule out hormone replacement therapy: liver disease, gall bladder problems, various hormone-dependent cancers, heart attacks, strokes, blood clots, thrombophlebitis, extreme obesity, et cetera. But there are ways to get around some of these considerations. Now that we have the estradiol patch and other nonoral methods to deliver estrogen *we can bypass the liver and gall bladder*. Someone with active liver disease should probably take no estrogen, but someone with reduced liver capacity could take the patch method, while avoiding the oral form. Someone with an old, low-grade history of thrombophlebitis might choose to take daily aspirin for prevention of blood clots and still take estrogen, progesterone, and testosterone. Or, she may decide to take just the testosterone, a little of which transforms into estrogen in the bloodstream. As mentioned earlier, the danger of clotting has been greatly reduced with the lower dosage used today, and other things are changing. Estrogen is now considered protective of your heart. In fact, very recently estrogen has actually been used to treat angina, by linguet (tablet under the tongue, just like nitroglycerine).

If you are in the gray zone, or just can't make up your mind, make a *really* educated choice: Try them. Try the patch for a few days or weeks. Then change to the pill for a month or two. See if it makes a valuable difference in your life. If you have side effects, switch around and try different methods until you are satisfied with one or convinced that ERT is not for you.

A short-term trial is not completely risk free, any more than aspirin or penicillin, but the potential for harm is minimal. To be extra safe, cover yourself by taking baby aspirin to prevent clotting problems.

For those who are diagnosed with cancer of some sort and told to stop their hormones immediately, weigh the pros and cons of stopping versus remaining on hormones until you have recovered from whatever surgery is involved. Discuss the matter with your doctor and then make the decision that is individually best for you. Some papers are coming out in the literature reevaluating this issue, and medical opinion may change on this point.

When You Absolutely Can't Take Estrogens

If you know ahead of time that you do not plan to take estrogens for menopause due to cancer, other medical reasons, or simply by personal choice, there are a number of measures you can take to prevent the more drastic effects of estrogen withdrawal. It would be misleading to suggest that these measures individually or cumulatively are equal to estrogen replacement, as is so often implied. It is not enough to take your calcium pill religiously and walk around the block, although it does help. On the other hand, these guidelines can markedly improve health, quality of life, and mood. Therefore, do take them to heart. Many of them are simply sound, commonsense measures, like good diet (low cholesterol), exercise, and avoiding alcohol and cigarettes.

Again, according to the nutritionist June Konopka:

"Diet can help alleviate menopausal symptoms in many women. A good diet includes a wide variety of unprocessed, unrefined foods (that means a minimal amount of white flour and white sugar products), green, leafy, and yellow vegetables (raw or lightly cooked), fresh fruit, whole grains, moderate amounts of unrefined vegetable oils, and a greater proportion of plant-based proteins compared to animal proteins with an emphasis on nuts, seeds, and legumes. Vegetables, pulses, and legumes (especially soybeans) contain phytoestrogens that can be beneficial. When estrogen levels are low in the body, phytoestrogens are able to exert some estrogenic activity; when estrogen levels are high, phytoestrogens reduce overall estrogenic activity by occupying estrogen receptor sites. This may explain, in part, why Japanese women have fewer menopausal symptoms and low breast cancer rates. Physical vitality is not only determined by what we choose to eat, but by what we choose to leave out of our diet. Alcohol, caffeine, and refined flour and sugar products significantly diminish our energy over time. Useful supplements during menopause might be evening primrose oil, flaxseed oil, bioflavinoids, and vitamin E. But remember, supplements cannot make up for an inadequate diet."

The estrogen replacement from certain soybean products and a form of ginseng show estrogen replacement potential, but any woman depending on "natural" measures alone should get blood tests for estrogens to insure adequate levels until more research has been performed establishing standardized, reliable dosing.

Whether or not a woman is on estrogens, weight bearing exercises—namely, exercise against resistance of some sort—become increasingly important as she ages. To compete with the onslaught of osteoporosis and other muscle and tissue deterioration, you must get your circulation going and your body moving. Exercise is also an antidepressant—not the least of its assets—and it raises your DHEA. Ten to fifteen minutes of exercise every morning will be the soundest investment you ever make. In addition to strengthening your bones and your muscles, it also helps your heart. So choose and follow your favorite cardiovascular fitness program.

And, speaking about exercise, don't forget your gymnastics in bed. Frequent sexual activity during perimenopause has been associated with higher natural estradiol levels and fewer hot flashes. Intercourse is superb exercise and can be weight bearing if you are on the bottom, though that is the wrong kind. You want the kind of weight bearing you get when you are on top. It's good for your thighs, buttock, hips, and arms.

As an aside, the most obvious genital symptoms of menopause—lack of lubrication—can be corrected artificially. Using vitamin E capsules is a nice technique: poke a hole in a capsule or two with a pin. Squirt the vitamin E into the vagina. This is not easy, but worth the effort. It will help to keep the tissues healthier and give you your daily dose at the same time.

Aren't Calcium Pills Enough?:   Calcium supplements alone have not been shown to substantially prevent osteoporosis, although high doses (1,500–2,000 mg/day) of calcium have some protective effects. However, calcium can increase the effectiveness of estrogen supplements, so if you are already on estrogen, you may be able to lower your dose of estrogen once you add calcium.

And, incidentally, the best insurance against bone loss after menopause is high *dietary* calcium intake when young and continuously thereafter. If it's too late for you, make sure that your daughters get enough calcium when they are young. It can only be efficiently stored and converted to bone growth until the early thirties, so these early years are strategic years.

Also, check with your doctor to ensure that you don't overdose on calcium. Too much can cause constipation, gas, kidney stones, and abnormal calcium deposits.

Just a few hours in the sun (without sunscreen) per week or fish in the diet can usually provide enough vitamin D, which is essential for bone formation and calcium absorption. A fine way to double dividends is to do your exercise outside in the morning sunshine. Exercise and sunshine are antidepressants as well and help to reset your circadian rhythm, enabling better sleep. So does early morning light.

Calcitonin: Calcitonin is a good alternative when estrogen is inadvisable. This is a peptide hormone secreted by the thyroid gland that can prevent the progression of osteoporosis by inhibiting bone reabsorption. It works for both men and women and is more widely used in Europe than in the U.S. It can also be used along with estrogen, allowing lower doses of estrogen for essentially the same effect, but with less breast cancer risk. Qualities shared by calcitonin and estrogen include appetite reduction and a certain degree of analgesia (pain reduction, numbing sensation) due to stimulation of beta-endorphin secretion.

There is also something brand-new and exciting on the horizon: A genetic treatment for bone and cartilage disease has been found to work that may make it possible to treat patients with osteoporosis with a procedure similar to transfusion but consisting of blood donor cells. It is like a bone marrow transplant, but using special cells—only the precursors of bone and cartilage that are found in marrow. A news release from Thomas Jefferson

University Hospital states that "one especially attractive feature of this type of therapy as described by the researchers is the ease with which the precursor cells can be isolated. Even though bone and cartilage precursors are an extremely small fraction of the cells of the marrow, they can be readily obtained from a donor with a needle and syringe, and in a few days, are ready to grow very rapidly in vitro. This makes them ideal for insertion into diseased patients." ("New Gene Treatment for Bone and Cartilage Disease Found," Thomas Jefferson University News Release, May 22, 1995.)

Partial Reprieve: Women with breast cancer are often placed on tamoxifen, an antiestrogen hormone. It was at first feared that this treatment would aggravate osteoporosis and all the other features of estrogen withdrawal. However, this turned out *not* to be the case. Tamoxifen actually helps to protect you against osteoporosis while preventing the progression of breast cancer. It also has *the opposite* effect on vaginal lubrication, promoting it, sometimes in abundance.

Don't Overlook Testosterone: For women who prefer not to take estrogen, testosterone therapy alone is sometimes an appropriate alternative. Strange as it seems, some testosterone will be transformed in the body to estradiol, potentially preventing some osteoporosis and skin changes in women who can't tolerate estrogen. The bulk of the testosterone replacement or supplement will give women who have had malignancies many of the benefits of testosterone therapy (see next section) without jeopardizing tumor recurrence by delivering much estrogen. (However, bear in mind that since a small amount does transform into estrogen, those women who want to totally eliminate estrogen can't even take testosterone.) Testosterone protects some tissues as mentioned earlier, and preserves sex drive. While testosterone alone won't compensate for the absence of estrogen—nothing will—it does fully compensate for the decline in testosterone, the benefits of which are worth considering.

In the future, for women who take estrogens, the addition of DHEA and perhaps tamoxifen or tamoxifenlike substances may fill the gap between what is available to her now and the quality of life and health with estrogen replacement therapy.

Perhaps one final point should be made here to round out the ERT issue. In fear of cancer risks, many women turn to over-the-counter or health store pharmacology. What they don't realize is that health store pharmacology is also experimental, with outcomes far less certain than when standard low-dose estrogen is prescribed by knowledgeable doctors who can assess both risks and benefits.

That said, there are numerous resources women are pursuing in the nontraditional mode that improve their QOL and health. It is important that women understand, however, that these alternatives will not provide osteoporosis and cardiac protection equivalent to estrogen.

PROGESTERONE REPLACEMENT: PROS/CONS AND PROTECTION

Originally, menopausal treatment consisted of estrogen replacement only. It was eventually discovered that in women undergoing natural menopause (those who still have their uterus and ovaries), giving estrogen alone can increase the risk of endometrial cancer. As mentioned earlier, without progesterone, estrogen alone can cause the uterine lining to hypertrophy (overgrow), which can lead to cancer. Adding progesterone to estrogen replacement therapy actually reduces this risk. (Progesterones are not as necessary in postmenopausal women who have had their uterus removed, because there is none left to protect, but may still be of some value.) With the combination of estrogen and low-dose progesterone replacement therapy, a woman can get the best of both worlds.

However, if a woman has had both her uterus and ovaries removed, it is important that she consider testosterone replacement therapy *in addition* to estrogen and progesterone, if she wishes to maintain certain qualities of her character including assertiveness, energetic sexual libido, and bright mood.

The New, Improved PMS and How to Avoid It

For all the benefit of this high-tech modern medicine, there is a hitch. Certain HRT strategies resurrect and perpetuate your period—and PMS—because if you use high doses of Provera (a progesterone) at the end of each cycle, you menstruate. It imitates, to some degree, your natural menstrual cycle. When progesterone is abruptly stopped for seven days, you get withdrawal bleeding much as you did prior to menopause.

Freedom from monthly misery is one of the few dividends of natural menopause, and some women won't take progesterone supplements for this reason alone, not to mention the renewed PMS and decreased sexual desire associated with relatively high doses (10 mg a day). But there are ways to get around it while still getting the right treatment.

Newer drug combinations have most of these problems licked. Now you can take Provera, not have any periods, and skip PMS, too. The solution is daily treatment with a combination of low-dose estrogen (.625 to 1.25 mg) and low-dose progesterone (2.5 mg), which has become quite popular with women and physicians for this reason. This is continuous treatment every day with no breaks, so you don't even have to count. Although there can be some breakthrough bleeding during the first four to twelve months, the use of lower doses of progestogen in continuous treatment (2.5 mg medroxyprogesterone or 1 mg or less of norethindrone) avoids most of this hormone's side effects and eventually prevents bleeding altogether, making the whole experience much happier, while still protecting the uterus.

TESTOSTERONE REPLACEMENT THERAPY FOR WOMEN

Although estrogen replacement therapy or a combination of estrogen-progesterone can solve some of the sexual symptoms of menopause, testosterone is often necessary to take care of the rest.

Physicians are appropriately reluctant to prescribe testosterone supplements for young women, but they often overlook the fact that surgical menopause is an exception to this general rule.

Testosterone replacement is an important consideration for natural menopause as well, for it can improve sex, mood, and quality of life. Yet testosterone supplementation for natural menopause is still frowned upon in the United States. In fact, testosterone supplements for menopause in the U.S. are rarely considered.

But research has found that, compared to placebo and estradiol-only groups, patients treated with a combination of testosterone and estradiol were less depressed, less anxious, less hostile, less tired, more clearheaded, more confident, and more energetic. Similar improvements occurred in the "estradiol-only" group, but were only half as pronounced as for those with the testosterone-estradiol combination. Testosterone behaves like an MAO inhibitor, similarly to estrogen, so it should have some antidepressant effect. Also, as we know, testosterone increases desire. (Conversely, decreased desire, activity, and mildly depressed states can decrease testosterone levels.)

Testosterone can be administered through injection, sublingually, orally, vaginally, or even by pellet implant, which is now available in the United States. We will also have testosterone patches soon.

Fluoxymesterone (5 to 20 mg a day) is frequently used for supplements in women. Oral methyltestosterone and fluoxymesterone are available in the U.S. Neither TU nor mesterolone are toxic to the liver but have not been approved for use in the U.S. Oral mesterolone (Proviron) is not metabolized to estrogens like other testosterone supplements, so it could be used in women where estrogen is not recommended. While sublingual testosterone (Testoral) is not as potent as other oral preparations, it can be advantageous for women because they do not usually require very high doses.

The pellet implant is placed under the skin, usually somewhere in the buttocks area through a very tiny incision. From there, it delivers testosterone until it melts away. There is no removal necessary, and it is done every two or three months. The oral route is less well absorbed, although sublingual administration goes directly into the bloodstream and is painless and convenient. On the other hand, the pellet avoids daily medication. Each has its own advantages.

Testosterone concentrations in the female genital tract (labia majora, clitoris, and pubic skin) decrease with age much more than in the genital tract of men. Testosterone cream (2 percent) applied to the vulva, vagina, and genital tissue, should be considered for maintaining tissue thickness and health. Cortisone creams are often prescribed but they thin the tissues even more. Estrogen helps somewhat by thickening subcutaneous tissue, protecting collagen, and promoting elasticity, but testosterone cream thickens and toughens these tissues. Sometimes "senile vaginitis" cannot be cured without it.

As we have mentioned in Chapter 5, testosterone also has the ability to raise blood pressure in women, cause or aggravate liver disease, increase cholesterol, and can cause weight gain both by increasing lean body mass and by causing edema. Acne is another possibility. Prostate cancer, of course, is not a worry.

Another word of caution: Among some elderly women, testosterone could aggravate aggressive behavior brought on by dementia and lead to destructive behavior.

In summary, while menopausal testosterone supplementation for either sex cannot be arbitrarily recommended because research is lacking, it remains the most promising pharmacological treatment for low sexual desire in menopausal women, particularly those resistant to estrogen treatment and/or sex therapy. For men, it promises to prevent or treat testopause as nothing else will.

Ironically, it has been overused for men by many of the same physicians who underutilize estrogen in women. In essence, then, these two hormones with relatively comparable side effects are used with curiously opposite patterns—too little for the woman who needs it, too much for the man who does not. Now that measurements of free and bound testosterone are readily available, testosterone function in both men and women warrants a second look.

FOR THE FUTURISTS AMONG US

With a look toward the future, other hormones and peptides we have discussed should be further evaluated from Zothara's point of view to prevent menopause and viropause. Indeed, we have done a nice but rudimentary job of reproducing the hormones we enjoyed at our prime to carry us into old age, but it is now time to do the fine tuning and include some of the other players in our standard treatment protocols, like oxytocin, vasopressin, growth hormone, and DHEA, which I am persuaded plays a major, but as yet unappreciated, role in menopause of both sexes.

Since both DHEA and growth hormone start dropping around age thirty in men and women, the ideal would be to replace them at their thirty-year-old values. DHEA would then theoretically continue fighting off cancer attacks, keeping your immune system strong, and sustaining your sex drive and scentual sex appeal, while growth hormone would keep you looking, feeling, and acting young and vital.

We originally learned about the potential rejuvenating features of growth hormone by treating short children for this deficiency. In young adults, the overall ratio of muscle to fat—an important index of conditioning, especially among athletes—improved by an average of almost 25 percent. (It is not surprising that growth hormone supplements are being abused by some athletes.) Now growth hormone is under consideration as a replacement hormone in the treatment of menopause and viropause because it shows such promise in returning or reviving strength, health, and stamina—an exciting substance to watch in the future as research data accumulates.

Of course, Zothara would have you install an LHRH pump along with sniffing vasopressin and oxytocin at regular intervals, in addition to these other methodologies. Such are the prospects for our future.

YOU, THE JURY

Altogether, your wonderful, savory sex soup is a well-balanced system that modern medicine has learned to manipulate. But it does change as you age. While we would like to have much more data than will be available to us in our lifetime, there is enough research to raise important questions and concerns over whether an important aspect of women's—and to some extent, men's—health care is being neglected or needlessly withheld when we address menopause and viropause.

You are the jury that counts. The verdict is in your hands. The only odd part about it is that you are also the defendant—whatever your decision is, it will be your life's sentence, and how wisely you make it will have a significant bearing on how long you live and your quality of life along the way.

If sexual considerations such as painful intercourse, lack of lubrication, bladder problems, and wanting to retain a sweeter smell, more sensitive touch, and more attractive appearance lead more postmenopausal women to use standard low-dose estrogen supplements, good sex will be both the catalyst and the dividend, bringing a sense of well-being and better overall health, but why wait till you deteriorate?

Nevertheless, even if estrogen treatment is not initiated at the start of menopause, estrogen replacement, at any time, will be valuable and can prevent the progression of osteoporosis, heart disease, and other symptoms. You can undo some of the past, and create a better future. *It is never too late to begin.*

The benefits of HRT-estrogen, progesterone, and testosterone—regardless of the route of administration—means an improvement of a whole collection of gynecological and urogenital symptoms, maintenance of healthy skin, prevention of emotional upheaval, and protection from arteriosclerosis and osteoporosis. That is a pretty attractive package. Weigh it against the risks.

We have explored three different approaches to menopause and viropause: letting nature take its course, treating the symptoms as they occur, and a futuristic look at preventing them altogether. Whichever approach you choose to take, as the years pass, other choices will present themselves to you. How you age and how long you live will depend not just on time, but on what choices you make.

*L*ASTING LONGER

There is a woman I know who is three or four decades older than she would like you to believe. Men still whistle at her legs, and she takes full advantage. She wears support hose with miniskirts, spike heels, and glittering bifocals. Her nails are perfection. Her skin has no pores.

She makes sure everyone knows how much she has had to sacrifice to stay fit— skipping fat, walking every day to see friends and family, and giving up her cigarettes cold turkey. She makes sure she gets her strokes, satisfying her skin hunger on a daily basis, getting petted. Whenever there is a crowd, she hogs all the attention with no apology.

She is actually an unabashed bisexual flirt, charming women and men alike, enthralling them, insisting they adore her. If they tell her she is pretty, she pretends she didn't hear it the first time to savor the encore, then she elegantly waves the compliment away as her just due.

Recently, she found out she had breast cancer, but took the news in stride and merely replaced both breasts with younger ones. A few years ago, she turned down a perfectly suitable offer of marriage, indicating that men on a daily basis were too great a chore. On another occasion, after a glass or two of champagne, she was taking her usual walk home. En route, she noticed a construction worker half in, half out of his pickup truck with his derriere tantalizingly exposed. She couldn't resist patting it on her way by. The construction worker, half, perhaps one third, her age was struck speechless.

Drawing from a well-lived life as a model, fashion designer, wife, and mother, she has outlived two virile husbands. With relentless self-discipline, she maintains a size four dress and is working on surviving her more lackadaisical daughter.

In fact, she claims to have taught me everything I know.

Here is a woman who has done instinctively all of her life what longevity scientists are only now discovering. She is taking her vitamins regularly and exercising every day. Her weight is below normal—lean and, she says proudly, mean. No question that she is stimulating her hormones and those of others, remaining alert and alive, refusing to equate chronological age with emotional age. She wholeheartedly agrees with Billie Burke, who lived to be a natty ninety-four, that age shouldn't matter unless you are a cheese. With attitude and fortitude, she has defied the odds—the traditional patterns of aging.

According to one study, sexually active seniors are the happiest men and women in America. Clearly, when you add sex to an otherwise healthy maturity, quality of life soars. But does sex perhaps have other dividends to offer individuals as they age, beyond the obvious? Most people do not just want to live longer, they want to enjoy life. Who wants to last longer unless they live better? That's where sex comes in.

This final chapter examines the relationship between hormones, love, lust, and longevity. I am going to take the understanding you have developed about hormones and stretch it. You now know the most current information about hormone replacement therapy—the consequences of doing it and the consequences of not doing it. You have learned how hormones manipulate us relentlessly from birth until death, how they dictate our stages, and influence our love lives. Wouldn't it be nice to be able to manipulate them right back and take charge again?

In this chapter I am going to tie all the various components together for you, as well as make specific recommendations so that you can use the information in this book to help you stay sexually fit for a lifetime. I will show you how to enjoy your hormones to the fullest, minimize the damage they can do, sabotage their power when necessary, and modulate your own stages so that you can even skip a few if you choose, lingering longer in the best ones. This chapter also discusses to what extent life-style, attitude, and medical intervention modifies our sex lives. Toward the end, I am going to take you back to Zothara's theme, into the future of marvelous possibilities that await. . . .

What about peaks, cycles, stages? We will again explore how they inter-connect, with particular attention to the dynamic differences between men and women. We've seen how hormones have the power to draw us together or drive us apart. I am going to suggest that they also exert forces to make us mutually interdependent. It's the same throughout the animal kingdom, with one very large exception: Humans, with knowledge, can alter their influence to suit our needs. Animals can't.

But first, let's find out: Does sex add a few years to your life or steal them away?

LASTING SEX

There is good reason to believe that sex lengthens your life span. Certainly there is extensive data to prove that frequent touching extends your life and improves its quality. As you have heard, older people become senile and die sooner if they are not touched. Those who have had heart attacks have better survival rates if they have a pet at home. The fact is, that having someone or something to touch keeps us alive longer. And sex is touching.

Touch exerts its power throughout your system by raising the substances that lengthen your life span—DHEA, oxytocin, endorphins, growth hormone —and lowering those that can shorten it, like cortisol and adrenaline. Touch lowers your pulse and heart rate, relieves stress, and increases your sense of well-being. Commonly when sex disappears from a relationship all touching comes to a halt. If intercourse is not possible for whatever reason, most people turn away from everything else. That is like skipping dinner just because you can't have dessert, and starving yourself.

If you practice enjoying other forms of sexual intimacy when you are younger, you will be much more resourceful and resilient sexually when you are older. I cannot emphasize this point enough. Health and longevity may not be the main reasons for enjoying an active sex life well into old age, but I call them a very nice dividend.

Many elderly couples stop having sex because one or both are ill or disabled to some degree. But touching between loving couples, as you now know, should not end no matter what—and sometimes one touch leads to another.

Will and Andrea may not be in the best of health, but sexually they are doing better than most healthy people.

Will had too much emphysema and too little energy from smoking heavily most of his life. Andrea had arthritis in her hands, knees, and hips—not crippling, but it made it hard for her to spread her legs, grip his penis in her hand, or get on top.

She snuggled up to him first thing in the morning.

"Your warm body wakes up my bones."

"Move over; let me sleep."

"Come on, a little loving would be good for your heart."

"Yeah, but when you kiss me, I can't breathe. What if I choke to death?"

"Can't think of a nicer way to go, can you? Besides, I'll kiss you other places instead. Meanwhile, get your oxygen—that's what we got this bedside unit with the nasal tubes for. I can still remember the expression on your doctor's face when we told him no face masks because they got in the way of tongue kissing. Almost fell out of his chair—as though it was against the law at our age."

"Well, now, perhaps it should be. I could use some protection."

"You know I can't get pregnant anymore."

"Hey, I thought you were going to use your mouth for something else," Will quipped.

Later . . . "That's pretty hard. Hold it for me while I find a comfortable position."

Side to side, they find their way to a familiar position their bodies don't object to.

"I wonder what your doctor would say to this kind of physical therapy?"

Andrea's endorphins are flowing, easing her pain. Warm blood is lubricating her joints. He is getting his cardiovascular workout—gently, resting on his side, with no respiratory distress. With oxygen flowing, he has enough peace of mind to enjoy himself. After a little thrusting, and a couple of orgasms for her, she takes him in her now limber hand. He indicates that ejaculation isn't on his agenda today, but tells her he expects a rematch later on.

She goes and gets the paper while he prepares juice and toast. They read in bed for a while.

Andrea said, "I went to see Emma in the hospital yesterday. I brought her a stack of Playgirl magazines to read and she laughed, asking me what on earth she was supposed to do with them. I told her she needed to get more blood to her brain —that's why she was there after all—a stroke."

"Where did you get your M.D. degree?" Will laughed.

"Oh, you know, it's just common sense. I thought it would be good for her circulation to get her sexually aroused."

"Should be a prescription item," said Will.

Will and Andrea kept their sense of humor and their sex life in working order by adapting to their health challenges and capitalizing on their resources. They not only enjoyed sex and intimacy, but got several additional dividends: the exercise, the antidepressant effect, and the erotic refreshment. They did not stop touching just because they ran into a few obstacles.

Ageless Sex

*During sex, the body is transformed. "One may become conscious of an increase in temperature in his own or the sexual partner's body surfaces," Alfred Kinsey wrote. "The identification of sexual arousal as a fever, a glow, a fire, heat or warmth testifies to the widespread understanding that there is this rise in surface temperature." At the same time, the skin begins to change color. What is known as the sex flush generally starts on the upper abdomen and face, then spreads to the breasts, neck, chest, thighs, arms, lower abdomen, buttocks and back, deepening in some cases from light to dark red or even to a rich reddish purple. A fine film of perspiration appears on various parts of the body. The eyes dilate and glisten with increased moisture. The blood that rushes to the surface alters the contours of various parts of the body, swelling the lips, thickening the nose, earlobes, enlarging the breasts. As the flesh around the eyes and mouth fills out, facial lines are reduced or erased, and years seem to fall away. (*Esquire, *May 1989)*

Without knowing it at the time, Alfred Kinsey could have been talking about vasopressin before it had yet been identified. The heat, the warmth, the flush as this peptide modulates our temperature. This is "vasopressin fever," and it can last throughout our lives, until we are very old indeed.

It used to be that old age was synonymous with the diseases of aging. With advancing age, the possibility of having to contend with one medical condition or another obviously increases. Actually, though, the pendulum is now swinging the other way. Age itself is considered by some to be a treatable disease. But how does one distinguish between the normal stages of aging and those imposed upon us by disease? It isn't that easy, although we're making great strides.

Many of the problems—including mental ones—we now associate with aging are actually due to arteriosclerosis and other chronic ailments. Weakness and fatigue are more often the result of heart disease, diabetes, or medication than pure old age. Less blood gets to strategic areas of the body or brain, and like a plant without water, they wither. Should we succeed in preventing these ailments, many conditions we now attribute to age would probably disappear, just as dowager's hump suddenly became almost obsolete with the effective treatment of osteoporosis.

I am not suggesting that if we diagnose and treat all disease we will live forever like Zothara, although that is theoretically possible and there are some scientists who subscribe to this theory. I do propose, however, that we are making great progress in separating the disease process from the aging process, and will continue to make advances along these lines, which should help us to modify our medical approach *from repairing problems to preventing them.*

None of these twenty-first century treatments will mean anything to the substance of our lives, however, if we don't change the regressive attitudes we have about intimacy, sexuality, and physical contact.

Concern for technique and performance has obscured the importance of emotional intimacy for old and young alike. In the quest for genital satisfactions, orgasms, orgasms, and more orgasms, often the caring and sharing is lost. The effects of these anachronistic attitudes are particularly pronounced as people get older—when the need for touching, companionship, love and intimacy become even more critical than at any other time of our lives. Even though the testosterone haze has lifted, old habits die hard. Nevertheless, we must reevaluate these old habits and goal-oriented approaches to sexuality, because there is no question about the health benefits of touch. No partner handy? Don't despair. Keep your favorite sexual partner alive in your mind. Bring him or her back in your fantasies by replaying the home movies of your memories, stirring up oxytocin, PEA, DHEA, and the rest of your sex soup.

Take advantage of self-repair and masturbate if you don't have a problem with it. Understand that if you choose not to masturbate and have no partner available, the consequences, for men, will probably be impotence, and for women, vaginal atrophy and other complications. The decision, of course, is yours. It's funny that so many men who masturbated with enthusiasm when they were teenagers won't masturbate at all as they get older, when they may really need to in order to prevent some of the irreversible sexual changes of aging. Jonathan is an exception. Although he is *psychologically* impotent, he didn't give up masturbation:

Jonathan, who had a lively sex life when he was younger, was already deep in the clutches of the adrenaline reflex before his relationship with Amy ended. Fears of performance and all the other trappings of psychological impotence, along with the upheaval of viropause, plagued him whenever he was in the presence of a woman, any woman—young, old, beautiful, ugly, skilled, or inexperienced—and nothing worked.

He was sentenced to his own company. "Rosy Palm," as he called her, was his only steady companion. But with her, he never faltered. Time after time, his penis responded just fine to masturbation, but let him down just as consistently every time he tried to function with a woman.

If Jonathan gets beyond his psychological problems, he will be sexually ready to work just fine for only one reason: he will have had sex regularly with someone who didn't make him nervous or feel pressure to perform— himself. Without realizing it, he continued to do the one thing that he could have done to protect himself from becoming biologically impotent as well. It happened to be "Rosy Palm," but any safe, trusted partner would have had the same beneficial effect. And when he wasn't worried (adrenaline reflex), funniest thing is, he worked.

Sexual Fitness

Both men and women who have sex often when they are younger retain the capacity to do so when they are older. However, after age sixty or so, if either one takes an extended intermission—even for just a few months—the physical capacity for sex rapidly fades. Think of how quickly your arm or leg can wither if kept in a cast for just a few weeks. The same is true for the rest of your body, your brain included. If you do not exercise them, they will atrophy. To keep sexually fit, the same principles apply. Having intercourse or masturbating with regular frequency is your best sexual insurance policy. But that's not all. There are some basic things you can do to become and remain sexually fit as long as you live:

All this will work—but, if like many older people, you are on any sex offender medications, you will need to consult with your doctor to change or adjust your medications so they don't sabotage your sex life.

Sex Offender Drugs

Some medications meddle with your peptides or your hormones and simply take the joy out of life. When prescription drugs sabotage your sex life, they do much more harm than most people realize. Twenty-five percent of impotence is the direct consequence of prescription drugs given for other reasons: heart disease, ulcers, depression, et cetera. I recommend that no patient accept a medication without appreciating its potential side effects—in particular the sexual ones. There are usually alternative medicines in each category of drug that can treat the medical problem equally well that are relatively "innocent" sexually.

In addition, beware of inadvertently combining medications that don't mix well. Some drug interactions not only can harm your sex life, they can trigger additional health problems. If you are being prescribed drugs by different doctors, make sure each knows what the other is doling out. Do not abandon any drug a physician has prescribed for you, and change them only under your doctor's supervision. But try to get by with as little as possible using all the nonprescription methods available to assist you—diet, exercise, and so forth.

EARTHWORM OR SPACE EXPLORER

If you have done all that you can to stay sexually fit it is time to contemplate the resources medicine now has to offer for making fit and healthy people better. But first, a disclaimer. There are certain advantages to growing older "naturally." Men and women blend together better as their biochemical features become more similar. I'm sure you have noticed that it is hard to tell newborn babies apart by sex in a hospital nursery. So it is with the elderly in convalescent homes. Our sexual distinctions are almost indistinguishable at the beginning of our life and then again at the end. Just think of all the trouble we managed to stir up in between. Much of it is the direct result of

THE FORMULA FOR SEXUAL FITNESS AS YOU AGE, THEN, GOES LIKE THIS:

- Sexual frequency of once a week or more, whether with a partner, with masturbation or both.
- Touch—a lot—throughout the day, during sex, on the way to sleep; if you don't have a partner, touch your friends, adopt a pet, cuddle your grandchildren. Find a way to get enough vitamin T to thrive.
- Diversify from genital sex. Don't limit yourself to intercourse. Savor the whole body and the broad smorgasbord of possibilities. Don't overlook manual stimulation, oral sex, touching, stroking, fondling, lips, tongue . . .
- Use your imagination; draw on fantasy; read novels that arouse you, look at pictures, rent movies, stimulate yourself. Play music that brings back erotic memories. Pin naked photos of your ideal lover on the refrigerator to help motivate you to keep slim and fit.
- Diet and exercise sensibly to keep your health, sexual pizzazz, and body image intact.
- Capitalize on your ability to use nonpharmacological methods of preserving your health—stress-reduction techniques, good nutrition, et cetera.
- Stop smoking.
- Cut down on or stop drinking alcohol.
- Avoid "sex offender" drugs (how I refer to prescription or over-the-counter drugs that cause sexual dysfunction)
- Keep an eye on the progress of hormone replacement strategies and longevity research.

these troublesome sex differences. Good riddance, perhaps, as Germaine Greer eloquently maintains. No more torturous love affairs, ill-defined longings, extreme misunderstandings, and mismatched temperaments.

As men and women age, their distinct characteristics begin to merge, biologically as well as psychologically. The enzymatic activity in men slowly shifts to that more characteristic of women (a return to the female mode from which males deviate during fetal growth). Similarly, the hormone profile of a postmenopausal woman not taking replacement therapy is closer to that of the male than it has ever been during her lifetime, largely due to the dramatic reduction of estrogen and along with it, oxytocin secretion.

As we've noted, some women prefer to age naturally because sex, to them, was a burden they feel well rid of. Surprisingly, some men share this philosophy about their own sexuality, but for a vastly different reason: they are convinced that the more often they ejaculate, the sooner they will die. There was a nematode study conducted some years back that lots of men identified with and worried about. These creatures did live longer the less sex they had. The more frequently they ejaculated, the sooner they died. This may be the source of the myth that men are born with a certain fixed number of ejaculations (like women with their eggs) and that once they are used up, so perhaps are you—a good example, I might add, of animal research leading us astray. Men with this worry wouldn't want testosterone therapy or anything else that promotes more orgasms.

Aging "naturally" is one perfectly legitimate course to follow, but it is not without risk. As you know, choosing against various hormone replacement therapies can subject you to depression, disease, and premature death. Then again, hormone replacement therapy as you also know is not risk free either. Now, though, I would like to take you with me to the cutting edge: Zothara's theme—reproducing the fountain of youth. However, the more we subscribe to the concept of young blood, the more gender distinct we remain. And there may be a hidden cost for men—shorter life spans. If you keep their testosterone levels up with replacement therapy, they may struggle with high cholesterol and heart disease. On the other hand, if you give them DHEA, it lengthens their life spans. For women, the big question is, will DHEA extend their lives or end them sooner? Some studies show that higher levels of DHEA in *women* correlate with shorter lives. Others don't. We just don't know yet. While the hormone replacement therapy usually offered today— estrogen and progesterone for the woman; nothing, or a little testosterone for the man—is a calculated risk, it is just the middle of the road. DHEA, oxytocin, growth hormone, and several others would ultimately have to be included to treat the deficiencies of age.

Let's explore what replacing these hormones, peptides, and neurotransmitters you have come to know so well can do for you. They already course through your body and will one day fade away unless you prevent it. Let's push the envelope a little, and play with the possibilities.

Longevity, Sex, Attitude, and Weight

Over the years while I was researching certain drugs for potential sexual side effects, I had identified more than a dozen that I felt warranted further research to investigate their sexual effectiveness. These were primarily pharmaceuticals that increased dopamine, testosterone, estrogen, or DHEA, or that lowered serotonin, prolactin, or progesterone. I'm sure that you can see why I targeted these mechanisms from reading about their profiles in previous chapters.

I noticed a recurring relationship between the drugs of interest to me and those receiving the most attention from longevity researchers. In fact, a striking percentage of these drugs all:

- increased sex drive;
- had an antidepressant effect;
- facilitated weight loss;
- were targeted for study by longevity researchers.

These features were sufficiently consistent that I began to suspect a link between sex and longevity, wondering if there was a causal connection. Does an active sex life extend your life? We don't know for certain, but many signposts point in that direction. Does good health promote good sex? That's plain common sense. Are there chemical common denominators between sex, happiness, fitness, and long life? I think so. But until we start looking with serious intent, we'll have to depend upon accidents of research for our learning—an inefficient and haphazard system.

Twenty-first-Century Alchemy

Now let's take a brief look at the hormones, peptides, and neurotransmitters that we have looked at from the sexual point of view to appreciate these additional dimensions: their impact on longevity, weight loss, and happiness.

Dopamine: Dopamine is essential to our happiness, our ability to pursue whatever we enjoy. Because of this, it is also at the core of most drug addictions in the sense that the drug transmits its rewarding effects through dopamine. It is also, as we've seen, responsible in large part, along with our endorphin response, for our romantic attachments and love addictions, causing us to *need* the other person once oxytocin, DHEA, and other factors have brought two people together.

While it may make you crave pleasure, love harder, and yearn intensely, it gives you more time to do so. For example:

Deprenyl is a prescription drug that works through promoting dopamine. According to *Medical World News*, a researcher named Jozsef Knoll claimed that deprenyl "can shift the life span of the human from one hundred fifteen years to one hundred forty-five years." And he has performed research on sexually sluggish old male rats in an attempt to prove it. His series of studies showed that the deprenyl-treated old rats recovered full sexual activity and maintained this activity for an extended life span. Placebo-treated rats continued to be sexually sluggish or became completely inactive. Further, the deprenyl rats reached peak sexual performance after eight months when the placebo rats were already dying. Pretty impressive, I'd say. He certainly succeeded in demonstrating that some of the same drugs that improve the sex life of various animals also cause them to live longer.

To paraphrase his report, the geriatric rat loses brain cells with age, just as we do, so that their dopamine response weakens. Also, there is more MAO activity, which can aggravate or cause depression. He concludes that the significant increase of depression in the elderly and the "age-dependent decline in male sexual vigor" might be due to decreased dopamine sensitivity and "trace amines," which are PEA derivatives.

Knoll proposes that deprenyl, which inhibits MAO and boosts dopamine activity, as well as PEA, might be the solution, improving "the quality of life in senescence." It will come as no surprise that deprenyl-treated rats were also thinner!

Other drugs that boost dopamine include L-dopa and Wellbutrin. The same holds true for them: they have antidepressant properties, tend to thin you down, and are of interest to longevity research.

Based on the sum total you have learned in the preceding chapters, I suspect that you would agree that what turns out to be fact for the rat in this case is a promising prospect for humans.

Oxytocin: Oxytocin plays a key role in our bonding, loving, nurturing, and parenting. It is involved in our orgasms and our pursuit of someone special to touch. It also makes male rats live longer: oxytocin injection three times per week extended the life span of male rats. Low doses were more effective than high doses.

Apparently, whether you raise oxytocin a few times a week with touch, nipple stimulation, orgasm, intercourse, or by injection, it is good for you. So good that you may actually live longer because of it. However, in contrast with vasopressin, it makes you forgetful, something that some people, surprisingly enough, consider a benefit. One of my elderly patients, when offered a "memory pill" by her neurologist, got annoyed with him, saying, "One of the few advantages of old age is that you can forget. Don't give me anything to make me remember things I no longer want to know." Oxytocin is also a mood elevator in the respect that it creates or promotes the euphoric state that results from holding someone you love. Does it make you thinner? I am not sure, but I bet it does. Wouldn't it be remarkable if constant physical contact with someone you love worked better than diet pills? Something to consider.

Vasopressin: Known to you now as our coping chemical, protecting us from emotional extremes, perhaps the "monogamy molecule" (at least for midwestern voles) and a player in orgasms, possibly essential for triggering them, vasopressin is also a favorite "smart" drug promoted by life extension books to help learning, attention, memory, and recall.

How vasopressin performs its miracles is unknown, but researcher Dr. David de Wied at the University of Utrecht in the Netherlands postulated that vasopressin causes changes within the central nervous system that help to transform the electrical impulses of learning into chemically encoded long-term memories. Durk Pearson, author and promoter of *Life Extension,* credits it with the ability to promote immunity. In this way, it fosters health and defers death.

While it is best known at the moment as an antidiuretic hormone (ADH), perhaps one of vasopressin's most captivating qualities is its influence on how we think. It focuses a person on the pragmatics of the "here and now," helping us to notice sexual cues, and to pay attention to what we are doing. With this special feature, vasopressin can help individuals avoid interference from distracting impulses, worrisome emotions (adrenaline reflex), and disturbing memories during sex.

For the same reasons, it may be a particularly good antidepressant/antianxiety drug. With its focus on the present, as opposed to the past or the future, it makes depression and anxiety difficult to sustain. Depression is manifested through a backward-looking state of mind. We get unhappy about things that have occurred in the distant or recent past. Anxiety is manifested as a forward-looking feeling. We worry about things that haven't happened yet. Vasopressin focuses us on the here and now—a therapy advance par excellence!

Here again is a drug, also a naturally occurring substance, that promotes sex, happiness, and longevity. If the pattern is consistent, we will also discover that it thins you down.

Melatonin: Melatonin is a hormone you have not met yet. It has just recently made its debut as a "miracle drug" and warrants a good look here.

This hormone has been little studied, and future research by pharmaceutical companies is unlikely since this substance is not patentable. Most of what follows is tantalizing, preliminary, and suggestive rather than established fact, but that said, here is what is being reported.

Various publications suggest that melatonin enhances sex, influences depression, and promotes longevity. Sound familiar? It also promotes a good night's sleep, cures jet lag, prevents pregnancy, and serves as a free radical scavenger and antioxidant. More extravagant claims state that melatonin is "the youth hormone," delays aging, boosts your immune system, and prevents and/or cures cancers of all sorts (especially breast cancer), heart disease, PMS, Alzheimer's disease, Parkinson's disease, and many more. It is also supposed to slow the growth of cataracts and improve diabetes.

This hormone already has a pretty impressive résumé. If only half these claims to fame are true, I'm excited. *But,* because it has such a potent effect on our other hormones, I would like to know considerably more before rushing to the store.

Melatonin is produced in darkness. Light turns it off. It derives from a pea-sized organ in our brain called the pineal gland. This remarkable hormone is best known for regulating our body clock (circadian rhythms) and treating insomnia. Just a little bit, 2 mg, of melatonin before bedtime can bring on restful sleep in the elderly, whom, as many of you know, often develop insomnia as they age. Melatonin deficiency seems to be the culprit. Replacing this hormone works better than sleeping pills, with no risk of addiction.

Melatonin peaks during childhood and declines gradually after puberty. Fifty- to seventy-year-olds have only trace amounts. In the future, melatonin may well be added, for obvious reasons, to the treatment and prevention of menopause and viropause once more is known about it.

"For every major hormone system that's been studied, we find age-related changes that suggest they could have meaning to the aging process," said Dr. Marc R. Blackman, chief of endocrinology at the Johns Hopkins Bayview Medical Center in Baltimore. (Jane Brody, "Restoring ebbing hormones may slow aging." *New York Times,* July 18, 1995.)

Melatonin inhibits estrogen and increases both prolactin and growth hormone. And there is a gender difference. Women are more sensitive to the cyclical fluctuations of melatonin than men for reasons we can't yet explain.

What does this wealth of information mean? Melatonin is a potent hormone that can influence your health, your attitude, and perhaps your sex life. One documented way to benefit from its virtues is simple: get more sleep *in the dark,* of course. Significant hormonal changes have been measured between those who get just eight hours versus those who get fourteen hours. Sleep is literally rejuvenating, and most of us are sleep deprived in this modern world, governed by artificial light rather than natural light.

Another way to benefit is to take melatonin supplements—if you are willing to be a part of a massive self-induced experiment. We don't know what the long-term effects might be, and the desirable short-term effects could backfire in serious ways. This is not just an idle caution that applies to any new drug, but a genuine concern specific to this drug. If it is strong enough to make women infertile, imagine what else it might do to our delicately tuned, rhythmic, cyclic endocrine systems.

Growth Hormone: Although we know less about the sexual effects of growth hormone than most of the other substances discussed in this book, anecdotal reports and the rare small study hint that this hormone too stimulates sex, improves attitude and mood, thins you down, and perhaps extends your life—a delightfully redundant theme. Babies deprived of touch fail to thrive partly because they produce less growth hormone than newborns that are touched often.

In one small study, twelve older men were placed on growth hormone. After taking it over six months, they averaged a 14 percent decrease in body fat while lean muscle mass increased by 9 percent. They also showed a 7.1 percent increase in skin thickness.

These preliminary findings gave rise to the rumor that human growth hormone could someday become a broad-spectrum anti-aging drug. According to Richard Cutler, Ph.D., a gerontologist at the National Institute of Aging at Bethesda, Maryland, "treatment with human growth hormone may enable people to retain a more youthful and vigorous body state as they get older. If we find that it also lengthens life span, that would be icing on the cake."

Some subjects receiving growth hormone injections three times a week over the course of six months felt that their bodies started moving backward through time. Their bulges receded and they regained the firm muscles of years ago. A seventy-two-year-old man told reporters that the wrinkles were disappearing from his face and hands. He could open jars with ease again and stride past younger people on the street. Older people receiving growth hormone comment on a strange but delightful feeling of robustness that seems to stay with them for some time even after injections stop.

Out of this research grew the concept that many of the emotional and physical changes seen in elderly patients could be attributed to a decrease in their ability to produce growth hormone.

These preliminary findings gave rise to headlines such as the one in *The New York Times*, "Human Growth Hormone Reverses Effects of Aging." Indeed, the theory of "young blood" would involve determining normal levels of growth hormone secretion in early adulthood, calculating the decline with aging and replacing the deficit accordingly. Welcome to Zothara's Land.

Growth hormone is not an innocent substance, however. It is potent stuff. Bear in mind that growth hormone makes things grow. That can include tumors. Also, too much growth hormone can be deadly, leading to enlargement of the heart and congestive heart failure. People who overproduce human growth hormone are prone to diabetes, arthritis, and a disfiguring condition called acromegaly, which causes bones of the feet, hands, fingers, nose and jaw, soft tissue of the nose, lips, forehead and tongue to grow to grotesque proportions while the rest of their features stay the same.

DHEA: You already know that DHEA promotes sex, weight loss, and happiness, so it should not surprise you that DHEA is one of the most promising potential longevity drugs under study. It is already popular in Europe as a treatment for the common problems of aging.

Time magazine reports:

> Dr. Etienne-Emile Baulieu, developer of the controversial RU-486 "abortion pill," had Paris in a tizzy last week. In a cover story in the French weekly *Le Point,* he touted the potential of an anti-aging pill based on a hormone that might ease many of the discomforts of the elderly. . . . The drug in question is known as DHEA. . . . Since Baulieu first encountered DHEA more than three decades ago, it has been scrutinized by many researchers, most extensively by Dr. Samuel Yen, an endocrinologist at the medical school of the University of California at San Diego. . . . Results of Yen's most recent trial, which involved giving elderly people small doses of DHEA daily, were published last June in the *Journal of Clinical Endocrinology and Metabolism.* They showed improved well-being, which Yen defines as "the ability to cope," increased mobility, less joint discomfort and sounder sleep. (Jaroff Leon, "New Age Therapy," *Time,* January 3, 1995, p. 52.)

The *Time* article went on to say that "Both proponents and naysayers agree that DHEA does not enhance sexual desire or performance." However, animal studies, our recent research at the Crenshaw Clinic, as well as that of Masters and Johnson, and at Sloan-Kettering, demonstrate that it does.

DHEA contributes to longevity in many ways. Higher DHEA levels reduce death from heart disease in both men and women as much as 48 percent and death from *any* cause by 36 percent (although other studies, while agreeing that DHEA reduces mortality in men, suggest that DHEA increases mortality in women). It also lowers cholesterol and raises the protective high-density lipoproteins (HDL). DHEA is reduced in autoimmune diseases like lupus erythematosus and rheumatoid arthritis. In addition, DHEA levels can predict which women will get breast cancer up to nine years in advance, and

GROWTH HORMONE PROFILE

MOST PEOPLE DON'T KNOW THAT GROWTH HORMONE:

- is being touted as the new fountain of youth
- increases bone density, bone mass, and lean body mass while reducing fat
- enhances the immune system
- increases skin thickness
- decreases cholesterol
- increases basal metabolic rate

GROWTH HORMONE:

- is produced in the anterior pituitary
- is pulsatile (six to eight pulses per day)
- pulse duration and amount released decreases with age
- stimulates skeletal and muscle growth
- can, in abnormal amounts, cause gigantism and acromegaly (disproportionate growth of skull, hands, and feet)

AS TO SEXUAL ROLES, GROWTH HORMONE:

- may increase sex drive
- may increase potency
- may increase responsiveness

AS TO BEHAVIOR, GROWTH HORMONE:

- increases energy and sense of well-being
- improves wound healing
- fosters lactation

GROWTH HORMONE HAS BEEN USED TO:

- treat short stature
- treat aging
- enhance athletic performance
- treat Turner's syndrome
- improve exercise tolerance
- improve wound healing, burns, grafts

HOW WE CAN INFLUENCE GROWTH HORMONE:

Increases growth hormone:

- clonidine
- desipramine
- strenuous activity
- sleep
- stress
- estrogen, testosterone, thyroid hormone
- dopamine
- bromocriptine
- opiates
- hypoglycemia
- anorexia nervosa

Decreases growth hormone:

- glucose
- menopause/aging
- glucocorticoids

have been found to be 50 percent lower in women with ovarian cancer. In fact, DHEA actually counteracts a well-known cancer-causing substance (12-0-tetradecanoyl-phorbol-13-acetate).

In addition to helping you live longer, slim down, and lower your cholesterol, DHEA improves your immune system, promotes bone growth, and protects against the toxic actions of high sugar and lipid levels, all encouraging a longer life span. DHEA also counteracts steroids like cortisol that rise in response to stress and disease. High cortisol can do harmful things to your body, whether these levels are raised through stress, drug abuse (i.e., athletes), or medication. DHEA polices your blood to watch over and guard against adverse health events. High levels of DHEA may also guard against Alzheimer's. In one study, 48 percent of the Alzheimer's patients had lower DHEA than healthy men and women of the same age.

In an attempt to discover whether low DHEA levels are bad for you, researchers have used DHEA supplements to evaluate obesity and osteoporosis respectively, with good results. Simply correlating DHEA levels with disease states and progression, however, doesn't tell us whether the relationship is cause or effect, or perhaps both. With further study, we should be able to determine whether high levels of DHEA are directly beneficial to health, adaptation, and recovery from disease, or just a consequence of it.

There is an unanswered question about whether DHEA's effects on longevity are equally beneficial in both sexes, or good for men and bad for women. It seems to extend life in men, while some studies suggest it *increases* mortality in females; others show the opposite effect. However, there is evidence that DHEA supplementation has greater effects in female animals than in males, and confounding matters, female animals, who usually live longer, have higher concentrations of DHEA in their brains. A study of rhesus monkeys has shown double the amount of DHEAS in lean females compared to spontaneously obese females, but no difference in DHEAS levels between obese and lean males. Until more research is performed, we won't know for sure, but this is definitely a molecule to monitor.

Beyond these drugs is a whole category of cognitive enhancers—better known as brain nutrients—that are boasted to boost your brain power, raise your grades, and improve your memory. They work through various neurotransmitters, among them dopamine and serotonin. These "neuronauts" will help us to decode the mysteries of our mind and meddle with it even more than we already do. There is even talk of "smart parties" replacing Tupperware parties someday. Most of these exciting new developments materialized out of research to treat diseases like Alzheimer's, Parkinson's, as well as strokes, brain injuries, comas, attention deficit disorder, and various psychiatric conditions, especially depression.

In Europe, cognitive-enhancing substances, called nootropics, are used primarily to treat the signs and symptoms of aging. We don't have a comparable category of drugs in the U.S. However, dopamine, DHEA, vasopressin, and growth hormone fall into this unnamed category. So do LH and LHRH. Estrogen may soon be invited to join this exclusive club. Along with demonstrating some cognitive-enhancing properties, they all appear to improve the general functioning of the central nervous system. Information acquisition and performance seem to be improved by normalizing certain disruptive processes in brain function associated with aging.

GENDER GAPS

Throughout this book, you have, I hope, come to appreciate the medical, biochemical, and sexual differences between men and women. This theme continues with regard to how we age and how long we live. The male has a higher mortality rate in almost all forms of life, from nematodes, insects, and spiders to birds, reptiles, fish, and mammals. "Whether you're a man or a mouse, fruit fly or an alligator," says one researcher, "females live longer." Eleven women in twelve outlive their husbands. And throughout the modern world, women outlive men by about seven years. Thirty-three percent more boy babies than girls die in the first year of life. If you look at the top ten or twelve causes of death, every single one kills more men.

Women are less vulnerable than men to life-threatening diseases but more susceptible to everyday sicknesses and pains—like arthritis, bunions, menstrual woes, and migraine headaches. Regarding mental health, women are more likely than men to be miserable daily, but men are more likely to have devastating psychological illnesses like schizophrenia. Women bend; men break.

Living longer is just one of many female features that represents the chemical and emotional distinctions between the sexes. Our brains represent another feature.

Sexing the Brain

Considering that we have different hormones and body parts, wouldn't it be odd if our brains were the same? They aren't, of course. And, not surprisingly, they don't age alike either.

According to University of Pennsylvania researchers, men lose their verbal abilities faster as they age. Using magnetic resonance imaging (MRI), they studied the brains of men and women between eighteen and eighty. They found men's brains deteriorated almost three times more quickly than women's, and that their left brain atrophied more so than the right. The change in women's brains was symmetrical, and the differences, "suggest that female sex hormones may protect the brain from atrophy associated with aging." They conclude that "women are less vulnerable to age-related changes in mental abilities, whereas men are particularly susceptible to aging effects on the left hemispheric functions."

Brain atrophy is a worry. Some scientists think that approximately fifty thousand neurons in the thinking part of the brain die every day and are never replaced. Perhaps more crucial is the natural aging of neurons, which like all cells, gradually deteriorate and lose efficiency. Some of their many dendrites (interconnecting branches) wither. More important, their production of neurotransmitters (dopamine, et cetera) decreases. Also the number of receptors decreases as well.

Because the receptors either decrease in number, deteriorate, or become less "receptive," it may be that counteracting the changes of aging through recreating "young blood" may require more of each hormone, neuropeptide, and neurotransmitter than mere *replacement*. *Supplementation* may be necessary, the rationale being that since the receptors are fewer and less sensitive, more than usual of all these substances may be necessary to achieve the same response that normal amounts produced in younger years when the receptors were healthy.

CYCLES AND PEAKS AGAIN

As we've discussed, all of our hormones fluctuate at some time to some degree, some considerably more than others. Oxytocin, for example, appears to be the most sensitive to inside and outside sources. A thought, a touch is all it takes. PEA is a close second, responding to a mere glance or a seductive smile. DHEA has a split personality. One form of the molecule sends and receives scents, ebbing and flowing accordingly. Another form of the molecule is one of the most stable and constant of all our hormones and hardly cycles at all. Estrogen also appears to be pretty stable. Although it cycles over the course of a month and a lifetime, it seems not to jump around on an hourly basis. Testosterone, on the other hand, flips out with the slightest provocation. When stress hits, testosterone drops and vasopressin goes up, perhaps in its role as a buffer or guardian, protecting us from going over the edge. It acts as one of our coping chemicals.

As we become more sophisticated about these cycles and peaks and the implications of these modulations—for our relationships, our sexuality, and our health—a fascinating picture emerges that is so dynamic it would overwhelm us, *except that this knowledge provides us with the power to intervene.*

Menstrual Cycling and Medication

Attention to body cycles and the timing of medication according to circadian rhythms is just beginning to get the serious attention of researchers. However, the relationship between rhythms, medication dosage, and sex-linked differences has barely gotten a glance. Dr. Margaret Jensvold emphasizes the importance of modifying the dose of psychotropic drugs according to a woman's menstrual cycle, suggesting that if her cycle isn't taken into consideration, blood levels of medication may be increased during the premenstruum and sustained at levels higher than necessary indefinitely.

Dr. Margaret Jensvold also advised that clinical interactions between oral contraceptives, hormone replacement therapy, and psychotropic medications be considered as well.

E. H. Ellinwood found that women taking oral contraceptives and a steady dose of diazepam (Valium) throughout the menstrual cycle were relatively "intoxicated" during the menses, showing that the same oral dose of benzodiazepine had a more potent effect during one part of the cycle than the others.

Premenstrually, one subgroup of women experienced water retention that could lead to unusually dilute levels of a drug. Some authors also suggest that the dopamine-lowering effect of estrogen may explain why young women are usually prescribed lower average doses of antipsychotics than young men. The side effects of antipsychotic drugs manifest differently between the sexes as well.

Also, the fact that these conditions change throughout a woman's menstrual cycle suggest that timing is an issue that deserves more attention in medicine than it has been receiving. In recent years, some controversial studies were published claiming that the time of the month a woman is scheduled for surgery in relation to her menstrual cycle has an important bearing on the outcome of the surgery, number of complications, et cetera. These researchers advised women not to schedule surgery when they were premenstrual, suggesting that it could affect the outcome of procedures like mastectomy. This recommendation was greeted with guffaws by most surgeons, and as yet these studies have not been confirmed. However, doesn't it stand to reason that you would feel a lot better and more resilient to face any physical or emotional challenge after menstruation is over rather than premenstrually? According to another study, even though the risk of clots is now low, to be extra safe during surgical procedures, oral contraceptives should be interrupted four weeks before surgery in order for blood clotting mechanisms to return to normal—an important but usually overlooked recommendation. For lack of a simple precautionary measure, how many women have undergone a minor, perhaps elective, surgical procedure only to die of a blood clot to the brain or lung?

Given these studies, it is clear that medications prescribed for women should be evaluated differently depending on whether that woman is estrogen normal or estrogen deprived. Much more research is needed to determine how to appropriately adjust the dosages of medication according to sex and our respective cycles. Since we are not aware of weekly or monthly cycles of key hormones in the male (perhaps because we have not looked carefully enough) there is not as much to take into consideration as of yet. A notable exception, however, are the frequent fluctuations of various hormones in the male during the day. As you learned in Chapter 5, giving constant doses of a drug like LH can have the exact opposite effect of giving the same dose in a pulsatile fashion. I suspect this is just the beginning of our education in this regard, and that not only our circadian rhythms, but our hormonal, peptide, and neurotransmitter rhythms will eventually have to be taken into account for proper drug dosing and rhythm of administration in both sexes.

THE NEWEST DRUG DELIVERY METHODS

Even once we have identified the optimal times to deliver medications based on cycles and peaks, we still need to create the best methods of delivering these drugs. As we've discussed, testosterone replacement, for example, would be more effective—more natural, if you will—when dispensed through a pump instead of by injection. Many other drugs have more effective impact when delivered to the body by one method rather than another.

There are a variety of new methods of drug delivery on the horizon. We are already expanding the patches. Since women and men have quite significant differences in subcutaneous tissue, which has a bearing on the dynamics of patch delivery, this distinction should be taken into consideration.

As you know, we already have vaginal delivery of estrogen. Delivering estrogen vaginally, rather than orally, allows for steady absorption rather than peaks and valleys, which is advantageous in the case of this hormone. Then there are pellet implants of estrogen, progesterone, and testosterone. Each one can be simply, surgically installed under the skin, where it is gradually absorbed into the system over the course of three to six months depending on the drug. Direct delivery of progesterone to the bloodstream through quickly dissolving tablets placed inside a cheek is right around the corner, and progesterone patches are already available in Britain. Both of these methods reduce the intensity of certain side effects, and bypass the liver.

Bioadhesive technology takes advantage of a specific molecule's (polycarbophil) ability to attach to mucous membranes. A hormone, or for that matter, just about any drug (even some forms of chemotherapy) is combined with this mucous-membrane-sensitive polymer. From there, for a very controlled, chemically engineered period of time—like sustained-release sinus capsules—this new polymer delivers the medicine trapped in its matrix to the bloodstream right through the mouth or the vagina. Hormones and anti-cancer medications are being considered for vaginal or oral distribution using this bioadhesive method, as well as a twelve-hour antacid and a three-hour breath freshener.

We can also expect a dramatic expansion of drugs delivered by nasal sprays and eye drops. Up to now, we have been used to nasal sprays that act only on the nose or the sinuses. But there is already a painkiller that is delivered this way. Butorphanol tartrate nasal spray (Stadol NS) is an opiate —a pain reliever you can spray in your nose for rapid relief. It has been shown to be effective for migraines and will undoubtedly find numerous other uses.

Synthetic peptides are largely destroyed in the gastrointestinal tract, so swallowing the drug is a less than ideal way of getting the desired effect. Frequent injections (which are usually necessary) are inconvenient to say the least, and virtually unmanageable for nonhospitalized patients. Alternative routes of administration have been developed, and both oxytocin and vasopressin have been available by nasal spray for some time. In addition, eye drops that administer synthetic peptides are being developed. This technique may circumvent some of the problems that occur with oral and intravenous drug delivery. Work is being done on "visual" versions of insulin, glucagon, vasopressin, and oxytocin, so keep your eye on eye drops.

Among the most exciting new methods are drugs designed to act directly on the brain and nowhere else (as discussed in Chapter 5). Carrier molecules transport them in an inactive state until they are released after they cross the blood/brain barrier. Many side effects of drugs can be avoided in this way, particularly as we develop the skill to target specific sites in the brain.

Of course, we still have to consider how men and women differ as they receive these drugs through new and more effective delivery systems. According to the Pharmaceutical Manufacturer's Association, "essential metabolic differences between men and women can create important differences in their ability to metabolize, absorb, and transport drugs to key areas of the body." Like the power of our knowledge about the hormones that reside and operate with in us—we are their neighborhood to cruise, so to speak. Unless we recognize that the problem of male/female drug delivery exists, we can hardly propose any solutions.

SEXUAL STAGES AND OUR HORMONES

Let us now take a look at how our various sexual stages affect us and can in turn be influenced by this new and extensive knowledge of how our hormones work. Some stages as you have seen are more enjoyable than others. A few can be downright miserable. In early stages, the challenge is learning to cope with too much of one hormone or another. Toward the other end of our lives, the big question is how to cope with too little. There are some simple guidelines that will work to reduce any problems and improve your relationship with yourself and others:

Don't let touching slip away. No matter what else is going on, hold hands, hold each other. Difficulties will disappear faster, you will feel happier, and chances are you will live longer than if you didn't keep on touching.

Don't let sexual problems persist for more than three months without seeking professional help. Some problems disappear on their own, but couples often give them too much time to do so without intervening, sometimes years.

Treat the one you love the best, not the worst. If the sexual chemistry has disappeared, contrary to popular belief, you can usually get it back! Loving consideration probably disappeared first. Get that back and the chemistry will follow.

Save prime time for sex. If you aren't spending the right kind of time together, start thinking of time-management techniques for lovers. Chances are, you are having sex when you are too tired to do anything else.

Have sex sometimes even when you are not in the mood. There will be times it is important to have sex even though you really don't feel like it. Don't force yourself if it hurts or you would hate it, but if you are just not in the mood, have sex at least once a week whether the urge moves you or not. You will be surprised to discover that more often than not, you enjoy it.

Make a choice to stay physically close and sexually intimate. When pregnancy, nursing, menopause, or other hormonal conditions mute you, compensate sexually for the emotional void by touching anyway—for your own sake as well as to protect the relationship, even when you are going through some rough times with your partner emotionally. You will have at least one bridge between you that isn't broken, and have a better chance of working the problem out.

Know the sexual consequences of any medication or combination of medications you are taking and advise your doctor that you want to be placed on the least sexually toxic medication available.

Make an intelligent, well-informed choice about hormone replacement therapy and remain current on new developments.

Respect the innate differences between men and women without being critical or disapproving. Try instead to listen to and learn from one another, and use these differences to the benefit of your relationship. Since the sexes so often have complementary skills, use your partner as an asset in areas where he or she has more strength. Develop your own skill further by asking your partner to teach you how to do something he or she does particularly well, perhaps naturally, like verbalizing feelings or setting them aside when necessary (compartmentalizing feelings). You will discover that many talents your sex does not innately possess can be acquired by using the right teacher. But much depends on the student.

Avoid situations that will trigger hormonal reactions that you aren't ready for yet or don't want. For example, don't put yourself into a sexual setting or a living situation with someone unless you are willing to take the consequences of your own hormones and feelings. If you don't want to become committed to someone, make it more difficult by not moving in with them.

Respect the power of your hormones over you, but at the same time, realize that unlike lower animals, the choices you make will determine how these hormones ultimately affect your behavior.

Speak up about your sexual needs and preferences before you are starved. Talk to your mate. Complaining to your best friend or your neighbor won't solve the problem.

Allow love. As discussed in Chapter 2, the experience of love smooths the bumps in the road for most relationships at any stage. I don't mean to suggest that it solves all the problems, but some actually do just inexplicably disappear. Others get easier to manage as a result of the goodwill and positive feelings between you.

Don't fight becoming interdependent, but maintain your individuality and balance.

In essence, there is a powerful relationship between sexual fitness, relationship happiness, and longevity that goes beyond coincidence to direct cause and effect. Each favorably influences the other through the neurochemicals and hormones they exercise and stimulate.

If you touch every day, address sexual problems before they become permanent, have intercourse, orgasms, or some kind of sex regularly, treat one another with love, romanticize and fantasize, laugh and play together, you will be stimulating all the molecules that keep you living longer while keeping your sexual chemistry alive.

MANIPULATING EACH OTHER'S HORMONES

Throughout my study of hormones, peptides, and neurotransmitters, one of the features that I could not ignore, and that I am sure has come through to you in the course of this reading, is how intimately and intensely we influence one another's chemistry. It is one thing to accept how our hormones affect our own feelings and behavior. It is a stretch to appreciate how the environment, the seasons, perhaps even lunar cycles affect our own. It is quite another to realize that another individual has the power to interfere with or regulate our body chemistry without our knowing, and, for that matter, without their knowing.

They do all this, and we do the same in return through sight, touch, scent, daily contact, sexual intercourse, and perhaps even orgasms. The net effect is that we become interdependent on one another for our health, our happiness, our well-being, and perhaps even our longevity. In which case, the reverse is also true—we can make each other extremely miserable, unhealthy, and perhaps even abbreviate our life span. For instance, Winnifred Berg Cutler wrote in *Love Cycles* that for peak fertility a woman should have intercourse contact with her mate at least once a week. Having intercourse regularly raises overall estrogen levels, changing a woman's hormonal chemistry, and to some extent her outlook on life. One researcher has suggested that the regular deposit of prostaglandins from the male ejaculate to the woman through intercourse is also a factor in her mood and state of mind. It has even been proposed that women who have sexual intercourse regularly delay menopause. Is the reverse also true? If you don't have a man in your life are you likely to go through menopause earlier?

One small study is a case in point:

"Salivary testosterone concentrations were measured in male and female members of four heterosexual couples on a total of eleven evenings before and after sexual intercourse and eleven evenings on which there was no intercourse. Testosterone increased across the evenings when there was intercourse and decreased when there was none. The pattern was the same for males and females. Early evening measures did not differ on the two kinds of days, suggesting that sexual activity affects testosterone more than initial testosterone affects sexual activity."

We need more research of this sort to fully address our questions: could it be that the emotional and physical bonding between a man and a woman is somehow central to our mental health and physical well-being? That sex isn't just elective, that frequency and continuity are not just for pleasure and reproduction but play a meaningful role in immunity, health, and longevity?

I'm sure you have read articles that claim married men live longer and single men die younger, and the reverse for women. You have probably known an elderly lifelong couple in which the survivor, who had seemed in good health, died shortly after the death of the spouse. All these questions lead to the bigger question of how dependent are we really upon a relationship? Are we as autonomous as we would like to believe? And if this interdependency proves true, how is it different for same-sex couples? Should these couples, or single men and women, have a selection of human pheromones that they dab on with prescribed frequency to remain in the optimum state of health?

Whatever the future holds as research unfolds, one thing is clear from the data available to us so far. A well-functioning relationship is good for our heart and our soul and our health. Good relationships are not accidents. There is, as you have seen, much you can do to influence them favorably throughout the sexual stages of your life.

THE VALUE OF VANITY

I am going to leave you with two thoughts to ponder—the value of vanity, and the chemistry of attitude. But before I do, get prepared to throw out all the advice your teachers, parents, and aunts gave you about beauty being only skin deep—a superficial frivolous feature. I propose instead that a little bit of vanity, along with "attitude," is great for both your health and your sex life. Think back on that flirtatious older woman I described in the beginning of this chapter.

Let's look at some other concrete examples. Various researchers have clearly established that thinner people live longer than obese people. In fact, body weight is among the most significant longevity factors we have so far identified. Men and women have different percentages of lean muscle mass to body fat and different distributions. Men are leaner with about 10 percent body fat. Maintaining a steady weight is easier too, because their metabolism is about 6 percent faster. They gulp more oxygen and throw off more heat, burning about twice as many calories as women.

Staying slim usually involves vigorous workouts that ensure physical fitness, not just looks. Physical fitness, added to weight-bearing exercise and the low-fat diets that reduce triglycerides and cholesterol, maintain bone strength and cardiovascular health. For women, a wasp waist and light weight pay numerous dividends, clearly assuming it doesn't extend to eating disorders.

Smoking and alcohol abuse are also incompatible with beauty, unless you want crepe-paper-thin skin full of wrinkles and crevices, a red nose, blotchy cheeks, a potbelly, and bad breath.

The Chemistry of Attitude

Vanity is not a bad thing from a psychodynamic point of view either. It's part of attitude, and attitude is the most important ingredient for counteracting "aging."

Separating emotional age from chronological age is the first way to liberate yourself. For women, it is especially important to discard the notion that men age well and women age poorly. Look at Vanessa Redgrave, Sophia Loren, and Princess Grace. Born natural beauties? Perhaps. But they clearly learned to preserve their gift and some have even bottled it. The point is, you can keep your natural beauty too, along with your health, unless you get careless and waste it. And perhaps you can influence your emotions through your motions.

It has always been assumed that if you feel pleasure, you smile; if you feel pain, you grimace, and so on. But according to some psychologists and neurologists, it might also work the other way around. They suggest that the chemistry of the smile, the blush, the frown, and other expressions could well influence the way we feel. An article in *Behavior Today* suggests that facial expressions such as a smile divert the blood flow momentarily from the brain to the face, which influences brain temperature, which stimulates the synthesis of endorphins in the brain. Dwelling on positive thoughts can also raise your endorphin levels. These are key chemicals for pain relief among other things and perhaps influence your emotions in other ways. If this is true, you might be able to smile yourself to health and happiness.

Remember Carolyn and Don? We left them having decaf at Ave's in St. Louis. Together now for three years, they continue to enjoy the special features of being with a contemporary. Carolyn, now sixty-four, keeps her youthful figure by pursuing her love of Rollerblade skating. Except for a hysterectomy in her fifties, she has remained in unusually good health. Hormone replacement with estrogen, progesterone, and testosterone has no doubt helped.

Don, seventy-two, is in pretty good shape too, but he developed a mild case of Parkinson's disease a year after they met and was placed on Eldepryl (deprenyl).

This turned out to be a blessing in disguise. He couldn't believe the boost it gave him. He felt almost like a young man again.

Retirement has eaten into their once abundant resources, but they still have enough in reserve to live modestly and travel if they aren't too extravagant. We catch up with them one May evening in the late afternoon on the Isle of Capri. While most of the Italian locals are still enjoying their afternoon rest, Don and Carolyn decide to rent a small boat to the Blue Grotto. They saw it for the first time earlier in the week and watched as young men and women dove overboard and played in the phosphorescent water. The kids took this older couple completely in stride, thinking perhaps of how much fun they were missing. However, the look in Carolyn's eye told a different story. Without words a plan took form between the two of them and settled in.

This afternoon turned out to be perfect. The tourist season had not yet gotten into full swing, but the weather was just right. The sea was calm and they could have the grotto to themselves.

Don and Carolyn put on their life vests and hopped into the boat. Despite having a one-and-a-half horsepower engine, they took turns rowing for the sheer fun of it until they were tired and then motored the rest of the way. As they approached the grotto they cut the engine, gliding quietly into the now-familiar wonder of this exotically romantic place. There was no one in sight. As though they had done it a thousand times before, with efficiency of movement, they slipped into the water marveling at the buoyancy of the salt-dense lift. The water was ink black but every time they moved it sparkled with fluorescent energy. They played like dolphins, frolicking in the surf.

Don swam up behind Carolyn and slipped his arms around her, crossing to cup each breast. "I hear these help you float, and I was having a sinking spell," he said.

She felt him press between her legs, and said, "I hope that's you. They don't have eels in here, do they?"

"Oh, yes," he said, "electric ones. They'll really turn you on."

Hindsight would have left their suits in the boat, but they managed to wriggle out of them. Don entered her from behind and then tried to stay united while they paddled around the grotto together with their best innocent look checking for intruders and savoring this new form of transportation.

The backstroke, however, was not such a success. Don couldn't keep his head above water. And the breaststroke, well, they were already doing that.

307

In her exuberance, Carolyn inhaled a little water, coughed and sneezed. "I think you fell out, hon."

"What do you mean, fell out! You blew me away."

"Now, that's an idea."

"No, don't even think about it, you'll drown."

"Here, face me and behave yourself for a change."

She covered his face with kisses, saying, "And the kids thought they were having fun."

He took both nipples in his mouth as she arched her back in a misty haze, mesmerized. She floated. He covered her.

The sun was setting and it was just beginning to get dark. Along with the changing light, the mood shifted from playful to passionate. Carolyn wrapped her legs around Don's waist. He grabbed on to her life vest for leverage. As their tempo was building, the water started to churn and sparkle. A distant motor sounded growing louder. Carolyn had hoped he would come before the visitors arrived and, just barely, he did. In relief she splashed her legs and arms and kicked and screamed and whooped exuberantly. Don said, "Some fireworks," as he watched the lights fly.

They scurried back into their suits and were just pulling themselves back into the boat looking nonchalant as the tourists pulled up. It wasn't until after Don had greeted them and exchanged a few pleasantries that he noticed that his swim trunks were not only on inside out, but twisted one leg up and one leg down. They laughed all the way back to the dock.

HUMAN OR ANIMAL: THE CHOICE IS YOURS

In closing, I come to what I think is the most important point of this chapter, and one that by now the information throughout this book has made abundantly clear: like other animals, we come under the spell of our hormones, but what sets us apart as human beings is that we need not remain helpless slaves whipped around senselessly by these molecular tyrants. We have a choice.

Just because testosterone is surging and you feel belligerent doesn't mean that you have to go out and shoot someone. Just because PEA and your pheromones are flooding doesn't mean that you have to have sex with the neanderthal you just met. You can exercise your power, not absolutely, but

quite extensively throughout your lives. In fact if you don't use this knowledge to your benefit every day, you won't have a chance against the force of
your hormones, because they never rest.

Finally, you can apply this newfound power to improving the quality of
your sexual stages. Your knowledge is your weapon. With it, you can rewrite
the script you would otherwise live out, courtesy of unopposed hormonal
influences, and enjoy a finer life, probably a longer life, and much better
relationships with your mate and the rest of those you love.

Carolyn and Don finally got it all together toward the end stages of their
lives. Had they had access to the knowledge contained in this book from the
time they were young, they wouldn't have had to wait so long.

SELECTED BIBLIOGRAPHY

AND REFERENCES

The following is a selected bibliography, organized by subject, related to the topics presented in the *Alchemy of Love and Lust*. This bibliography ends with a general list of selected reading for those interested in further exploration of some of the broader topics I have introduced in this book.

For a more comprehensive bibliography and extensive discussion of hormones, please refer to *Sexual Pharmacology*, written by Theresa L. Crenshaw, M.D., and James Goldberg, Ph.D., published by W. W. Norton, 1996.

AGGRESSION

Albert, D. J., E. M. Dyson, M. L. Walsh, and R. Wong. 1988. Defensive aggression and testosterone-dependent intermale social aggression are each elicited by food competition. *Physiol. Behav.* 43:21–28.

Albert, D. J., M. L. Walsh, and R. H. Jonik. 1993. Aggression in humans: what is its biological foundation? *Neurosci. Biobehav. Rev.* 17:405–25.

Angier, N. 1995. Does testosterone equal aggression? Maybe not. *New York Times,* 20 June 1995, A1–C3.

Archer, J. 1991. The influence of testosterone on human aggression. *Br. J. Psychol.* 82:1–28.

Bagatell, C. J., J. R. Heiman, J. E. Rivier, and W. J. Bremner. 1994. Effects of endogenous testosterone and estradiol on sexual behavior in normal young men [published erratum appears in *J. Clin. Endocrinol. Metab.* 1994. 78(6):1520]. *J. Clin. Endocrinol. Metab.* 78:711–16.

Booth, A., G. Shelley, A. Mazur, G. Tharp, and R. Kittok. 1989. Testosterone, and winning and losing in human competition. *Horm. Behav.* 23:556–71.

Buchanan, C. P., E. M. Shrier, and W. L. Hill. 1994. Time-dependent effects of PCPA on social aggression in chicks. *Pharmacol. Biochem. Behav.* 49:483–88.

Butovskaia, M. L., M. A. Deriagina, A. M. Chirkov, V. G. Startsev, and V. G. Chalian. 1985. [Effect of stress-inducing factors on the behavior of monkeys. I. The behavior of monkeys under acute emotional stress] Vliianie stressogennykh faktotov na povedenie obez'ian. I. Povedenie obez'ian v usloviiakh ostrogo emotsional'nogo stressa. *Biologicheskie Nauki.* 67–74.

Christiansen, K., and R. Knussmann. 1987. Androgen levels and components of aggressive behavior in men. *Horm. Behav.* 21:170–80.

Constantino, J. N., D. Grosz, P. Saenger, D. W. Chandler, R. Nandi, and F. J. Earls. 1993. Testosterone and aggression in children. *J. Am. Acad. Child Adolesc. Psychiatry* 32:1217–22.

Cooper, A. J., D. Baxter, W. Wong, and S. Losztyn. 1987. Sadistic homosexual pedophilia treatment with cyproterone acetate [letter] [published erratum appears in *Can. J. Psychiatry* 1988 Feb; 33(1):77]. *Can. J. Psychiatry* 32:738–40.

Cooper, A. J., S. Losztyn, N. C. Russell, and Z. Cernovsky. 1990. Medroxyprogesterone acetate, nocturnal penile tumescence, laboratory arousal, and sexual acting out in a male with schizophrenia. *Arch. Sex. Behav.* 19:361–72.

Ferguson, J., S. Henriksen, H. Cohen, G. Mitchell, J. Barchas, and W. Dement. 1970. "Hypersexuality" and behavioral changes in cats caused by administration of p-chlorophenylalanine. *Science* 168:499–501.

Fishbein, D. H. 1992. The psychobiology of female aggression. *Criminal Justice and Behavior* 19:99–126.

Gladue, B. A., M. Boechler, and K. D. McCaul. 1989. Hormonal response to competition in human males. *Aggressive Behavior* 15:409–22.

Hellhammer, D. H., W. Hubert, and T. Schurmeyer. 1985. Changes in saliva testosterone after psychological stimulation in men. *Psychoneuroendocrinology* 10:77–81.

Jeffcoate, W. J., N. B. Lincoln, C. Selby, and M. Herbert. 1986. Correlation between anxiety and serum prolactin in humans. *J. Psychosom. Res.* 30:217–22.

Keverne, E. B. 1978. Sexual and aggressive behaviour in social groups of talapoin monkeys. *Ciba. Found. Symp.* 62:271–97.

Kiersch, T. A. 1990. Treatment of sex offenders with Depo-Provera. *Bull. Am. Acad. Psychiatry Law.* 18:179–87.

Linn, G. S., and H. D. Steklis. 1990. The effects of depo-medroxyprogesterone acetate (DMPA) on copulation-related and agonistic behaviors in an island colony of stumptail macaques (Macaca arctoides). *Physiol. Behav.* 47:403–8.

Loosen, P. T., S. E. Purdon, and S. N. Pavlou. 1994. Effects on behavior of modulation of gonadal function in men with gonadotropin-releasing hormone antagonists. *Am. J. Psychiatry* 151:271–73.

Marti-Carbonell, M. A., S. Darbra, A. Garau, and F. Balada. 1992. [Hormones and aggression] Hormonas y agresion. *Arch. Neurobiol. (Madr.)* 55:162–74.

Mendelson, J. H., N. K. Mello, S. K. Teoh, S. E. Lukas, W. Phipps, J. Ellingboe, S. L. Palmieri, and I. Schiff. 1992. Human studies of the biological basis of reinforcement. In *Addictive states.* C. P. O'Brien and J. Jaffe, editors. New York: Raven Press. 131–55.

Meyer, W. J., C. Cole, and E. Emory. 1992. Depo-Provera treatment for sex offending behavior: an evaluation of outcome. *Bull. Am. Acad. Psychiatry Law.* 20:249–59.

Murray, J. B. 1988. Psychopharmacological therapy of deviant sexual behavior. *J. Gen. Psychol.* 115:101–10.

Olweus, D., A. Mattsson, D. Schalling, and H. Low. 1988. Circulating testosterone levels and aggression in adolescent males: a causal analysis. *Psychosom. Med.* 50:261–72.

Packer, C., D. A. Collins, A. Sindimwo, and J. Goodall. 1995. Reproductive constraints on aggressive competition in female baboons. *Nature* 373:60–63.

Piacente, G. J. 1986. Aggression. *Psychiatr. Clin. North Am.* 9:329–39.

Prescott, J. W. 1990. Affectional bonding for the prevention of violent behaviors: Neurobiological, psychological and religious/spiritual determinants. In *Violent Behavior Vol. I: Assessment and Intervention.* L. J. Hertzberg et al., editors. New York: PMA Publishing Corp. 110–42.

Raboch, J., H. Cerna, and P. Zemek. 1987. Sexual aggressivity and androgens. *Br. J. Psychiatry* 151:398–400.

Robel, P., E. Bourreau, C. Corpechot, D. C. Dang, F. Halberg, C. Clarke, M. Haug, M. L. Schlegel, M. Synguelakis, and C. Vourch. 1987. Neuro-steroids: 3 beta-hydroxy-delta 5-derivatives in rat and monkey brain. *J. Steroid Biochem.* 27:649–55.

Sanchez, C., and J. Hyttel. 1994. Isolation-induced aggression in mice: effects of 5-hydroxytryptamine uptake inhibitors and involvement of postsynaptic 5-HT1A receptors. *Eur. J. Pharmacol.* 264:241–47.

Schiavi, R. C., A. Theilgaard, D. R. Owen, and D. White. 1984. Sex chromosome anomalies, hormones, and aggressivity. *Arch. Gen. Psychiatry* 41:93–99.

Schmitz, A. 1993. A shot in the dark. (Hormone injection to treat sex offenders.) *Health* 7:22(2).

Shively, C. A., S. B. Manuck, J. R. Kaplan, and D. R. Koritnik. 1990. Oral contraceptive administration, interfemale relationships, and sexual behavior in Macaca fascicularis. *Arch. Sex. Behav.* 19:101–17.

Thibaut, F., and L. Colonna. 1992. Cyproterone acetate in the treatment of aggression [letter]. *Am. J. Psychiatry* 149:411.

Thibaut, F., J. M. Kuhn, and L. Colonna. 1991. A possible antiaggressive effect of cyproterone acetate [letter]. [See comments.] *Br. J. Psychiatry* 159:298–99.

Uzych, L. 1992. Anabolic-androgenic steroids and psychiatric-related effects: a review. *Can. J. Psychiatry* 37:23–28.

Van de Poll, N. E., and S. H. Van Goozen. 1992. Hypothalamic involvement in sexuality and hostility: comparative psychological aspects. *Prog. Brain Res.* 93:343–61.

Virkkunen, M., E. Kallio, R. Rawlings, R. Tokola, R. E. Poland, A. Guidotti, C. Nemeroff, G. Bissette, K. Kalogeras, and S. L. Karonen. 1994. Personality profiles and state aggressiveness in Finnish alcoholics, violent offenders, fire setters, and healthy volunteers. *Arch. Gen. Psychiatry* 51:28–33.

Virkkunen, M., and M. Linnoila. 1993. Brain serotonin, type II alcoholism and impulsive violence. *J. Stud. Alcohol Suppl.* 11:163–69.

Weiner, M. F., M. Denke, K. Williams, and R. Guzman. 1992. Intramuscular medroxyprogesterone acetate for sexual aggression in elderly men [letter]. *Lancet* 339:1121–22.

Worthman, C. M., and M. J. Konner. 1987. Testosterone levels change with subsistence hunting effort in !Kung San men. *Psychoneuroendocrinology* 12:449–58.

BREAST CANCER

Bakker, G. H., B. Setyono-Han, H. Portengen, F. H. De Jong, J. A. Foekens, and J. G. Klijn. 1990. Treatment of breast cancer with different antiprogestins: preclinical and clinical studies. *J. Steroid Biochem. Mol. Biol.* 37:789–94.

Belchetz, P. E. 1994. Hormonal treatment of postmenopausal women. *N. Engl. J. Med.* 330:1062–71.

Bird, C. E., V. Masters, E. E. Sterns, and A. F. Clark. 1985. Effects of tamoxifen on testosterone metabolism in postmenopausal women with breast cancer. *Clin. Invest. Med.* 8:97–102.

Cobleigh, M. A., R. F. Berris, T. Bush, N. E. Davidson, N. J. Robert, J. A. Sparano, D. C. Tormey, and W. C. Wood. 1994. Estrogen replacement therapy in breast cancer survivors. A time for change. Breast Cancer Committees of the Eastern Cooperative Oncology Group. *JAMA* 272:540–45.

Colditz, G. A., K. M. Egan, and M. J. Stampfer. 1993. Hormone replacement therapy and risk of breast cancer: results from epidemiologic studies. *Am. J. Obstet. Gynecol.* 168:1473–80.

Ebeling, P., and V. A. Koivisto. 1994. Physiological importance of dehydroepiandrosterone. *Lancet* 343:1479–81.

Gambrell, R. D. 1993. Estrogen replacement therapy and breast cancer risk: a new look at the data. *The Female Patient* 18:50+.

Grady, D., S. M. Rubin, D. B. Petitti, C. S. Fox, D. Black, B. Ettinger, V. L. Ernster, and S. R. Cummings. 1992. Hormone therapy to prevent disease and prolong life in postmenopausal women. *Ann. Intern. Med.* 117:1016–37.

Hamed, H., M. Caleffi, I. S. Fentiman, B. Thomas, and R. D. Bulbrook. 1991. Steroid hormones in lymph and blood from women with early breast cancer. *Eur. J. Cancer* 27:42–44.

Henrich, J. B. 1992. The postmenopausal estrogen/breast cancer controversy [see comments]. *JAMA* 268:1900–1902.

Kaplan, H. S. 1992. A neglected issue: the sexual side effects of current treatments for breast cancer. *J. Sex Marital Ther.* 18:3–19.

Kaufman, D. W., J. R. Palmer, J. de Mouzon, L. Rosenberg, P. D. Stolley, M. E. Warshauer, A. G. Zauber, and S. Shapiro. 1991. Estrogen replacement therapy and the risk of breast cancer: results from the case-control surveillance study. *Am. J. Epidemiol.* 134:1375–85; discussion 1396+.

Palmer, J. R., L. Rosenberg, E. A. Clarke, D. R. Miller, and S. Shapiro. 1991. Breast cancer risk after estrogen replacement therapy: results from the Toronto Breast Cancer Study. *Am. J. Epidemiol.* 134:1386–95; discussion 1396+.

Skegg, D. C., E. A. Noonan, C. Paul, G. F. Spears, O. Meirik, and D. B. Thomas. 1995. Depot medroxyprogesterone acetate and breast cancer. A pooled analysis of the World Health Organization and New Zealand studies. *JAMA* 273:799–804.

Strickland, D. M., R. D. Gambrell, C. A. Butzin, and K. Strickland. 1992. The relationship between breast cancer survival and prior postmenopausal estrogen use. *Obstet. Gynecol.* 80:400–404.

CONTRACEPTIVES

Adams, D. B., A. R. Gold, and A. D. Burt. 1978. Rise in female-initiated sexual activity at ovulation and its suppression by oral contraceptives. *N. Engl. J. Med.* 299:1145–50.

Bancroft, J., and N. Sartorius. 1990. The effects of oral contraceptives on well-being and sexuality. *Oxf. Rev. Reprod. Biol.* 12:57–92.

Bancroft, J., B. B. Sherwin, G. M. Alexander, D. W. Davidson, and A. Walker. 1991. Oral contraceptives, androgens, and the sexuality of young women: I. A comparison of sexual experience, sexual attitudes, and gender role in oral contraceptive users and nonusers. *Arch. Sex. Behav.* 20:105–20.

————. 1991. Oral contraceptives, androgens, and the sexuality of young women: II. The role of androgens. *Arch. Sex. Behav.* 20:121–35.

Board of Trustees, A. M. A. 1992. Requirements or incentives by government for the use of long-acting contraceptives. Board of Trustees, American Medical Association [published erratum appears in *JAMA* 1992 Nov 11; 268(18):2518]. *JAMA* 267:1818–21.

Chi, I. 1993. The safety and efficacy issues of progestin-only oral contraceptives—an epidemiologic perspective. *Contraception* 47:1–21.

Ellinwood, E. H., M. E. Easler, M. Linnoila, D. W. Molter, D. G. Heatherly, and T. D. Bjornsson. 1984. Effects of oral contraceptives on diazepam-induced psychomotor impairment. *Clin. Pharmacol. Ther.* 35:360–66.

Ellis, D. J., and R. J. Hewat. 1985. Mothers' postpartum perceptions of spousal relationships. *J. Obstet. Gynecol. Neonatal. Nurs.* 14:140–46.

Gaspard, U. J., M. Dubois, D. Gillain, P. Franchimont, and J. Duvivier. 1984. Ovarian function is effectively inhibited by a low-dose triphasic oral contraceptive containing ethinylestradiol and levonorgestrel. *Contraception* 29:305–18.

Glick, I. D., and S. E. Bennett. 1981. Psychiatric complications of progesterone and oral contraceptives. *J. Clin. Psychopharmacol.* 1:350–67.

Hankinson, S. E., G. A. Colditz, D. J. Hunter, T. L. Spencer, B. Rosner, and M. J. Stampfer. 1992. A quantitative assessment of oral contraceptive use and risk of ovarian cancer. *Obstet. Gynecol.* 80:708–14.

He, C. H., Y. E. Shi, J. Q. Xu, and P. F. Van Look. 1991. A multicenter clinical study on two types of levonorgestrel tablets administered for postcoital contraception. *Int. J. Gynaecol. Obstet.* 36:43–48.

Runnebaum, B. 1992. The androgenicity of oral contraceptives: the young patient's concerns. *Int. J. Fertil.* 37(suppl.)4:211–17.

Shaaban, M. M. 1993. Experience with Norplant in Egypt. *Ann. Med.* 25:167–69.

Vermeulen, A., and M. Thiery. 1982. Metabolic effects of the triphasic oral contraceptive Trigynon. *Contraception* 26:505–12.

Wang, I. Y., and I. S. Fraser. 1994. Reproductive function and contraception in the postpartum period. *Obstet. Gynecol. Surv.* 49:56–63.

DHEA

Abraham, G. E., Z. H. Chakmakjian, J. E. Buster, and J. R. Marshall. 1974. Effect of exogenous conjugated estrogen on plasma gonadotropins and ovarian steroids during the menstrual cycle. *Obstet. Gynecol.* 43:676–84.

Barrett-Connor, E., K. T. Khaw, and S. S. Yen. 1986. A prospective study of dehydroepiandrosterone sulfate, mortality, and cardiovascular disease. *N. Engl. J. Med.* 315:1519–24.

Bouloux, P. M., P. Munroe, J. Kirk, and G. M. Besser. 1992. Sex and smell—an enigma resolved. *J. Endocrinol.* 133:323–26.

Carette, B., and P. Poulain. 1984. Excitatory effect of dehydroepiandrosterone, its sulphate ester and pregnenolone sulphate, applied by iontophoresis and pressure, on single neurones in the septo-preoptic area of the guinea pig. *Neurosci. Lett.* 45:205–10.

Cleary, M. P. 1991. The antiobesity effect of dehydroepiandrosterone in rats. *Proc. Soc. Exp. Biol. Med.* 196:8–16.

Cleary, M. P., A. Shepherd, and B. Jenks. 1984. Effect of dehydroepiandrosterone on growth in lean and obese Zucker rats. *J. Nutr.* 114:1242–51.

Crenshaw, T. L., J. P. Goldberg, and W. C. Stern. 1987. Pharmacologic modification of psychosexual dysfunction. *J. Sex Marital Ther.* 13:239–52.

Cumming, D. C., R. W. Rebar, B. R. Hopper, and S. S. Yen. 1982. Evidence for an influence of the ovary on circulating dehydroepiandrosterone sulfate levels. *J. Clin. Endocrinol. Metab.* 54:1069–71.

Dallo, J., T. T. Yen, I. Farago, and J. Knoll. 1988. The aphrodisiac effect of (−) deprenyl in non-copulator male rats. *Pharmacol. Res. Commun.* 20:(suppl.)25–26.

Deslypere, J. P., L. Verdonck, and A. Vermeulen. 1985. Fat tissue: a steroid reservoir and site of steroid metabolism. *J. Clin. Endocrinol. Metab.* 61:564–70.

Feher, T., K. S. Szalay, and G. Szilagyi. 1985. Effect of ACTH and prolactin on dehydroepiandrosterone, its sulfate ester and cortisol production by normal and tumorous human adrenocortical cells. *J. Steroid Biochem.* 23:153–57.

Glaser, J., L. J. Brind, L. J. Vogelman, H. M. Eisner, and others. 1992. Elevated serum dehydroepiandrosterone sulfate levels in practitioners of the Transcendental Meditation (TM) and TM-Sidhi programs. *Journal of Behavioral Medicine* 15:327–41.

Key, T. J., M. C. Pike, D. Y. Wang, and J. W. Moore. 1990. Long term effects of a first pregnancy on serum concentrations of dehydroepiandrosterone sulfate and dehydroepiandrosterone. *J. Clin. Endocrinol. Metab.* 70:1651–53.

Klove, K. L., R. Subir, and R. A. Lobo. 1984. The effect of different contraceptive treatments on the serum concentration of dehydroepiandrosterone sulfate. *Contraception* 29:319–24.

Lobo, R. A. 1988. Endocrine therapy of hyperandrogenism. In *Reproductive endocrine therapeutics*. R. Barbieri and I. Schiff, editors. New York: Alan R. Liss. 101–26.

Morales, A. J., J. J. Nolan, J. C. Nelson, and S. S. Yen. 1994. Effects of replacement dose of dehydroepiandrosterone in men and women of advancing age. *J. Clin. Endocrinol. Metab.* 78:1360–67.

Parker, L., J. Eugene, D. Farber, E. Lifrak, M. Lai, and G. Juler. 1985. Dissociation of adrenal androgen and cortisol levels in acute stress. *Horm. Metab. Res.* 17:209–12.

Parker, L. N., E. R. Levin, and E. T. Lifrak. 1985. Evidence for adrenocortical adaptation to severe illness. *J. Clin. Endocrinol. Metab.* 60:947–52.

Regelson, W., R. Loria, and M. Kalimi. 1988. Hormonal intervention: "Buffer hormones" or "state dependency": The role of dehydroepiandrosterone (DHEA), thyroid hormone, estrogen and hypoplysectomy in aging. In *Neuroimmunomodulation: interventions in aging and cancer.* W. Pierpaoli and N. H. Spector, editors. New York: New York Academy of Sciences. 260–73.

Robel, P., Y. Akwa, C. Corpechot, H. Zhony-Yi, I. Jung-Testas, K. Kabbadj, C. Le Goascogne, R. Morfin, C. Vourc'h, J. Young, and E. E. Baulieu. 1991. Neurosteroids: Biosynthesis and function of pregnenolone and dehydroepiandrosterone in the brain. In *Brain endocrinology.* M. Motta, editor. New York: Raven Press. 105–30.

Robel, P., C. Corpechot, M. Sunguelakis, A. Groyer, C. Clarke, M. L. Schlegel, P. Brazeau, and E. E. Baulieu. 1984. Pregnenolone, dehydroepiandrosterone, and their sulfate esters in the rat brain. In *Metabolism of hormonal steroids in the neuroendocrine structures.* F. Celotti, L. Naftolin, and L. Martini, editors. New York: Raven Press. 185–93.

Rozenberg, S., H. Ham, D. Bosson, A. Peretz, and C. Robyn. 1990. Age, steroids and bone mineral content. *Maturitas* 12:137–143.

Rozenberg, S., H. Ham, A. Caufriez, D. Bosson, A. Peretz, and C. Robyn. 1990. Sex and adrenal steroids in female in- and out-patients. In *Multidisciplinary perspectives on menopause.* M. Flint and F. Kronenberg, editors. New York: New York Academy of Sciences. 466–68.

Schneider, R. H., P. J. Mills, W. Schramm, K. G. Walton, M. C. Dillbeck, and R. K. Wallace. 1989. Dehydroepiandrosterone sulfate (DHEA) levels in type A behavior and the transcendental meditation program [abstract]. *Psychosom. Med.* 51:256.

Schwartz, A. G., G. C. Hard, L. L. Pashko, M. Abou-Gharbia, and D. Swern. 1981 Dehydroepiandrosterone: an anti-obesity and anti-carcinogenic agent. *Nutr. Cancer* 3: 46–53.

Tagliaferro, A. R., J. R. Davis, S. Truchon, and N. Van Hamont. 1986. Effects of dehydroepiandrosterone acetate on metabolism, body weight and composition of male and female rats. *J. Nutr.* 116:1977–83.

DOPAMINE

Argiolas, A. 1989. Central dopamine-oxytocin link in the control of sexual behavior in male rats. *Abstracts of the 12th annual meeting of the European Neuroscience Association* 12: (abstr.).

Argiolas, A., M. R. Melis, and G. L. Gessa. 1988. Yawning and penile erection: central dopamine-oxytocin-adrenocorticotropin connection. *Ann. N. Y. Acad. Sci.* 525:330–37.

Austin, M. C., and P. W. Kalivas. 1991. Dopaminergic involvement in locomotion elicited from the ventral pallidum/substantia innominata. *Brain Res.* 542:123–31.

Carlsson, M., and A. Carlsson. 1988. A regional study of sex differences in rat brain serotonin. *Prog. Neuro-psychopharmacol. Biol. Psychiatry* 12:53–61.

Dackis, C. A., and M. S. Gold. 1985. New concepts in cocaine addiction: the dopamine depletion hypothesis. *Neurosci. Biobehav. Rev.* 9:469–77.

Dallo, J., N. Lekka, and J. Knoll. 1986. The ejaculatory behavior of sexually sluggish male rats treated with (−) deprenyl, apomorphine, bromocriptine and amphetamine. *Pol. J. Pharmacol. Pharm.* 38:251–55.

DiPaolo, T., S. Masson, M. Daigle, and A. Belanger. 1987. Effect of a combined physiological dose of 17B-estradiol and progesterone on male and female rat striatum dopamine and serotonin metabolism. *J. Neurochemistry* 48:(suppl.)5126.

Foreman, M. M., and J. L. Hall. 1987. Effects of D2-dopaminergic receptor stimulation on male rat sexual behavior. *J. Neural Transm.* 68:153–70.

Gessa, G. L., and L. Napoli-Farris. 1983. Dopamine receptors and premature ejaculation. In *Psychopharmacology and sexual disorders.* D. Wheatley, editor. New York: Oxford University Press. 15–21.

Gessa, G. L., and A. Tagliamonte. 1974. Possible role of brain serotonin and dopamine in controlling male sexual behavior. In *Serotonin: new vistas.* E. Costa, M. Sandler, and G. L. Gessa, editors. New York: Raven Press. 217–28.

————. 1975. Role of brain serotonin and dopamine in male sexual behavior. In *Sexual Behavior: Pharmacology and biochemistry.* M. Sandler and G. L. Gessa, editors. New York: Raven Press. 117–28.

Gessa, G. L., A. Tagliamonte, and B. B. Brodie. 1970. Essential role of testosterone in the sexual stimulation induced by p-chlorophenylalanine in male animals. *Nature* 227:616–17.

Harris, G. C., and G. Aston-Jones. 1994. Involvement of D2 dopamine receptors in the nucleus accumbens in the opiate withdrawal syndrome. *Nature* 371:155–57.

Jones, G. H., D. B. Neill, and J. B. Justice. 1990. Nucleus accumbens, dopamine, and anticipatory behavior. *Society for Neuroscience, 20th Meeting* 16:(Pt. 1)437.

Knoll, J. 1985. The facilitation of dopaminergic activity in the aged brain by (−) deprenyl. A proposal for a strategy to improve the quality of life in senescence. *Mech. Ageing Dev.* 30:109–22.

Knoll, J., J. Dallo, and T. T. Yen. 1989. Striatal dopamine, sexual activity and lifespan. Longevity of rats treated with (−) deprenyl. *Life Sci.* 45:525–31.

Lanca, A. J. 1994. Reduction of voluntary alcohol intake in the rat by modulation of the dopaminergic mesolimbic system: transplantation of ventral mesencephalic cell suspensions. *Neuroscience* 58:359–69.

Mas, M., J. L. Gonzalez-Mora, A. Louilot, C. Sole, and T. Guadalupe. 1990. Increased dopamine release in the nucleus accumbens of copulating male rats as evidenced by in vivo voltammetry. *Neurosci. Lett.* 110:303–8.

Melis, M. R., A. Argiolas, R. Stancampiano, and G. L. Gessa. 1990. Effect of apomorphine on oxytocin concentrations in different brain areas and plasma of male rats. *Eur. J. Pharmacol.* 182:101–7.

Mitchell, J. B., and J. Stewart. 1989. Effects of castration, steroid replacement, and sexual experience on mesolimbic dopamine and sexual behaviors in the male rat. *Brain Res.* 491:116–27.

Napoli-Farris, L., W. Fratta, and G. L. Gessa. 1984. Stimulation of dopamine autoreceptors elicits "premature ejaculation" in rats. *Pharmacol. Biochem. Behav.* 20:69–72.

Pfaus, J. G., and A. G. Phillips. 1991. Role of dopamine in anticipatory and consummatory aspects of sexual behavior in the male rat. *Behav. Neurosci.* 105:727–43.

Pomerantz, S. M. 1991. Quinelorane (LY163502), a D2 dopamine receptor agonist, acts centrally to facilitate penile erections of male rhesus monkeys. *Pharmacol. Biochem. Behav.* 39:123–28.

————. 1992. Dopaminergic influences on male sexual behavior of rhesus monkeys: effects of dopamine agonists. *Pharmacol. Biochem. Behav.* 41:511–17.

Rosen, R. C., and A. K. Ashton. 1993. Prosexual drugs: empirical status of the "new aphrodisiacs." *Arch. Sex. Behav.* 22:521–43.

Zini, D., C. Carani, A. Baldini, C. Cavicchioli, D. Piccinini, and P. Marrama. 1986. Further acquisitions on gonadal function in bromocriptine treated hyperprolactinemic male patients. *Pharmacol. Res. Commun.* 18:601–9.

ESTROGEN

Abraham, G. E., Z. H. Chakmakjian, J. E. Buster, and J. R. Marshall. 1974. Effect of exogenous conjugated estrogen on plasma gonadotropins and ovarian steroids during the menstrual cycle. *Obstet. Gynecol.* 43:676–84.

Bachmann, G. A. 1993. Estrogen-androgen therapy for sexual and emotional well-being. *The Female Patient* 18:35+.

Barrett-Connor, E., and T. L. Bush. 1991. Estrogen and coronary heart disease in women. *JAMA* 265:1861–67.

Barrett-Connor, E., and D. Kritz-Silverstein. 1993. Estrogen replacement therapy and cognitive function in older women. *JAMA* 269:2637–41.

Beach, F. A. 1976. Sexual attractivity, proceptivity, and receptivity in female mammals. *Horm. Behav.* 7:105–38.

Becker, U., C. Gluud, P. Bennett, S. Micic, B. Svenstrup, K. Winkler, N. J. Christensen, and F. Hardt. 1988. Effect of alcohol and glucose infusion on pituitary-gonadal hormones in normal females. *Drug Alcohol Depend.* 22:141–49.

Corson, S. L. 1993. A decade of experience with transdermal estrogen replacement therapy: overview of key pharmacologic and clinical findings. *Int. J. Fertil.* 38:79–91.

Cumming, D. C., R. W. Rebar, B. R. Hopper, and S. S. Yen. 1982. Evidence for an influence of the ovary on circulating dehydroepiandrosterone sulfate levels. *J. Clin. Endocrinol. Metab.* 54:1069–71.

Ditkoff, E. C., W. G. Crary, M. Cristo, and R. A. Lobo. 1991. Estrogen improves psychological function in asymptomatic postmenopausal women. *Obstet. Gynecol.* 78:991–95.

Fahrback, S. E., J. I. Morrelli, and D. W. Pfaff. 1985. Role of oxytocin in the onset of estrogen-facilitated maternal behavior. In *Oxytocin: clinical and laboratory studies: proceedings of the Second International Conference on Oxytocin, Lac Beauport, Quebec, Canada, June 29 to July 1, 1984.* J. A. Amico and A. G. Robinson, editors. New York: Elsevier Science Pub. Co. 372–88.

Gambrell, R. D. 1993. Estrogen replacement therapy and breast cancer risk: a new look at the data. *The Female Patient* 18:50+.

Hafner, H., A. Riecher-Rossler, W. An Der Heiden, K. Maurer, B. Fatkenheuer, and W. Loffler. 1993. Generating and testing a causal explanation of the gender difference in age at first onset of schizophrenia. *Psychol. Med.* 23:925–40.

Henderson, B. E., A. Paganini-Hill, and R. K. Ross. 1991. Decreased mortality in users of estrogen replacement therapy. *Arch. Intern. Med.* 151:75–78.

Henderson, V. W., A. Paganini-Hill, C. K. Emanuel, M. E. Dunn, and J. G. Buckwalter. 1994. Estrogen replacement therapy in older women. Comparisons between Alzheimer's disease cases and nondemented control subjects. *Arch. Neurol.* 51:896–900.

Johnson, D. F., and C. H. Phoenix. 1976. Hormonal control of female sexual attractiveness proceptivity, and receptivity in rhesus monkeys. *J. Comp. Physiol. Psychol.* 90:473–83.

Kampen, D. L., and B. B. Sherwin. 1994. Estrogen use and verbal memory in healthy postmenopausal women. *Obstet. Gynecol.* 83:979–83.

Lobo, R. A. 1987. Absorption and metabolic effects of different types of estrogens and progestogens. *Obstet. Gynecol. Clin. North Am.* 14:143–67.

Nabulsi, A. A., A. R. Folsom, A. White, W. Patsch, G. Heiss, K. K. Wu, and M. Szklo. 1993. Association of hormone-replacement therapy with various cardiovascular risk factors in postmenopausal women. The Atherosclerosis Risk in Communities Study Investigators [see comments]. *N. Engl. J. Med.* 328:1069–75.

Nachtigall, L. E. 1994. Comparative study: Replens versus local estrogen in menopausal women. *Fertil. Steril.* 61:178–80.

Palinkas, L. A., and E. Barrett-Connor. 1992. Estrogen use and depressive symptoms in postmenopausal women. *Obstet. Gynecol.* 80:30–36.

Petitti, D. 1994. Hormone-replacement therapy: risk and benefits. *The Female Patient* 19:63+.

Pfaff, D. W. 1973. Luteinizing hormone-releasing factor potentiates lordosis behavior in hypophysectomized ovariectomized female rats. *Science* 182:1148–49.

Robinson, D., L. Friedman, R. Marcus, J. Tinklenberg, and J. Yesavage. 1994. Estrogen replacement therapy and memory in older women. *J. Am. Geriatr. Soc.* 42:919–22.

Schiff, I. 1993. Keys to balancing the risks and benefits of estrogen therapy for postmenopausal women. *Modern Medicine* 61:72–93.

Slotes, F. 1994. The power of estrogen. *The Tennessean* November 15 1D–4D.

Udry, J. R., and L. M. Talbert. 1988. Sex hormone effects on personality at puberty. *J. Pers. Soc. Psychol.* 54:291–95.

Unknown. 1992. Transdermal estrogen patch shown to preserve bone mineral density. *Geriatrics* 47:17(1).

Zubialde, J. P., F. Lawler, and N. Clemenson. 1993. Estimated gains in life expectancy with use of postmenopausal estrogen therapy: a decision analysis. *J. Fam. Pract.* 36:271–80.

GROWTH HORMONE

Clopper, R. R., J. M. Adelson, and J. Money. 1976. Postpubertal psychosexual function in male hypopituitarism without hypogonadotropinism after growth hormone therapy. *Journal of Sex Research* 12:14–32.

Dinan, T. G., L. N. Yatham, V. O'Keane, and S. Barry. 1991. Blunting of noradrenergic-stimulated growth hormone release in mania. *Am. J. Psychiatry* 148:936–38.

Ikeda, Y., I. Tanaka, Y. Oki, H. Morita, K. Komatsu, and T. Yoshimi. 1995. Testosterone or GnRH treatment improves impaired response of plasma vasopressin stimuli in men with hypogonadism. *Endocrine Society 72nd Meeting Abstracts* (abstr.).

Pace, J. N., J. L. Miller, and L. I. Rose. 1991. GnRH agonists: gonadorelin, leuprolide and nafarelin. *Am. Fam. Physician* 44:1777–82.

Schwartz, A. G., G. C. Hard, L. L. Pashko, M. Abou-Gharbia, and D. Swern. 1981. Dehydroepiandrosterone: an anti-obesity and anti-carcinogenic agent. *Nutr. Cancer* 3:46–53.

Shaw, R. W. 1992. The role of GnRH analogues in the treatment of endometriosis. *Br. J. Obstet. Gynaecol.* 99 (suppl.) 7:9–12.

Spark, R. F., R. A. White, and P. B. Connolly. 1980. Impotence is not always psychogenic. Newer insights into hypothalamic-pituitary-gonadal dysfunction. *JAMA* 243:750–55.

HORMONE REPLACEMENT THERAPY

Appleby, L., and J. Montgomery. 1987. Effect of combined implants of oestradiol and testosterone on libido in postmenopausal women [letter]. *Br. Med. J. (Clin. Res. Ed.)* 294:1417–18.

Barrett-Connor, E., and T. L. Bush. 1991. Estrogen and coronary heart disease in women. *JAMA* 265:1861–67.

Barrett-Connor, E., and D. Kritz-Silverstein. 1993. Estrogen replacement therapy and cognitive function in older women. *JAMA* 269:2637–41.

Burger, H., J. Hailes, J. Nelson, and M. Menelaus. 1987. Effect of combined implants of oestradiol and testosterone on libido in postmenopausal women. *Br. Med. J. (Clin. Res. Ed.)* 294:936–37.

Burris, A. S., S. M. Banks, C. S. Carter, J. M. Davidson, and R. J. Sherins. 1992. A long-term, prospective study of the physiologic and behavioral effects of hormone replacement in untreated hypogonadal men. *J. Androl.* 13:297–304.

Burris, A. S., R. H. Gracely, C. S. Carter, R. J. Sherins, and J. M. Davidson. 1991. Testosterone therapy is associated with reduced tactile sensitivity in human males. *Horm. Behav.* 25:195–205.

Carani, C., D. Zini, A. Baldini, L. Della Casa, A. Ghizzani, and P. Marrama. 1990. Effects of androgen treatment in impotent men with normal and low levels of free testosterone. *Arch. Sex. Behav.* 19:223–34.

Clopper, R. R., M. L. Voorhess, M. H. MacGillivray, P. A. Lee, and B. Mills. 1993. Psychosexual behavior in hypopituitary men: a controlled comparison of gonadotropin and testosterone replacement. *Psychoneuroendocrinology* 18:149–61.

Corson, S. L. 1993. A decade of experience with transdermal estrogen replacement therapy: overview of key pharmacologic and clinical findings. *Int. J. Fertil.* 38:79–91.

Deems, D. A., R. L. Doty, R. G. Settle, V. Moore-Gillon, P. Shaman, A. F. Mester, C. P. Kimmelman, V. J. Brightman, and J. B. Snow. 1991. Smell and taste disorders, a study of 750 patients from the University of Pennsylvania Smell and Taste Center. *Arch. Otolaryngol. Head. Neck Surg.* 117:519–28.

Henderson, B. E., A. Paganini-Hill, and R. K. Ross. 1991. Decreased mortality in users of estrogen replacement therapy. *Arch. Intern. Med.* 151:75–78.

Kampen, D. L., and B. B. Sherwin. 1994. Estrogen use and verbal memory in healthy postmenopausal women. *Obstet. Gynecol.* 83:979–83.

Labrie, F., A. Dupont, A. Belanger, R. St-Arnaud, M. Giguere, Y. Lacourciere, J. Emond, and G. Monfette. 1986. Treatment of prostate cancer with gonadotropin-releasing hormone agonists. *Endocr. Rev.* 7:67–74.

Nachtigall, L. E. 1994. Comparative study: Replens versus local estrogen in menopausal women. *Fertil. Steril.* 61:178–80.

Phillips, G. B., T. Y. Jing, L. M. Resnick, M. Barbagallo, J. H. Laragh, and J. E. Sealey. 1993. Sex hormones and hemostatic risk factors for coronary heart disease in men with hypertension. *J. Hypertens.* 11:699–702.

Robinson, D., L. Friedman, R. Marcus, J. Tinklenberg, and J. Yesavage. 1994. Estrogen replacement therapy and memory in older women. *J. Am. Geriatr. Soc.* 42:919–22.

Sherwin, B. B. 1988. Estrogen and/or androgen replacement therapy and cognitive functioning in surgically menopausal women. *Psychoneuroendocrinology* 13:345–57.

———. 1991. The impact of different doses of estrogen and progestin on mood and sexual behavior in postmenopausal women. *J. Clin. Endocrinol. Metab.* 72:336–43.

LHRH

Doelle, G. C., A. N. Alexander, R. M. Evans, R. Linde, J. Rivier, W. Vale, and D. Rabin. 1983. Combined treatment with an LHRH agonist and testosterone in man. Reversible oligozoospermia without impotence. *J. Androl.* 4:298–302.

Ehrensing, R. H., and A. J. Kastin. 1976. Clinical investigations for emotional effects of neuropeptide hormones. *Pharmacol. Biochem. Behav.* 5:89–93.

Kendrick, K. M., and A. F. Dixson. 1985. Luteinizing hormone releasing hormone enhances proceptivity in a primate. *Neuroendocrinology* 41:449–53.

Labrie, F. 1993. Mechanism of action and pure antiandrogenic properties of flutamide. *Cancer* 72:3816–27.

Labrie, F., A. Belanger, L. Cusan, C. Seguin, G. Pelletier, P. A. Kelly, F. A. Lefebvre, A. Lemay, and J. P. Raynaud. 1980. Antifertility effects of LHRH agonists in the male. *J. Androl.* 1:209–28.

Labrie, F., A. Dupont, A. Belanger, L. Cusan, Y. Lacourciere, G. Monfette, J. G. Laberge, J. P. Emond, A. T. Fazekas, J. P. Raynaud, and J. M. Husson. 1982. New hormonal therapy in prostatic carcinoma: combined treatment with an LHRH agonist and an antiandrogen. *Clin. Invest. Med.* 5:267–75.

Labrie, F., A. Dupont, L. Cusan, J. L. Gomez, and P. Diamond. 1993. Major advantages of "early" administration of endocrine combination therapy in advanced prostate cancer. *Clin. Invest. Med.* 16:493–98.

Sakuma, Y., and D. W. Pfaff. 1980. LH-RH in the mesencephalic central grey can potentiate lordosis reflex of female rats. *Nature* 283:566–67.

LONGEVITY

Barrett-Connor, E. 1990. Smoking and endogenous sex hormones in men and women. In *Smoking and hormone-related disorders.* N. Wald and J. Baron, editors. New York: Oxford University Press. 183–96.

Birkmayer, W., and P. Riederer. 1984. Deprenyl prolongs the therapeutic efficacy of combined L-DOPA in Parkinson's disease. *Adv. Neurol.* 40:475–81.

Brody, J. E. 1995. Restoring ebbing hormones may slow aging; researchers hope supplements may have rejuvenating effects. (nine teams researching hormone replacement for people over 60 years old focus on growth hormone and other 'trophic factors') (Science Times Pages). *New York Times* 144:B5(N)–C1(L).

Camilleri, J. P. 1992. [Structural approach of vascular aging] Approche structurale du vieillissement vasculaire. *Presse Med.* 21:1184–87.

Curtis, R. H. 1986. Mind and mood: understanding and controlling your emotions. New York: Scribner.

Glaser, J. L., J. L. Brind, J. H. Vogelman, M. J. Eisner, et al. 1992. Elevated serum dehydroepiandrosterone sulfate levels in practitioners of the Transcendental Meditation (TM) and TM-Sidhi programs. *Journal of Behavioral Medicine* 15:327–41.

Gregori, A., L. Marchionni, P. Ossanna, and D. Pozza. 1993. [Life habits and impotence] Abitudini di vita ed impotenza. *Minerva. Med.* 84:191–94.

Grow, D. R., and H. P. Wiczyk. 1994. Mifepristone (RU 486) and other anti-progestins: a review of their clinical applications. *Assisted Reproductive Reviews* 4:101–9.

Heath, R. G. 1972. Pleasure and brain activity in man. Deep and surface electroencephalograms during orgasm. *J. Nerv. Ment. Dis.* 154:3–18.

Jaroff, L. 1995. New age therapy. (research on anti-aging pill) *Time* 145:52.

Kemnitz, J. W., R. W. Goy, T. J. Flitsch, J. J. Lohmiller, and J. A. Robinson. 1989. Obesity in male and female rhesus monkeys: fat distribution, glucoregulation, and serum androgen levels. *J. Clin. Endocrinol. Metab.* 69:287–93.

Knoll, J. 1988. Extension of life span of rats by long-term (–)deprenyl treatment. *Mt. Sinai. J. Med.* 55:67–74.

Knoll, J., T. T. Yen, and J. Dallo. 1983. Long-lasting, true aphrodisiac effect of (–)-deprenyl in sexually sluggish old male rats. *Mod. Probl. Pharmacopsychiatry* 19:135–53.

Kopera, H. 1983. Sex hormones and the brain. In *Sex and the brain*. J. Durden-Smith and D. deSimone, editors. New York: Arbor House.

Labrie, C., J. Simard, H. F. Zhao, A. Belanger, G. Pelletier, and F. Labrie. 1990. Stimulation of androgen-dependent gene expression by the adrenal precursors dehydroepiandrosterone and androstenedione in the rat ventral prostate. In *Steroid formation, degradation, and action in peripheral tissues*. L. Castagnetta, S. D'Aquino, F. Labrie, and H. L. Bradlow, editors. New York: New York Academy of Sciences. 395–98.

Li, B. H., and G. C. Chiou. 1992. Systemic administration of calcitonin through ocular route. *Life Sci.* 50:349–54.

Mazurek, M. F., M. F. Beal, E. D. Bird, and J. B. Martin. 1987. Oxytocin in Alzheimer's disease: postmortem brain levels. *Neurology* 37:1001–3.

McGaugh, J., editor. 1985. *Neuropeptides, neurotransmitters and memory: Sociosexual aspects*. New York: Elsevier Science Pub. Co.

Pearson, D., and S. Shaw. 1982. *Life extension: a practical scientific approach*. New York: Warner Books.

Pelton, R. 1986. *Mind food and smart pills: nutrients and drugs that increase intelligence and prevent brain aging*. Poway, CA: T & R Publishers.

Roberts, E. 1986. Guides through the labyrinth of AD: Dehydroepiandrosterone, potassium channels, and the C4 component of complement. In *Treatment strategies for Alzheimer's disease*. T. Crook, R. T. Bartus, S. Ferris, and S. Sershon, editors. Madison: Mark Powley. 173–200.

Stone, N. N., W. R. Fair, and J. Fishman. 1986. Estrogen formation in human prostatic tissue from patients with and without benign prostatic hyperplasia. *Prostate* 9:311–18.

Sunderland, T., C. R. Merril, M. G. Harrington, B. A. Lawlor, S. E. Molchan, R. Martinez, and D. L. Murphy. 1989. Reduced plasma dehydroepiandrosterone concentrations in Alzheimer's disease [letter]. *Lancet* 2:570.

Terra, R. 1993. Working molecules: progress toward nanotechnology. *Analog Science Fiction & Fact* 113:44(20)

Toone, B. K. 1992. Sexual dysfunction. In *Drug-induced dysfunction in psychiatry*. M. S. Keshavan and J. S. Kennedy, editors. New York: Hemisphere Publ. Corp. 273–80.

Van Voorhies, W. A. 1992. Production of sperm reduces nematode lifespan. *Nature* 360:456–58.

Winn, R. L., and N. Newton. 1982. Sexuality in aging: a study of 106 cultures. *Arch. Sex. Behav.* 11:(4).

Witelson, S. F. 1991. Sex differences in neuroanatomical changes with aging [letter]. *N. Engl. J. Med.* 325:211–12.

Woodward, K. L. 1992. Better than a gold watch; sex after 60 turns out to be good and plentiful. (survey results) *Newsweek* 120:71.

Zajonc, R. B., S. T. Murphy, and M. Inglehart. 1989. Feeling and facial efference: implications of the vascular theory of emotion. *Psychol. Rev.* 96:395–416.

Zubialde, J. P., F. Lawler, and N. Clemenson. 1993. Estimated gains in life expectancy with use of postmenopausal estrogen therapy: a decision analysis. *J. Fam. Pract.* 36:271–80.

MELATONIN

Brody, J. 1995. Restoring ebbing hormones may slow aging. *New York Times*, 18 July 1995.

Brugger, P., W. Marktl, and M. Herold. 1995. Impaired nocturnal secretion of melatonin in coronary heart disease. *Lancet* 345:1408.

Haimov, I., P. Lavie, M. Laudon, P. Herer, C. Vigder, and N. Zisapel. 1995. Melatonin replacement therapy of elderly insomniacs. *Sleep* 18:598–603.

Huyghe, P. 1994. The hormone whose time has come. *Hippocrates* 7/8:22–25.

Pierpaoli, W., and W. Regelson. *The melatonin miracle: nature's age-reversing, disease-fighting, sex-enhancing hormone.* New York: Simon & Schuster.

―――. 1994. Pineal control of aging: effect of melatonin and pineal grafting on aging mice. *Proc. Nat. Acad. Sci. U.S.A.* 91:787–791.

Puig-Domingo, M., S. M. Webb, J. Serrano, et al. 1992. Brief report: melatonin-related hypogonadotropic hypogonadism. *New Engl. J. Med.* 327:1356–1359.

Utiger, R.D. 1992. Melatonin—the hormone of darkness [letter]. *New Engl. J. Med.* 327:1377–79.

MENOPAUSE

Abrahamsson, G., P. O. Janson, and S. Kullander. 1990. An in vitro perfusion method for metabolic studies on human postmenopausal ovaries. *Acta Obstet. Gynecol. Scand.* 69:527–32.

Ajabor, L. N., C. C. Tsai, P. Vela, and S. S. Yen. 1972. Effect of exogenous estrogen on carbohydrate metabolism in postmenopausal women. *Am. J. Obstet. Gynecol.* 113:383–87.

al-Azzawi, F. 1992. Endocrinological aspects of the menopause. *Br. Med. Bull.* 48:262–75.

Appleby, L., and J. Montgomery. 1987. Effect of combined implants of oestradiol and testosterone on libido in postmenopausal women [letter]. *Br. Med. J. (Clin. Res. Ed.)* 294:1417–18.

Bachmann, G. A. 1990. Sexual issues at menopause. *Ann. N. Y. Acad. Sci.* 592:87–94; discussion 123–33.

―――. 1991. Sexual dysfunction in the older woman. *Medical Aspects of Human Sexuality* 42+.

―――. 1993. Estrogen-androgen therapy for sexual and emotional well-being. *The Female Patient* 18:35+.

Ballinger, S. 1990. Stress as a factor in lowered estrogen levels in the early postmenopause. *Ann. N. Y. Acad. Sci.* 592:95–113; discussion 123–33.

Booher, D. L. 1990. Estrogen supplements in menopause. *Cleve. Clin. J. Med.* 57:154–60.

Brooks, T. R. 1993. Sexuality in the aging woman. *The Female Patient* 18:27+.

Burger, H., J. Hailes, J. Nelson, and M. Menelaus. 1987. Effect of combined implants of oestradiol and testosterone on libido in postmenopausal women. *Br. Med. J. (Clin. Res. Ed.)* 294:936–37.

Burger, H. G. 1994. Diagnostic role of follicle-stimulating hormone (FSH) measurements during the menopausal transition—an analysis of FSH, oestradiol and inhibin. *Eur. J. Endocrinol.* 130:38–42.

Carlson, K. J., B. A. Miller, and F. J. Fowler. 1994. The Maine Women's Health Study: I. Outcomes of hysterectomy. *Obstet. Gynecol.* 83:556–65.

Carlstrom, K., S. Brody, N. O. Lunell, A. Lagrelius, G. Mollerstrom, A. Pousette, G. Rannevik, R. Stege, and B. von Schoultz. 1988. Dehydroepiandrosterone sulphate and dehydroepiandrosterone in serum: differences related to age and sex. *Maturitas* 10:297–306.

Cumming, D. C., R. W. Rebar, B. R. Hopper, and S. S. Yen. 1982. Evidence for an influence of the ovary on circulating dehydroepiandrosterone sulfate levels. *J. Clin. Endocrinol. Metab.* 54:1069–71.

Ettinger, B. 1991. Estrogen supplements in menopause [letter; comment]. *Cleve. Clin. J. Med.* 58:366.

Grady, D., S. M. Rubin, D. B. Petitti, C. S. Fox, D. Black, B. Ettinger, V. L. Ernster, and S. R. Cummings. 1992. Hormone therapy to prevent disease and prolong life in postmenopausal women. *Ann. Intern. Med.* 117:1016–37.

Greer, G. 1992. *The change: women, aging, and the menopause.* Distributed by Random House, New York.

Kampen, D. L., and B. B. Sherwin. 1994. Estrogen use and verbal memory in healthy postmenopausal women. *Obstet. Gynecol.* 83:979–83.

Khaw, K. T. 1992. Epidemiology of the menopause. *Br. Med. Bull.* 48:249–61.

Krummel, D., T. D. Etherton, S. Peterson, and P. M. Kris-Etherton. 1993. Effects of exercise on plasma lipids and lipoproteins of women. *Proc. Soc. Exp. Biol. Med.* 204:123–37.

Kvale, J. N., and J. K. Kvale. 1993. Common gynecologic problems after age 75. *Postgrad. Med.* 93:263–68, 271–72.

Martin, K. A., and M. W. Freeman. 1993. Postmenopausal hormone-replacement therapy [editorial; comment]. *N. Engl. J. Med.* 328:1115–17.

Martinez-Jordan, N. 1993. [Relation between psychological symptoms and gynecological, hormonal and perimenopausal disorders] Relacion entre sintomas psicologicos, alteraciones ginecologico-hormonales y periodo perimenopausico. I y II parte. *Actas. Luso. Esp. Neurol. Psiquiatr. Cienc. Afines.* 21:131–42.

McBride, P. A., H. Tierney, M. DeMeo, J. S. Chen, and J. J. Mann. 1990. Effects of age and gender on CNS serotonergic responsivity in normal adults. *Biol. Psychiatry* 27:1143–55.

Morrell, M. J., J. M. Dixen, C. S. Carter, and J. M. Davidson. 1984. The influence of age and cycling status on sexual arousability in women. *Am. J. Obstet. Gynecol.* 148:66–71.

Nabulsi, A. A., A. R. Folsom, A. White, W. Patsch, G. Heiss, K. K. Wu, and M. Szklo. 1993. Association of hormone-replacement therapy with various cardiovascular risk factors in postmenopausal women. The Atherosclerosis Risk in Communities Study Investigators [see comments]. *N. Engl. J. Med.* 328:1069–75.

Nielsen, N. M., P. von der Recke, M. A. Hansen, K. Overgaard, and C. Christiansen. 1994. Estimation of the effect of salmon calcitonin in established osteoporosis by biochemical bone markers. *Calcif. Tissue Int.* 55:8–11.

O'Neil, P. M., and K. S. Calhoun. 1975. Sensory deficits and behavioral deterioration in senescence. *J. Abnorm. Psychol.* 84:579–82.

Overgaard, K. 1994. Effect of intranasal salmon calcitonin therapy on bone mass and bone turnover in early postmenopausal women: a dose-response study. *Calcif. Tissue Int.* 55:82–86.

Paganini-Hill, A., and V. W. Henderson. 1994. Estrogen deficiency and risk of Alzheimer's disease in women. *Am. J. Epidemiol.* 140:256–61.

Persky, H., L. Dreisbach, W. R. Miller, C. P. O'Brien, M. A. Khan, H. I. Lief, N. Charney, and D. Strauss. 1982. The relation of plasma androgen levels to sexual behaviors and attitudes of women. *Psychosom. Med.* 44:305–19.

Petitti, D. 1994. Hormone-replacement therapy: risk and benefits. *The Female Patient* 19:63+.

Randall, T. 1993. Women need more and better information on menopause from their physicians, says survey [news]. *JAMA* 270:1664.

Raz, R. and W. E. Stamm. 1993. A controlled trial of intravaginal estriol in postmenopausal women with recurrent urinary tract infections [see comments]. *N. Engl. J. Med.* 329:753–56.

Regelson, W., R. Loria, and M. Kalimi. 1988. Hormonal intervention: "Buffer hormones" or "state dependency": the role of dehydroepiandrosterone (DHEA), thyroid hormone, estrogen and hypoplysectomy in aging. In *Neuroimmunomodulation: interventions in aging and cancer.* W. Pierpaoli and N. H. Spector, editors. New York: New York Academy of Sciences. 260–73.

Renard, E., J. Bringer, and C. Jaffiol. 1993. [Sex steroids. Effects on the carbohydrate metabolism before and after menopause] Steroides sexuels. Effets sur le metabolisme hydrocarbone avant et après la ménopause. *Presse Med.* 22:431–35.

Rozenberg, S., H. Ham, D. Bosson, A. Peretz, and C. Robyn. 1990. Age, steroids and bone mineral content. *Maturitas* 12:137–43.

Rozenberg, S., H. Ham, A. Caufriez, D. Bosson, A. Peretz, and C. Robyn. 1990. Sex and adrenal steroids in female in- and out-patients. In *Multidisciplinary perspectives on menopause.* M. Flint and F. Kronenberg, editors. New York: New York Academy of Sciences. 466–68.

Sands, R., and J. Studd. 1995. Exogenous androgens in postmenopausal women. *Am. J. Med.* 98:76S–79S.

Sarrel, P. M. 1990. Sexuality and menopause. *Obstet. Gynecol.* 75:26S–30S; discussion 31S–35S.

Sarrel, P. M., and M. I. Whitehead. 1985. Sex and menopause: defining the issues. *Maturitas* 7:217–24.

Schiff, I. 1993. Keys to balancing the risks and benefits of estrogen therapy for postmenopausal women. *Modern Medicine* 61:72–93.

Sherman, C. 1993. Recommends early estrogen therapy for menopause. *Family Practice News* 26. Dr. Thomas E. Nolan, of the Medical College of Georgia, Augusta, remarks at the annual meeting of the South Carolina Academy of Family Physicians, ". . . More and more we are coming to view menopause as not a natural process . . . but as organ failure . . . as an endocrinopathy."

Sherman, S. S. 1993. Gender, health, and responsible research. *Clin. Geriatr. Med.* 9:261–69.

Sherwin, B. B. 1988. Estrogen and/or androgen replacement therapy and cognitive functioning in surgically menopausal women. *Psychoneuroendocrinology* 13:345–57.

———. 1991. The impact of different doses of estrogen and progestin on mood and sexual behavior in postmenopausal women. *J. Clin. Endocrinol. Metab.* 72:336–43.

Sherwin, B. B., and M. M. Gelfand. 1987. The role of androgen in the maintenance of sexual functioning in oophorectomized women. *Psychosom. Med.* 49:397–409.

Sherwin, B. B., M. M. Gelfand, and W. Brender. 1985. Androgen enhances sexual motivation in females: a prospective, crossover study of sex steroid administration in the surgical menopause. *Psychosom. Med.* 47:339–51.

Sherwin, B. B., and S. Phillips. 1990. Estrogen and cognitive functioning in surgically menopausal women. In *Multidisciplinary perspectives on menopause.* M. Flint and F. Kronenberg, editors. New York: New York Academy of Sciences. 474–75.

Siddle, N., P. Sarrel, and M. Whitehead. 1987. The effect of hysterectomy on the age at ovarian failure: identification of a subgroup of women with premature loss of ovarian function and literature review. *Fertil. Steril.* 47:94–100.

Slotes, F. 1994. The power of estrogen. *The Tennessean* November 15 1D–4D.

Upton, G. V. 1987. Contraception for the perimenopausal patient. *Obstet. Gynecol. Clin. North Am.* 14:207–27.

Vagenakis, A. G. 1989. Endocrine aspects of menopause. *Clin. Rheumatol.* 8 (suppl.) 2:48–51.

Voda, A. M. 1994. Risks and benefits associated with hormonal and surgical therapies for healthy midlife women. *West. J. Nurs. Res.* 16:507–23.

MENSTRUAL CYCLE

Abplanalp, J. M., A. F. Donnelly, and R. M. Rose. 1979. Psychoendocrinology of the menstrual cycle: I. Enjoyment of daily activities and moods. *Psychosom. Med.* 41:587–604.

Abplanalp, J. M., L. Livingston, R. M. Rose, and D. Sandwisch. 1977. Cortisol and growth hormone responses to psychological stress during the menstrual cycle. *Psychosom. Med.* 39:158–77.

Abplanalp, J. M., R. M. Rose, A. F. Donnelly, and L. Livingston-Vaughan. 1979. Psychoendocrinology of the menstrual cycle: II. The relationship between enjoyment of activities, moods, and reproductive hormones. *Psychosom. Med.* 41:605–15.

Abraham, G. E., G. B. Maroulis, and J. R. Marshall. 1974. Evaluation of ovulation and corpus luteum function using measurements of plasma progesterone. *Obstet. Gynecol.* 44:522–25.

Apodaca, L., and L. Fink. 1989. Pre-menstrual syndrome in the courts. In *Representing . . . battered women who kill.* S. L. Johann and F. Osanka, editors. Springfield, IL: Charles C. Thomas. 113–27.

Backstrom, T., D. Sanders, R. Leask, D. Davidson, P. Warner, and J. Bancroft. 1983. Mood, sexuality, hormones, and the menstrual cycle. II. Hormone levels and their relationship to the premenstrual syndrome. *Psychosom. Med.* 45:503–7.

Bancroft, J. H., D. Sanders, D. Davidson, and P. Warner. 1983. Mood, sexuality, hormones and the menstrual cycle. III. Sexuality and the role of androgens. *Psychosom. Med.* 45:509–16.

Constant, M., C. A. Abrams, and F. I. Chasalow. 1993. Gonadotropin-associated psychosis in perimenstrual behavior disorder. *Horm. Res.* 40:141–44.

Cutler, W. B., G. Preti, A. Krieger, G. R. Huggins, C. R. Garcia, and H. J. Lawley. 1986. Human axillary secretions influence women's menstrual cycles: the role of donor extract from men. *Horm. Behav.* 20:463–73.

Dabbs, J. M., and D. de La Rue. 1991. Salivary testosterone measurements among women: relative magnitude of circadian and menstrual cycles. *Horm. Res.* 35:182–84.

Dennerstein, L., J. B. Brown, G. Gotts, C. A. Morse, T. M. Farley, and A. Pinol. 1993. Menstrual cycle hormonal profiles of women with and without premenstrual syndrome. *J. Psychosom. Obstet. Gynaecol.* 14:259–68.

Dennerstein, L., G. Gotts, J. B. Brown, C. A. Morse, T. M. Farley, and A. Pinol. 1994. The relationship between the menstrual cycle and female sexual interest in women with PMS complaints and volunteers. *Psychoneuroendocrinology* 19:293–304.

Donaldson, R., et al. 1995. Premenstrual peanut butter cravings [letter]. *Cortlandt Forum* 55.

Fiser, D., and S. Radic. 1973. [Changes of olfactory acuity during normal menstrual cycle] Varijacije olfakcijske ostrine u normalnom menstrulnom ciklusu. *Med. Pregl.* 26:61–64.

Freeman, E., K. Rickels, S. J. Sondheimer, and M. Polansky. 1990. Ineffectiveness of progesterone suppository treatment for premenstrual syndrome. *JAMA* 264:349–53.

Genazzani, A. R., M. DeVoto, C. Cianchetti, C. Pintor, F. Facchinetti, A. Mangoni, and P. Fioretti. 1978. Possible correlation between plasma androgen variations during the menstrual cycle and sexual behavior in the human female. In *Clinical psychoneuroendocrinology in reproduction.* L. Carenza, P. Pancheri, and L. Zichella, editors. New York: Academic Press. 419–36.

Graham, C. A., and W. C. McGrew. 1980. Menstrual synchrony in female undergraduates living on a coeducational campus. *Psychoneuroendocrinology* 5:245–52.

Harvey, S. M. 1987. Female sexual behavior: fluctuations during the menstrual cycle. *J. Psychosom. Res.* 31:101–10.

Hoon, P. W., K. Bruce, and B. Kinchloe. 1982. Does the menstrual cycle play a role in sexual arousal? *Psychophysiology* 19:21–27.

Jarett, L. R. 1984. Psychosocial and biological influences on menstruation: synchrony, cycle length, and regularity. *Psychoneuroendocrinology* 9:21–28.

Jensvold, M. F. 1993. Psychiatric aspects of the menstrual cycle. In *Psychological aspects of women's health care: The interface between psychiatry and obstetrics and gynecology.* Washington, D. C.: American Psychiatric Press. 165–92.

Kudrow, L. 1993. Migraine in women: recognizing hormonal influences. *Female Patient* 18:33–38.

Ladisich, W. 1977. Influence of progesterone on serotonin metabolism: a possible causal factor for mood changes. *Psychoneuroendocrinology* 2:257–66.

McClintock, M. K. 1971. Menstrual synchrony and suppression. *Nature* 229:244–45.

Mendoza, S. P. 1984. The psychobiology of social relationships. In *Social cohesion: Essays toward a sociophysiological perspective*. P. R. Barchas and S. P. Mendoza, editors. Westport, Conn: Greenwood, pp. 3–29.

Michael, R., and D. Zumpe. 1990. Sexual preferences during artificial menstrual cycles in social groups of rhesus monkeys (Macaca mulatta). *Primates* 31:225–41.

Morrell, M. J., J. M. Dixen, C. S. Carter, and J. M. Davidson. 1984. The influence of age and cycling status on sexual arousability in women. *Am. J. Obstet. Gynecol.* 148:66–71.

Parazzini, F., L. Tozzi, R. Mezzopane, L. Luchini, M. Marchini, and L. Fedele. 1994. Cigarette smoking, alcohol consumption, and risk of primary dysmenorrhea. *Epidemiology* 5:469–72.

Persky, H., C. P. O'Brien, and M. A. Kahn. 1976. Reproductive hormone levels, sexual activity and moods during the menstrual cycle. *Psychosom. Med.* 38: 62–63.

Preti, G., W. B. Cutler, C. R. Garcia, G. R. Huggins, and H. J. Lawley. 1986. Human axillary secretions influence women's menstrual cycles: the role of donor extract of females. *Horm. Behav.* 20:474–82.

Rossmanith, W. G. 1993. Ultradian and circadian patterns in luteinizing hormone secretion during reproductive life in women. *Hum. Reprod.* 8 (suppl.) 2:77–83.

Sanders, D., and J. Bancroft. 1982. Hormones and the sexuality of women—the menstrual cycle. *Clin. Endocrinol. Metab.* 11:639–59.

Sanders, D., P. Warner, T. Backstrom, and J. Bancroft. 1983. Mood, sexuality, hormones and the menstrual cycle. I. Changes in mood and physical state: description of subjects and method. *Psychosom. Med.* 45:487–501.

Satinoff, E. 1982. Are there similarities between thermoregulation and sexual behavior? In *The physiological mechanisms of motivation*. D. W. Pfaff, editor. New York: Springer-Verlag. 217–51.

Schreiner-Engel, P., R. C. Schiavi, H. Smith, and D. White. 1981. Sexual arousability and the menstrual cycle. *Psychosom. Med.* 43:199–214.

Serrander, A. M., and K. E. Peek. 1993. Changes in contact lens comfort related to the menstrual cycle and menopause. A review of articles. *J. Am. Optom. Assoc.* 64:162–66.

Sherwin, B. B. 1988. A comparative analysis of the role of androgens in human male and female sexual behavior: behavioral specificity, critical thresholds and sensitivity. *Psychobiology* 16: 416–25.

Silberstein, S. D., and G. R. Merriam. 1991. Menstrual migraine. *Del. Med. J.* 63:477–84.

Spiegel, A. D. 1988. Temporary insanity and premenstrual syndrome: Medical testimony in an 1865 murder trial. *New York State Journal of Medicine* 88: 482–92.

Tessman, I. 1979. Female sexual activity at ovulation [letter]. *N. Engl. J. Med.* 300:626–27.

Udry, J. R., and N. M. Morris. 1968. Distribution of coitus in the menstrual cycle. *Nature* 220:593–96.

———. 1977. The distribution of events in the human menstrual cycle. *J. Reprod. Fertil.* 51:419–25.

Vierling, J. S., and J. Rock. 1967. Variations in olfactory sensitivity to exaltolide during the menstrual cycle. *J. Appl. Physiol.* 22:311–15.

Weller, L., and A. Weller. 1993. Human menstrual synchrony: a critical assessment. *Neurosci. Biobehav. Rev.* 17:427–39.

Yen, S. S., and R. B. Jaffe. 1986. *Reproductive endocrinology: physiology, pathophysiology and clinical management.* Philadelphia: Harcourt Brace Jovanovich.

OXYTOCIN

Albeck, D., T. Smock, S. Arnold, K. Raese, K. Paynter, and S. Colaprete. 1991. Peptidergic transmission in the brain. IV. Sex hormone dependence in the vasopressin/oxytocin system. *Peptides* 12:53–56.

Angier, N. 1991. A potent peptide prompts an urge to cuddle. (oxytocin) (Science Times pages) *New York Times* ed. 140th. C1–C10.

Ansseau, M., J. J. Legros, C. Mormont, J. L. Cerfontaine, P. Papart, V. Geenen, F. Adam, and G. Franck. 1987. Intranasal oxytocin in obsessive-compulsive disorder. *Psychoneuroendocrinology* 12:231–36.

Argiolas, A. 1992. Oxytocin stimulation of penile erection. Pharmacology, site, and mechanism of action. *Ann. N. Y. Acad. Sci.* 652:194–203.

Argiolas, A., and G. L. Gessa. 1987. Oxytocin: A powerful stimulant of penile erection and yawning in male rats. In *Hypothalamic dysfunction in neuropsychiatric disorders*. D. Nerozzi, F. K. Goodwin, and E. Costa, editors. New York: Raven Press. 153–63.

Arletti, R., C. Bazzani, M. Castelli, and A. Bertolini. 1985. Oxytocin improves male copulatory performance in rats. *Horm. Behav.* 19:14–20.

Arletti, R., A. Benelli, and A. Bertolini. 1990. Sexual behavior of aging male rats is stimulated by oxytocin. *Eur. J. Pharmacol.* 179:377–81.

Arletti, R., A. Benelli, C. Luppi, C. Caroni, and A. Bertolini. 1990. Testosterone modulation of the effect of oxytocin on male sexual behavior. *Neuroscience* 39:(suppl.).

Bicknell, R. J., and G. Leng. 1982. Endogenous opiates regulate oxytocin but not vasopressin secretion from the neurohypophysis. *Nature* 298:161–62.

Bohus, B., I. Urban, T. B. van Wimersma Greidanus, and D. de Wied. 1978. Opposite effects of oxytocin and vasopressin on avoidance behaviour and hippocampal theta rhythm in the rat. *Neuropharmacology* 17:239–47.

Caldwell, J. D., A. J. Prange, and C. A. Pedersen. 1986. Oxytocin facilitates the sexual receptivity of estrogen-treated female rats. *Neuropeptides* 7:175–89.

Carmichael, M. S., R. Humbert, J. Dixen, G. Palmisano, W. Greenleaf, and J. M. Davidson. 1987. Plasma oxytocin increases in the human sexual response. *J. Clin. Endocrinol. Metab.* 64:27–31.

DeVries, G. J., R. M. Buijs, F. W. Van Leeuwen, A. R. Caffe, and D. F. Swaab. 1985. The vasopressinergic innervation of the brain in normal and castrated rats. *J. Comp. Neurol.* 233:236–54.

Elands, J., E. van Doremalen, B. Spruyt, and R. Kloet. 1991. Oxytocin receptors in the rat hypothalamic ventromedial nucleus: a study of possible mediators of female sexual behavior. In *Proceedings / Actes / de la troisième conference internationale sur la vasopressine, Le Corum, Montpellier, France, 5–10 août 1990.* S. Jard and R. L. Jamison, editors. London: J. Libbey Eurotext. 311–19.

Else J. G. E., and P. C. L. Lee, editors. 1984. *Postpartum sexual behavior of American women as a function of absence or frequency of breast feeding: a preliminary communication.* Cambridge; New York: Cambridge University Press.

Eriksson, M., and K. Uvnas-Moberg. 1990. Plasma levels of vasoactive intestinal polypeptide and oxytocin in response to suckling, electrical stimulation of the mammary nerve and oxytocin infusion in rats. *Neuroendocrinology* 51:237–40.

Ferrier, B. M., D. J. Kennett, and M. C. Devlin. 1980. Influence of oxytocin on human memory processes. *Life Sci.* 27:2311–17.

Field, T. M., S. M. Schanberg, F. Scafidi, C. R. Bauer, N. Vega-Lahr, R. Garcia, J. Nystrom, and C. M. Kuhn. 1986. Tactile/kinesthetic stimulation effects on preterm neonates. *Pediatrics* 77:654–58.

Fox, C. A., and G. S. Knaggs. 1969. Milk-ejection activity (oxytocin) in peripheral venous blood in man during lactation and in association with coitus. *J. Endocrinol.* 45:145–46.

Hermes, M. L., R. M. Buijs, M. Masson-Pevet, and P. Pevet. 1988. Oxytocinergic innervation of the brain of the garden dormouse (Eliomys quercinus L.). *J. Comp. Neurol.* 273:252–62.

Hughes, A. M., B. J. Everitt, S. L. Lightman, and K. Todd. 1987. Oxytocin in the central nervous system and sexual behavior in male rats. *Brain Res.* 414:133–37.

Insel, T. R. 1992. Oxytocin—a neuropeptide for affiliation: evidence from behavioral, receptor autoradiographic, and comparative studies. *Psychoneuroendocrinology* 17:3–35.

Insel, T. R., and L. E. Shapiro. 1992. Oxytocin receptors and maternal behavior. *Ann. N. Y. Acad. Sci.* 652:122–41.

Jirikowski, G. F., J. D. Caldwell, C. A. Pedersen, and W. E. Stumpf. 1988. Estradiol influences oxytocin-immunoreactive brain systems. *Neuroscience* 25:237–48.

Johnson, A. E., G. F. Ball, H. Coirini, C. R. Harbaugh, B. S. McEwen, and T. R. Insel. 1989. Time course of the estradiol-dependent induction of oxytocin receptor binding in the ventromedial hypothalamic nucleus of the rat. *Endocrinology* 125:1414–19.

Kim, Y. M., N. Tejani, B. Chayen, and U. L. Verma. 1986. Management of the third stage of labor with nipple stimulation. *J. Reprod. Med.* 31:1033–34.

Legros, J. J., P. Chiodera, V. Geenen, S. Smitz, and R. von Frenckell. 1984. Dose-response relationship between plasma oxytocin and cortisol and adrenocorticotropin concentrations during oxytocin infusion in normal men. *J. Clin. Endocrinol. Metab.* 58:105–9.

Leng, G., and J. A. Russell. 1988. Opioid control of oxytocin secretion. In *Recent progress in posterior pituitary hormones 1988: proceedings of the Satellite Symposium on Posterior Pituitary Hormones, Hakone, Japan, 14–16 July 1988.* S. Yoshida and L. Share, editors. New York: Elsevier Science Pub. Co. 89–96.

Mashini, I. S., L. D. Devoe, J. S. McKenzie, H. A. Hadi, and D. M. Sherline. 1987. Comparison of uterine activity induced by nipple stimulation and oxytocin. *Obstet. Gynecol.* 69:74–78.

McNeilly, A. S., and H. A. Ducker. 1972. Blood levels of oxytocin in the female goat during coitus and in response to stimuli associated with mating. *J. Endocrinol.* 54:399–406.

Melis, M. R., A. Argiolas, R. Stancampiano, and G. L. Gessa. 1988. Oxytocin-induced penile erection and yawning: structure-activity relationship studies. *Pharmacol. Res. Commun.* 20:1117–18.

Melis, M. R., R. Stancampiano, G. L. Gessa, and A. Argiolas. 1992. Prevention by morphine of apomorphine-

and oxytocin-induced penile erection and yawning: site of action in the brain. *Neuropsychopharmacology* 6:17–21.

Murphy, M. R., S. A. Checkley, J. R. Seckl, and S. L. Lightman. 1990. Naloxone inhibits oxytocin release at orgasm in man. *J. Clin. Endocrinol. Metab.* 71:1056–58.

Newton, N. 1978. The role of the oxytocin reflexes in three interpersonal reproductive acts: Coitus, birth and breast-feeding. In *Clinical psychoneuroendocrinology in reproduction*. L. Carenza, P. Pancheri, and L. Zichella, editors. New York: Academic Press. 411–18.

Newton, N., and C. Modahl. 1989. New frontiers of oxytocin research. In *Free woman: women's health in the 1990's*. E. V. Van Hall and W. Everard, editors. Park Ridge, NJ: Parthenon Publ. Group.

Page, S. R., V. T. Ang, R. Jackson, A. White, S. S. Nussey, and J. S. Jenkins. 1990. The effect of oxytocin infusion on adenohypophyseal function in man. *Clin. Endocrinol. (Oxf.)* 32:307–13.

Pedersen, C. A., J. D. Caldwell, G. Peterson, C. H. Walker, and G. A. Mason. 1992. Oxytocin activation of maternal behavior in the rat. *Ann. N. Y. Acad. Sci.* 652:58–69.

Pedersen, C. 1992. *Oxytocin in maternal, sexual, and social behaviors*. New York: New York Academy of Sciences.

Pepe, F., G. Garozzo, N. Rotolo, V. Cali, E. Chirico, V. Leanza, S. Di Mauro, and P. Pepe. 1991. [Breast feeding and pleasure] Allattamento materno e piacere. *Minerva. Ginecol.* 43:115–18.

Schumacher, M., H. Coirini, M. Frankfurt, and B. S. McEwen. 1989. Localized actions of progesterone in hypothalamus involve oxytocin. *Proc. Natl. Acad. Sci. U. S. A.* 86:6798–6801.

Silber, M., O. Almkvist, B. Larsson, and K. Uvnas-Moberg. 1990. Temporary peripartal impairment in memory and attention and its possible relation to oxytocin concentration. *Life Sci.* 47:57–65.

Silber, M., B. Larsson, and K. Uvnas-Moberg. 1991. Oxytocin, somatostatin, insulin and gastrin concentrations vis-a-vis late pregnancy, breastfeeding and oral contraceptives. *Acta Obstet. Gynecol. Scand.* 70:283–89.

Stern, J. M. 1986. Licking, touching, and suckling: contact stimulation and maternal psychobiology in rats and women. *Ann. N. Y. Acad. Sci.* 474:95–107.

Stoneham, M. D., B. J. Everitt, S. Hansen, S. L. Lightman, and K. Todd. 1985. Oxytocin and sexual behavior in the male rat and rabbit. *J. Endocrinol.* 107:97–106.

Suh, B. Y., J. H. Liu, D. D. Rasmussen, D. M. Gibbs, J. Steinberg, and S. S. Yen. 1986. Role of oxytocin in the modulation of ACTH release in women. *Neuroendocrinology* 44:309–13.

PEA

Birkmayer, W., P. Riederer, W. Linauer, and J. Knoll. 1984. L-deprenyl plus L-phenylalanine in the treatment of depression. *J. Neural Transm.* 59:81–87.

DeLisi, L. E., D. L. Murphy, et al. 1984. Phenylethylanine excretion in depression. *Psychiatry Research* 13:193–201.

Ornstein, R., and D. Sobel. 1989. *Healthy Pleasures*. Reading, PA: Addison Wesley.

Reite, M. 1984. Touch, attachment and health: is there a relationship? In *The many facets of touch*. C. C. Brown, editor. Skillman, NJ: Johnson & Johnson.

Romano, M., L. Diomede, G. Guiso, S. Caccia, C. Perego, and M. Salmona. 1990. Plasma and brain kinetics of large neutral amino acids and of striatum monoamines in rats given aspartame. *Food Chem. Toxicol.* 28:317–21.

Schuman, M., M. J. Gitlin, and L. Fairbanks. 1987. Sweets, chocolate, and atypical depressive traits. *J. Nerv. Ment. Dis.* 175:491–95.

Shah, A., and M. Sircar. 1991. Postcoital asthma and rhinitis. *Chest* 100:1039–41.

Weil, A., and W. Rosen. 1983. *Chocolate to morphine: understanding mind-active drugs*. Boston: Houghton-Mifflin.

PHEROMONES

Baskin, A. 1994. Aggressive mail. *Omni* 17:(3)86.

Berliner, D. L., C. Jennings-White, and R. M. Lavker. 1991. The human skin: fragrances and pheromones. *J. Steroid Biochem. Mol. Biol.* 39:671–79.

Bower, B. 1989. Garter snakes yield sexual chemistry. *Science News* 136:55.

Cohn, B. A. 1994. In search of human skin pheromones. *Arch. Dermatol.* 130:1048–51.

Cotterill, J. A. 1992. Dog days and antiandrogens [letter]. *Lancet* 340:986.

Deems, D. A., R. L. Doty, R. G. Settle, V. Moore-Gillon, P. Shaman, A. F. Mester, C. P. Kimmelman, V. J. Brightman, and J. B. Snow. 1991. Smell and taste disorders, a study of 750 patients from the University of Pennsylvania Smell and Taste Center. *Arch. Otolaryngol. Head. Neck Surg.* 117:519–28.

Douglas-Wilson, I. 1971. A human pheromone? [editorial]. *Lancet* 1:279–80.

Dranov, P. 1995. Making sense of scents (a surprisingly scary update!). *Cosmopolitan* 204–7.

Gibbons, B. 1986. The intimate sense of smell. *National Geographic* 324–60.

Gower, D. B., and B. A. Ruparelia. 1993. Olfaction in humans with special reference to odorous 16-androstenes: their occurrence, perception and possible social, psychological and sexual impact. *J. Endocrinol.* 137:167–87.

Hudson, R., G. Gonzalez-Mariscal, and C. Beyer. 1990. Chin marking behavior, sexual receptivity, and pheromone emission in steroid-treated, ovariectomized rabbits. *Horm. Behav.* 24:1–13.

Krantz, M. 1994. Two scents' worth: a new fragrance company takes advantage of pheromones. *Omni* 16:24(1).

Landolt, P. J., and R. R. Heath. 1990. Sexual role reversal in mate-finding strategies of the cabbage looper moth. *Science* 249:1026(3).

Levine, A. 1993. Love potion no. 10: the search for instant sex appeal, or seduction in a can. (man tries unsuccessfully to attract women by wearing scents advertised as aphrodisiacs) *Los Angeles Times* 112:MAG39.

Lothstein, L. M. 1995. An olfactory perversion (osmophilia) successfully treated with Depo-provera and psychotherapy. *Sexual Addiction and Compulsivity* 2:40–53.

Mason, R. T., H. M. Fales, T. H. Jones, L. K. Pannell, J. W. Chinn, and D. Crews. 1989. Sex pheromones in snakes. *Science* 245:290–93.

Ponte, L. 1982. Secret scents that affect behavior. *Reader's Digest* 121–25.

Roelofs, W. L. 1995. Chemistry of sex attraction. *Proc. Natl. Acad. Sci. U. S. A.* 92:44–49.

Schal, C., X. Gu, E. L. Burns, and G. J. Blomquist. 1994. Patterns of biosynthesis and accumulation of hydrocarbons and contact sex pheromone in the female German cockroach, Blattella germanica. *Arch. Insect Biochem. Physiol.* 25:375–91.

Stevens, J. C., and A. D. Dadarwala. 1993. Variability of olfactory threshold and its role in assessment of aging [published erratum appears in Percept Psychophys 1993 Oct; 54(4):562]. *Percept. Psychophys.* 54:296–302.

Udry, J. R. 1988. Biological predispositions and social control in adolescent sexual behavior. *Am. Soc. Rev.* 53:709–22.

Warrick, P. 1993. Call it animal magnetism. (why certain mammals mate for life; includes related article.) *Los Angeles Times* ed. 113th. E1–E5.

PROGESTERONE

Cooper, A. J. 1986. Progestogens in the treatment of male sex offenders: a review. *Can. J. Psychiatry* 31:73–79.

Cooper, A. J., S. Sandhu, S. Losztyn, and Z. Cernovsky. 1992. A double-blind placebo controlled trial of medroxyprogesterone acetate and cyproterone acetate with seven pedophiles. *Can. J. Psychiatry* 37:687–93.

Cordoba, O. A., and J. L. Chapel. 1983. Medroxyprogesterone acetate antiandrogen treatment of hypersexuality in a pedophiliac sex offender. *Am. J. Psychiatry* 140:1036–39.

Feder, H. H., and B. L. Marrone. 1977. Progesterone: its role in the central nervous system as a facilitator and inhibitor of sexual behavior and gonadotropin release. *Ann. N. Y. Acad. Sci.* 286:331–54.

Freeman, E. W., L. Weinstock, K. Rickels, S. J. Sondheimer, and C. Coutifaris. 1992. A placebo-controlled study of effects of oral progesterone on performance and mood. *Br. J. Clin. Pharmacol.* 33:293–98.

Lanthier, A., and V. V. Patwardhan. 1986. Sex steroids and 5-en-3 beta-hydroxysteroids in specific regions of the human brain and cranial nerves. *J. Steroid Biochem.* 25:445–49.

Lovejoy, J., and K. Wallen. 1990. Adrenal suppression and sexual initiation in group-living female rhesus monkeys. *Horm. Behav.* 24:256–69.

Nadler, R. D. 1970. A biphasic influence of progesterone on sexual receptivity of spayed female rats. *Physiol. Behav.* 5:95–97.

Stoll, R., and N. Faucounau. 1991. [Compared action of norethindrone and various steroids on Muller ducts of the female chick embryo] Action comparée de la norethindrone et de divers steroides sur les canaux de Muller de l'embryon femelle de Poulet. *C. R. Seances. Soc. Biol. Fil.* 185:206–10.

van der Schoot, P., and R. Baumgarten. 1992. Interactions between oestradiol and the progesterone antagonist RU-486 in establishing and maintaining female rats' sexual responsiveness: central versus peripheral effects. *Behav. Brain Res.* 47:105–12.

Walker, P. A., and W. J. Meyer III. 1981. Medroxyprogesterone acetate treatment for paraphiliac sex offenders. In *Violence and the violent individual: proceedings of the twelfth annual symposium, Texas Research Institute of Mental Sciences, Houston, Texas, November 1–3, 1979*. T. K. Roberts, K. S. Solway, L. Feldman, and J. R. Hays, editors. New York: Spectrum Publications.

PROLACTIN

Baker, E. R., R. S. Mathur, R. F. Kirk, S. C. Landgrebe, L. O. Moody, and H. O. Williamson. 1982. Plasma gonadotropins, prolactin, and steroid hormone concentrations in female runners immediately after a long-distance run. *Fertil. Steril.* 38:38–41.

Carter, J. N., J. E. Tyson, G. Tolis, S. Van Vliet, C. Faiman, and H. G. Friesen. 1978. Prolactin-screening tumors and hypogonadism in 22 men. *N. Engl. J. Med.* 299:847–52.

Felicetta, J. V. 1991. Determining the cause of hyperprolactinemia. *Family Practice Recertification* 13:(11)74–89.

Forster, C., S. Abraham, A. Taylor, and D. Llewellyn-Jones. 1994. Psychological and sexual changes after the cessation of breast-feeding. *Obstet. Gynecol.* 84:872–76.

Freed, G. L., S. J. Clark, J. Sorenson, J. A. Lohr, R. Cefalo, and P. Curtis. 1995. National assessment of physicians' breast-feeding knowledge, attitudes, training, and experience. *JAMA* 273:472–76.

Mockel, M., L. Rocker, T. Stork, J. Vollert, O. Danne, H. Eichstadt, R. Muller, and H. Hochrein. 1994. Immediate physiological responses of healthy volunteers to different types of music: cardiovascular, hormonal and mental changes. *Eur. J. Appl. Physiol.* 68:451–59.

Newton, N. 1973. Interrelationships between sexual responsiveness, birth and breast feeding. In *Contemporary sexual behaviors: critical issues in the 1970s.* J. Money and J. Zubin, editors. Baltimore: Johns Hopkins University Press. 77–98.

Segraves, R. T., H. W. Schoenberg, and J. Ivanoff. 1983. Serum testosterone and prolactin levels in erectile dysfunction. *J. Sex Marital Ther.* 9:19–26.

Stern, J. M. 1991. Nursing posture is elicited rapidly in maternally naive, haloperidol-treated female and male rats in response to ventral trunk stimulation from active pups. *Horm. Behav.* 25:504–17.

Stern, J. M., M. Konner, T. N. Herman, and S. Reichlin. 1986. Nursing behaviour, prolactin and postpartum amenorrhoea during prolonged lactation in American and !Kung mothers. *Clin. Endocrinol. (Oxf.)* 25:247–58.

Tordjman, G. 1982. Prolactin, sexuality and bromocriptine. In *Sexology: sexual biology, behavior and therapy: selected papers of the 5th World Congress of Sexology, Jerusalem, Israel, June 21–26, 1981.* Z. Hoch and H. Lief, editors. New York: Elsevier. 72–76.

Uvnas-Moberg, K., A. M. Widstrom, S. Werner, A. S. Matthiesen, and J. Winberg. 1990. Oxytocin and prolactin levels in breast-feeding women. Correlation with milk yield and duration of breast-feeding. *Acta Obstet. Gynecol. Scand.* 69:301–6.

Vermesh, M., G. T. Fossum, and O. A. Kletzky. 1988. Vaginal bromocriptine: pharmacology and effect on serum prolactin in normal women. *Obstet. Gynecol.* 72:693–98.

Werner, O. R., R. K. Wallace, B. Charles, G. Janssen, T. Stryker, and R. A. Chalmers. 1986. Long-term endocrinologic changes in subjects practicing the Transcendental Meditation and TM-Sidhi program. *Psychosom. Med.* 48:59–66.

SEROTONIN

Balogh, S., S. E. Hendricks, and J. Kang. 1992. Treatment of fluoxetine-induced anorgasmia with amantadine [letter]. *J. Clin. Psychiatry* 53:212–13.

Baumgarten, H. G., and H. G. Schlossberger. 1984. Anatomy and function of central serotonergic neurons. In *Progress in tryptophan and serotonin research: proceedings.* H. G. Schlossberger and A. Butenandt, editors. New York: W. de Gruyter. 173–88.

Cohen, A. J. 1992. Fluoxetine-induced yawning and anorgasmia reversed by cyproheptadine treatment. *J. Clin. Psychiatry* 53:174.

Costa, E., G. L. Gessa, and M. Sandler. 1974. *Serotonin; new vistas.* New York: Raven Press.

Datla, K. P., S. K. Mitra, and S. K. Bhattacharya. 1991. Serotonergic modulation of footshock induced aggression in paired rats. *Indian J. Exp. Biol.* 29:631–35.

Gessa, G. L., and A. Tagliamonte. 1974. Role of brain monoamines in male sexual behavior. *Life Sci.* 14:425–36.

McCormick, S., J. Olin, and A. W. Brotman. 1990. Reversal of fluoxetine-induced anorgasmia by cyproheptadine in two patients. *J. Clin. Psychiatry* 51:383–84.

Patterson, W. M. 1993. Fluoxetine-induced sexual dysfunction [letter]. *J. Clin. Psychiatry* 54:71

Swenson, J. R. 1993. Fluoxetine and sexual dysfunction [letter]. *Can. J. Psychiatry* 38:297

Tagliamonte, A., P. Tagliamonte, G. L. Gessa, and B. B. Brodie. 1969. Compulsive sexual activity induced by p-chlorophenylalanine in normal and pinealectomized male rats. *Science* 166:1433–35.

Walker, P. W., J. O. Cole, E. A. Gardner, A. R. Hughes, J. A. Johnston, S. R. Batey, and C. G. Lineberry. 1993. Improvement in fluoxetine-associated sexual dysfunction in patients switched to bupropion. *J. Clin. Psychiatry* 54:459–65.

TESTOSTERONE

Angier, N. 1995. Does testosterone equal aggression? Maybe not. *New York Times* A1–C3.

Bachmann, G. A. 1993. Estrogen-androgen therapy for sexual and emotional well-being. *The Female Patient* 18:35+.

Booth, A., and J. M. Dabbs. 1993. Testosterone and men's marriages. *Social Forces* 72:463–77.

Booth, A., G. Shelley, A. Mazur, G. Tharp, et al. 1989. Testosterone, and winning and losing in human competition. *Hormones & Behavior* 23:556–71

Brisson, G. R., A. Quirion, M. Ledoux, D. Rajotte, and J. Pellerin-Massicotte. 1984. Influence of long-distance swimming on serum androgens in males. *Horm. Metab. Res.* 16:160.

Buena, F., R. S. Swerdloff, B. S. Steiner, P. Lutchmansingh, M. A. Peterson, M. R. Pandian, M. Galmarini, and S. Bhasin. 1993. Sexual function does not change when serum testosterone levels are pharmacologically varied within the normal male range. *Fertil. Steril.* 59:1118–23.

Burris, A. S., S. M. Banks, C. S. Carter, J. M. Davidson, and R. J. Sherins. 1992. A long-term, prospective study of the physiologic and behavioral effects of hormone replacement in untreated hypogonadal men. *J. Androl.* 13:297–304.

Burris, A. S., R. H. Gracely, C. S. Carter, R. J. Sherins, and J. M. Davidson. 1991. Testosterone therapy is associated with reduced tactile sensitivity in human males. *Horm. Behav.* 25:195–205.

Carani, C., D. Zini, A. Baldini, L. Della Casa, A. Ghizzani, and P. Marrama. 1990. Effects of androgen treatment in impotent men with normal and low levels of free testosterone. *Arch. Sex. Behav.* 19:223–34.

Clopper, R. R., M. L. Voorhess, M. H. MacGillivray, P. A. Lee, and B. Mills. 1993. Psychosexual behavior in hypopituitary men: a controlled comparison of gonadotropin and testosterone replacement. *Psychoneuroendocrinology* 18:149–61.

Dabbs, J. M. 1993. Salivary testosterone measurements in behavioral studies. *Ann. N. Y. Acad. Sci.* 694:177–83.

Dabbs, J. M., D. de La Rue, and P. M. Williams. 1990. Testosterone and occupational choice: actors, ministers, and other men. *J. Pers. Soc. Psychol.* 59:1261–65.

Dabbs, J. M., R. L. Frady, T. S. Carr, and N. F. Besch. 1987. Saliva testosterone and criminal violence in young adult prison inmates. *Psychosom. Med.* 49:174–82.

Dabbs, J. M., G. J. Jurkovic, and R. L. Frady. 1991. Salivary testosterone and cortisol among late adolescent male offenders. *J. Abnorm. Child Psychol.* 19:469–78.

Dabbs, J. M., Jr. 1993. Salivary testosterone measurements in behavioral studies. *Saliva as a diagnostic fluid. Annals of the New York Academy of Sciences.* 177–83.

Davidson, J. M., M. L. Stefanick, B. D. Sachs, and E. R. Smith. 1978. Role of androgen in sexual reflexes of the male rat. *Physiol. Behav.* 21:141–46.

Davies, R. H., B. Harris, D. R. Thomas, N. Cook, G. Read, and D. Riad-Fahmy. 1992. Salivary testosterone levels and major depressive illness in men [see comments]. *Br. J. Psychiatry* 161:629–32.

Driscoll, R., and C. Thompson. 1993. Salivary testosterone levels and major depressive illness in men [letter]. *Br. J. Psychiatry* 163:122–23.

Ehlers, C. L., K. C. Rickler, and J. E. Hovey. 1980. A possible relationship between plasma testosterone and aggressive behavior in a female outpatient population. In *Limbic epilepsy and the dyscontrol syndrome: proceedings of the 1st International Symposium on Limbric Epilepsy and the Dyscontrol Syndrome, held in Sydney, Australia, February 6–9, 1980.* L. G. Kiloh and M. Girgis, editors. New York: Elsevier North Holland. 183–94.

Fox, C. A., A. A. Ismail, D. N. Love, K. E. Kirkham, and J. A. Loraine. 1972. Studies on the relationship between plasma testosterone levels and human sexual activity. *J. Endocrinol.* 52:51–58.

Gessa, G. L., and A. Tagliamonte. 1974. Role of brain monoamines in male sexual behavior. *Life Sci.* 14:425–36.

Gilbert, R. N., C. W. Graham, and J. B. Regan. 1992. Double-blind, cross-over trial of testosterone enanthate in impotent men with low or low-normal serum testosterone levels. *J. Urology* 147:264A(abstr.)

Gordon, T. P., I. S. Bernstein, and R. M. Rose. 1978. Social and seasonal influences on testosterone secretion in the male rhesus monkey. *Physiol. Behav.* 21:623–27.

Gouchie, C., and D. Kimura. 1991. The relationship between testosterone levels and cognitive ability patterns. *Psychoneuroendocrinology.* 16:323–34.

Hassler, M. 1991. Testosterone and artistic talents. *Int. J. Neurosci.* 56:25–38.

———. 1991. Testosterone and musical talent. *Exp. Clin. Endocrinol.* 98:89–98.

———. 1992. Creative musical behavior and sex hormones: musical talent and spatial ability in the two sexes. *Psychoneuroendocrinology* 17:55–70.

Hassler, M., D. Gupta, and H. Wollmann. 1992. Testosterone, estradiol, ACTH and musical, spatial and verbal performance. *Int. J. Neurosci.* 65:45–60.

Johnson, A. R., and J. P. Jarow. 1992. Is routine endocrine testing of impotent men necessary? *J. Urol.* 147:1542–43; discussion 1543–44.

Kaplan, H. S., and T. Owett. 1993. The female androgen deficiency syndrome. *J. Sex. Marital Ther.* 19:3–24.

Kemper, T. D. 1990. Social structure and testosterone: explorations of the socio-bio-social chain. New Brunswick: Rutgers University Press.

Levitt, A. J., and R. T. Joffe. 1988. Total and free testosterone in depressed men. *Acta Psychiatr. Scand.* 77:346–48.

Margolick, D. 1990. Among lawyers, do litigators have some kind of hormonal edge? Science may deliver the answer. (James M. Dabbs Jr. testing hormonal levels of trial lawyers) (column) *New York Times* 140:B15(N)–B5(L).

Mendoza, S. P., C. L. Coe, E. L. Lowe, and S. Levine. 1978. The physiological response to group formation in adult male squirrel monkeys. *Psychoneuroendocrinology* 3:221–29.

Persky, H., L. Dreisbach, W. R. Miller, C. P. O'Brien, M. A. Khan, H. I. Lief, N. Charney, and D. Strauss. 1982. The relation of plasma androgen levels to sexual behaviors and attitudes of women. *Psychosom. Med.* 44:305–19.

Persky, H., H. I. Lief, D. Strauss, W. R. Miller, and C. P. O'Brien. 1978. Plasma testosterone level and sexual behavior of couples. *Arch. Sex. Behav.* 7:157–73.

Phillips, G. B. 1992. The variability of the serum estradiol level in men: effect of stress (college examinations), cigarette smoking, and coffee drinking on the serum sex hormone and other hormone levels. *Steroids* 57:135–41.

Rose, R. M., I. S. Berstein, and T. P. Gordon. 1975. Consequences of social conflict on plasma testosterone levels in rhesus monkeys. *Psychosom. Med.* 37:50–61.

Rosenfield, R. L., and A. W. Lucky. 1993. Acne, hirsutism, and alopecia in adolescent girls. Clinical expressions of androgen excess. *Endocrinol. Metab. Clin. North Am.* 22:507–32.

Salvador, A., V. Simon, F. Suay, and L. Llorens. 1987. Testosterone and cortisol responses to competitive fighting in human males: A pilot study. *Aggressive Behavior* 13:9–13.

Shoham, S. G., J. J. Askenasy, G. Rahav, F. Chard, A. Addi, and M. Addad. 1988. Violent prisoners. *Med. Law.* 7:247–67.

Toth, I., and I. Faredin. 1985. Steroids excreted by human skin. II. C19-steroid sulphates in human axillary sweat. *Acta Med. Hung.* 42:21–28.

Unknown. 1992. Testosterone replacement looks promising. (Includes related article on Estracombi estrogen-progesterone patch for menopausal symptoms.) *Medical World News* 33:16(1).

Virkkunen, M., R. Rawlings, R. Tokola, R. E. Poland, A. Guidotti, C. Nemeroff, G. Bissette, K. Kalogeras, S. L. Karonen, and M. Linnoila. 1994. CSF biochemistries, glucose metabolism, and diurnal activity rhythms in alcoholic, violent offenders, fire setters, and healthy volunteers. *Arch. Gen. Psychiatry* 51:20–27.

Wincze, J. P., S. Bansal, and M. Malamud. 1986. Effects of medroxyprogesterone acetate on subjective arousal, arousal to erotic stimulation, and nocturnal penile tumescence in male sex offenders. *Arch. Sex. Behav.* 15:293–305.

VASOPRESSIN

Beckwith, B. E., T. V. Petros, D. I. Couk, and T. P. Tinius. 1990. The effects of vasopressin on memory in healthy young adult volunteers. Theoretical and methodological issues. *Ann. N.Y. Acad. Sci.* 579:215–26.

Danguir, J. 1983. Sleep deficits in rats with hereditary diabetes insipidus. *Nature* 304:163–64.

de Vries, G. J., R. M. Buijs, and A. A. Sluiter. 1984. Gonadal hormone actions on the morphology of the vasopressinergic innervation of the adult rat brain. *Brain Res.* 298:141–45.

de Wied, D., O. Gaffori, J. M. van Ree, and W. de Jong. 1984. Central target for the behavioural effects of vasopressin neuropeptides. *Nature* 308:276–78.

Fehm-Wolfsdorf, G., G. Bachholz, J. Born, K. Voigt, and H. L. Fehm. 1988. Vasopressin but not oxytocin enhances cortical arousal: an integrative hypothesis on behavioral effects of neurohypophyseal hormones. *Psychopharmacology (Berl.)* 94:496–500.

Ferkin, M. H. 1992. Time course of androgenic modulation of odor preferences and odor cues in male meadow voles, Microtus pennsylvanicus. *Horm. Behav.* 26:512–21.

Ferkin, M. H., and M. R. Gorman. 1992. Photoperiod and gonadal hormones influence odor preferences of the male meadow vole, Microtus pennsylvanicus. *Physiol. Behav.* 51:1087–91.

Ferkin, M. H., M. R. Gorman, and I. Zucker. 1991. Ovarian hormones influence odor cues emitted by female meadow voles, Microtus pennsylvanicus. *Horm. Behav.* 25:572–81.

———. 1992. Influence of gonadal hormones on odours emitted by male meadow voles (Microtus pennsylvanicus). *J. Reprod. Fertil.* 95:729–36.

Ferkin, M. H., E. S. Sorokin, M. W. Renfroe, and R. E. Johnston. 1994. Attractiveness of male odors to females varies directly with plasma testosterone concentration in meadow voles. *Physiol. Behav.* 55:347–53.

Ferris, C. F., J. Pollock, H. E. Albers, and S. E. Leeman. 1985. Inhibition of flank-marking behavior in golden hamsters by microinjection of a vasopressin antagonist into the hypothalamus. *Neurosci. Lett.* 55:239–43.

Hermes, M. L., R. M. Buijs, M. Masson-Pevet, T. P. van der Woude, P. Pevet, R. Brenkle, and R. Kirsch. 1989. Hibernation requires the absence of central vasopressin release. *European J. Neuroscience* 1:(suppl. 2)287.

———. 1989. Central vasopressin infusion prevents hibernation in the European hamster (Cricetus cricetus). *Proc. Natl. Acad. Sci. U.S.A.* 86:6408–11.

Ikeda, Y., I. Tanaka, Y. Oki, R. Gemmma, H. Morita, K. Komatsu, and T. Yoshimi. 1993. Testosterone normalizes plasma vasopressin response to osmotic stimuli in men with hypogonadism. *Endocr. J.* 40:387–92.

McGaugh, J., editor. 1985. *Neuropeptides, neurotransmitters and memory: Sociosexual aspects.* New York: Elsevier Science Pub. Co.

Moltz, H. 1990. E-series prostaglandins and arginine vasopressin in the modulation of male sexual behavior. *Neurosci. Biobehav. Rev.* 14:109–15.

Murphy, M. R., J. R. Seckl, S. Burton, S. A. Checkley, and S. L. Lightman. 1987. Changes in oxytocin and vasopressin secretion during sexual activity in men. *J. Clin. Endocrinol. Metab.* 65:738–41.

Naylor, A. M., W. D. Ruwe, and W. L. Veale. 1986. Thermoregulatory actions of centrally-administered vasopressin in the rat. *Neuropharmacology* 25:787–94.

Onaka, T., K. Yagi, and M. Hamamura. 1982. Vasopressin: Physical stress potentiates but emotional stress suppresses its secretion. In *Neuroendocrinology of vasopressin, corticoliberin and opiomelanocortins.* A. J. Baert-schi and J. J. Dreifuss, editors. New York: Academic Press. 221–25.

Peck, E. J., A. E. Boyd, and A. M. Gotto, editors. 1979. *Sexual function and brain peptides.* New York: Elsevier North-Holland.

Pitman, R. K., S. P. Orr, and N. B. Lasko. 1993. Effects of intranasal vasopressin and oxytocin on physiologic responding during personal combat imagery in Vietnam veterans with posttraumatic stress disorder. *Psychiatry Res.* 48:107–17.

Sodersten, P., G. J. de Vries, R. M. Buijs, and P. Melin. 1985. A daily rhythm in behavioral vasopressin sensitivity and brain vasopressin concentrations. *Neurosci. Lett.* 58:37–41.

Sodersten, P., M. Henning, P. Melin, and S. Ludin. 1983. Vasopressin alters female sexual behaviour by acting on the brain independently of alterations in blood pressure. *Nature* 301:608–10.

Steele, B. T. 1993. Nocturnal enuresis. Treatment options. *Can. Fam. Physician* 39:877–80.

Sukhai, R. N. 1993. Enuresis nocturna: long term use and safety aspects of minrin (desmopressin) spray. *Regul. Pept.* 45:309–10.

Van den Hooff, P., and I. J. Urban. 1990. Vasopressin facilitates excitatory transmission in slices of the rat dorso-lateral septum. *Synapse* 5:201–6.

Weingartner, H., P. Gold, J. C. Ballenger, S. A. Smallberg, R. Summers, D. R. Rubinow, R. M. Post, and F. K. Goodwin. 1981. Effects of vasopressin on human memory functions. *Science* 211:601–3.

Zorgniotti, A. W., B. Rossman, and M. Claire. 1987. Possible role of chronic use of nasal vasoconstrictors in impotence [letter]. *Urology* 30:594

VIROPAUSE

Aoki, M., and Y. Kumamoto. 1990. [Analyses of factors contributing to the erectile dysfunction with aging by nocturnal penile tumescence, penile blood pressure index and papaverine test]. *Nippon. Hinyokika. Gakkai. Zasshi.* 81:1633–41.

Atala, A., M. Amin, and J. I. Harty. 1992. Diethylstilbestrol in treatment of postorchiectomy vasomotor symptoms and its relationship with serum follicle-stimulating hormone, luteinizing hormone, and testosterone. *Urology* 39:108–10.

Belander, A., B. Candas, A. Dupont, L. Cusan, P. Diamond, J. L. Gomez, and F. Labrie. 1994. Changes in serum concentrations of conjugated and unconjugated steroids in 40- to 80-year-old men. *J. Clin. Endocrinol. Metab.* 79:1086–91.

Chen, I. 1994. Hormone replacement for men? *Hippocrates* 22–25.

Dornan, W. A., and C. W. Malsbury. 1989. Neuropeptides and male sexual behavior. *Neurosci. Biobehav. Rev.* 13:1–15.

Doty, R. L. 1991. Olfactory capacities in aging and Alzheimer's disease. Psychophysical and anatomic considerations. *Ann. N. Y. Acad. Sci.* 640:20–27.

Feldman, H. A., I. Goldstein, D. G. Hatzichristou, R. J. Krane, and J. B. McKinlay. 1994. Impotence and its medical and psychosocial correlates: results of the Massachusetts Male Aging Study. *J. Urol.* 151:54–61.

Gray, A., H. A. Feldman, J. B. McKinlay, and C. Longcope. 1991. Age, disease, and changing sex hormone levels in middle-aged men: results of the Massachusetts Male Aging Study. *J. Clin. Endocrinol. Metab.* 73:1016–25.

Hill, A. M. 1993. *Viropause/andropause: the male menopause: emotional and physical changes mid-life men experience.* Far Hills, NJ: New Horizon Press.

Kaiser, F. E., and J. E. Morley. 1994. Gonadotropins, testosterone, and the aging male. *Neurobiol. Aging* 15:559–63.

Kaufman, J. M., F. D. Borges, W. P. Fitch, R. A. Geller, M. B. Gruber, J. G. Hubbard, D. L. McKay, J. P. Tuttle, and F. R. Witten. 1993. Evaluation of erectile dysfunction by dynamic infusion cavernosometry and cavernosography (DICC). Multi-institutional study. *Urology* 41:445–51.

Labrie, F., A. Dupont, A. Belanger, R. St-Arnaud, M. Giguere, Y. Lacourciere, J. Emond, and G. Monfette. 1986. Treatment of prostate cancer with gonadotropin-releasing hormone agonists. *Endocr. Rev.* 7:67–74.

Longcope, C., S. R. Goldfield, D. J. Brambilla, and J. McKinlay. 1990. Androgens, estrogens, and sex hormone-binding globulin in middle-aged men. *J. Clin. Endocrinol. Metab.* 71:1442–46.

Loprinzi, C. L., R. M. Goldberg, J. R. O'Fallon, S. K. Quella, A. W. Miser, L. A. Mynderse, L. D. Brown, L. K. Tschetter, M. B. Wilwerding, and M. Dose. 1994. Transdermal clonidine for ameliorating post-orchiectomy hot flashes. *J. Urol.* 151:634–36.

McLeod, D. G. 1993. Antiandrogenic drugs. *Cancer* 71:1046–49.

Morales, A. J., J. J. Nolan, J. C. Nelson, and S. S. Yen. 1994. Effects of replacement dose of dehydroepiandrosterone in men and women of advancing age. *J. Clin. Endocrinol. Metab.* 78:1360–67.

Nofzinger, E. A., C. F. Reynolds, M. E. Thase, E. Frank, J. R. Jennings, A. L. Fasiczka, L. R. Sullivan, and D. J. Kupfer. 1995. REM sleep enhancement by bupropion in depressed men. *Am. J. Psychiatry* 152:274–76.

Phillips, G. B., T. Y. Jing, L. M. Resnick, M. Barbagallo, J. H. Laragh, and J. E. Sealey. 1993. Sex hormones and hemostatic risk factors for coronary heart disease in men with hypertension. *J. Hypertens.* 11:699–702.

Radlmaier, A., K. Bormacher, and F. Neumann. 1990. Hot flushes: mechanism and prevention. *Prog. Clin. Biol. Res.* 359:131–40; discussion 141–5.

Redmond, D. E., T. R. Kosten, and M. F. Reiser. 1983. Spontaneous ejaculation associated with anxiety: psychophysiological considerations. *Am. J. Psychiatry* 140:1163–66.

Solstad, K., and K. Garde. 1992. Middle-aged Danish men's ideas of a male climacteric—and of the female climacteric. *Maturitas* 15:7–16.

Steiger, A., U. von Bardeleben, J. Guldner, C. Lauer, B. Rothe, and F. Holsboer. 1993. The sleep EEG and nocturnal hormonal secretion studies on changes during the course of depression and on effects of CNS-active drugs. *Prog. Neuropsychopharmacol. Biol. Psychiatry* 17:125–37.

Szarvas, F. 1992. [The male climacteric from the practical viewpoint] Mannliches Klimakterium aus praktischer Sicht. *Wien. Med. Wochenschr.* 142:100–103.

Tsitouras, P. D., C. E. Martin, and S. M. Harman. 1982. Relationship of serum testosterone to sexual activity in healthy elderly men. *J. Gerontol.* 37:288–93.

ADDITIONAL READING

Ackerman, D. 1990. *A natural history of the senses.* New York: Random House.

Barnhart, E. R., and B. B. Huff. 1994. *Physicians' desk reference: PDR.* Oradell, NJ: Medical Economics.

Beck, A. M., and A. H. Katcher. 1984. *Between pets and people: the importance of animal companionship.* New York: Perigee Books.

Berscheid, E., and E. Hatfield. 1978. *Interpersonal attraction.* Reading, PA: Addison-Wesley Pub. Co.

Bowlby, J. 1969. *Attachment and loss.* New York: Basic Books.

———. 1988. *A secure base: clinical applications of attachment theory.* London: Routledge.

Brownmiller, S. 1975. *Against our will: men, women and rape.* New York: Simon and Schuster.

Chopra, D. 1993. *Ageless body, timeless mind: the quantum alternative to growing old.* New York: Harmony Books.

Cohen, S. S. 1987. *The magic of touch.* New York: Harper & Row.

Colbern, D. L., and W. H. Gispen, editors. 1988. *Neural mechanisms and biological significance of grooming behavior.* New York: New York Academy of Sciences. 525.

Crenshaw, T. L. 1983. *Bedside manners: your guide to better sex.* New York: McGraw-Hill.

Crenshaw, T. L., and J. Goldberg. 1996. *Sexual Pharmacology.* New York: Norton.

Cutler, W. B. 1991. *Love cycles: the science of intimacy.* New York: Villard Books.

DeCherney, A. H., and M. L. Pernoll. 1994. *Current Obstetric and Gynecological Diagnosis and Treatment.* East Norwalk: Appleton & Lange.

Eibl-Eibesfeldt, I. 1974. *Love and hate; the natural history of behavior patterns.* New York: Schocken Books.

Fisher, H. E. 1992. *Anatomy of Love: The Natural History of Monogamy, Adultery, and Divorce*. New York: Norton.

Graedon, J., and T. Graedon. 1991. *Graedon's best medicine: from herbal remedies to high-tech RX breakthroughs*. New York: Bantam Books.

Greer, G. 1992. *The change: women, aging, and the menopause*. New York: Distributed by Random House.

Harlow, H. F., and C. Mears. 1979. *The human model: primate perspectives*. New York: Wiley.

Hill, A. M. 1993. *Viropause/andropause: the male menopause: emotional and physical changes mid-life men experience*. Far Hills, NJ: New Horizon Press.

Hooper, A. 1980. *The body electric: a unique account of sex therapy for women*. London, England: n.p.

Janiger, O., and P. Goldberg. 1993. *A different kind of healing: doctors speak candidly about their successes with alternative medicine*. New York: Putnam.

Kaplan, H. S. 1987. *Sexual aversion, sexual phobias, and panic disorder*. New York: Brunner/Mazel.

Keeton, K., and Y. Baskin. 1985. *Woman of tomorrow*. New York: St. Martin's/Marek.

Keirsey, D., and M. Bates. 1984. *Please understand me: character & temperament types*. Del Mar, CA: Prometheus Nemesis Book Company.

Kinsey, A. C., W. B. Pomeroy, and C. E. Martin. 1948. *Sexual behavior in the human male*. Philadelphia: W. B. Saunders Co.

Krieger, D. 1987. *Living the therapeutic touch: healing as a lifestyle*. New York: Dodd, Mead.

Laumann, E., R. Michael, and S. Michaels. 1994. *The social organization of sexuality: sexual practices in the United States*. Chicago: University of Chicago Press.

Leiblum, S. R., and R. Rosen. 1988. *Sexual desire disorders*. New York: Guilford.

Liebowitz, M. R. 1983. *The chemistry of love*. Boston: Little, Brown.

Lynch, J. J. 1979. *The broken heart: the medical consequences of loneliness*. New York: Basic Books.

Masters, W. H., and V. E. Johnson. 1966. *Human sexual response*. Boston: Little, Brown.

Montagu, A. 1986. *Touching: the human significance of the skin*. New York: Harper & Row.

Moore, T. 1994. *Soul mates: honoring the mysteries of love and relationship*. New York: HarperCollins Publishers.

Morris, D. 1967. *The naked ape: a zoologist's study of the human animal*. New York: McGraw-Hill.

————. 1971. *Intimate behaviour*. New York: Bantam Books.

————. 1985. *Bodywatching: a field guide to the human species*. London: Cape.

Motta, M. 1991. *Brain endocrinology*. New York: Raven Press.

Newton, N. 1955. *Maternal emotions; a study of women's feelings toward menstruation, pregnancy, childbirth, breast feeding, infant care, and other aspects of their femininity*. New York: P. B. Hoeber.

————. 1990. *Newton on birth and women: selected works of Niles Newton, both classic and current*. Seattle: Birth & Life Bookstore.

Older, J. 1982. *Touching is healing*. New York: Stein and Day.

Ornstein, R., and D. Sobel. 1989. *Healthy pleasures*. Reading, PA: Addison Wesley.

Parker, L. N. 1989. *Adrenal androgens in clinical medicine*. San Diego: Academic Press.

Patterson, M. L., J. L. Powell, and M. G. Lenihan. 1986. Touch, compliance, and interpersonal affect. *Journal of Nonverbal Behavior* 10:41–50.

Peck, M. S. 1978. *The road less traveled: a new psychology of love, traditional values and spiritual growth*. New York: Simon & Schuster.

Pelton, R. 1986. *Mind food and smart pills: nutrients and drugs that increase intelligence and prevent brain aging*. Poway, CA: T & R Publishers.

Person, E. S. 1988. *Dreams of love and fateful encounters: the power of romantic passion*. New York: Norton.

Pisano, M. D., S. M. Wall, and A. Foster. 1986. Perceptions of nonreciprocal touch in romantic relationships. *Journal of Nonverbal Behavior* 10:29–39.

Raines, H. 1993. *Fly fishing through the midlife crisis*. New York: Morrow.

Reite, M. 1984. Touch, attachment and health: is there a relationship? In *The many facets of touch*. C. C. Brown, editor. Skillman, NJ: Johnson & Johnson.

Reynolds, M. 1990. *Erotica: women's writing from Sappho to Margaret Atwood*. New York: Fawcett Columbine.

Rountree, C. 1993. *On women turning 50: celebrating mid-life discoveries*. New York: HarperCollins.

Sandler, M., and G. L. Gessa, editors. 1975. *Sexual behavior: pharmacology and biochemistry*. New York: Raven Press.

Sheehy, G. 1993. *The silent passage: menopause*. New York: Pocket Books.

Solomon, M. F. 1994. *Lean on me: the power of positive dependency in intimate relationships*. New York: Simon & Schuster.

Spitz, R. A. 1957. *No and yes: on the genesis of human communication*. New York: International Universities Press.

————. 1965. *The first year of life: a psychoanalytic study of normal and deviant development of object relations [by] Rene A. Spitz, in collaboration with W. Godfrey Cobliner*. New York: International Universities Press.

Tannen, D. 1990. *You just don't understand: women and men in conversation.* New York: Ballantine Books.

Turek, F. W. 1992. Biological rhythms in reproductive processes. *Horm. Res.* 37 (suppl.) 3:93–98.

Walton, A. H. 1958. *Aphrodisiacs: from legend to prescription: a study of aphrodisiacs throughout the ages, with sections on suitable food, glandular extracts, hormone stimulation and rejuvenation.* Westport, CT: Associated Booksellers.

Wood, E. J., and F. W. Wood. 1992. *She said, he said: what men and women really think about money, sex, politics and other issues of essence.* Detroit: Visible Ink Press.

Woody, J. D. 1992. *Treating sexual distress: integrative systems therapy.* Newbury Park, CA: Sage Publications.

Yen, S. S. C., and R. B. Jaffe. 1991. *Reproductive endocrinology: physiology, pathophysiology, and clinical management.* Philadelphia: Saunders.

Zilbergeld, B. 1992. *The new male sexuality.* New York: Bantam.

INDEX

ACTH, 77
addictions
    dopamine and, 7
    to touch, 91
adolescence
    estrogen during, 31–32
    sexual development during, 27–34
    sexual intercourse during, 33–34
    testosterone during, 29–31
    touch during, 32
adrenal glands, 77, 79, 122, 225
adrenaline reflex, 217–18, 240–41, 282
afterglow, and oxytocin, 114
age differences, 20–24
aggression. *See also* sex drive, aggressive
    aggressive sex drive and, 140–42
    serotonin levels and, 6
    testosterone and, 5, 30–31, 140–51
    of women, 184–85
    women's response to, 184–85
aging. *See also* longevity
    attraction to younger partner and, 20–22
    deprenyl and, 162
    DHEA and, 11, 84, 211–12, 229, 273–74, 292, 294–95
    dopamine and, 287
    growth hormone and, 273
    and hormonal changes, 228–29
    and hormonal patterns, 11
    and lack of touch, 92
    melatonin and, 289, 290
    and normal male sex changes, 218
    normal stages of, versus disease, 281
    sexual distinctions and, 283, 284–85
    sexual fitness and, 282–83
    testosterone and, 211–12
    as treatable, 280–81
    vasopressin and, 102
    viropause and, 239–40

alcohol
    DHEA and, 84
    oxytocin and, 117
    serotonin and, 132
Alzheimer's disease, melatonin and, 289
androgens
    DHEA as, 77, 79
    progesterone as, 175
andropause. *See* viropause
androsterone, 68
antidepressant(s)
    estrogen as, 253–54
    PEA levels and, 5
    progesterone as, 177–78, 183
    testosterone as, 6, 124, 144–46, 156, 272
antioxidant, melatonin as, 289
aphrodisiacs, 54–89
    DHEA as, 77, 78–83
    pheromones as (*see* pheromones)
aromatherapy, 69
arthritis, and sex, 279
attitude, and longevity, 306–8
aversive drive, 169, 174

Baulieu, Etienne-Emile, 292
Berliner, David, 65
Blackman, Marc R., 290
bioadhesive technology, 300
biofeedback, and DHEA levels, 86
birth control pills, 237, 258
    low-dose, 248–49
    progesterone and, 181–84
birth process
    oxytocin and, 98
    vasopressin and, 102–3
body heat, 112
body weight. *See* weight
bonding
    DHEA and, 11

mother-infant, 100–101
oxytocin and, 4, 37, 44, 93, 97–98
touch and, 90–91
vasopressin and, 93–94
Booth, Alan, 148
brain
anti-aging supplements and, 297
DHEA and, 77, 81–82, 103, 295
facial expressions and, 306
female sex hormones and, 296–97
limbic activity, 86, 103
sex differences in aging of, 296–97
vasopressin and, 102–3, 288–89
breast cancer
estrogen replacement therapy and, 260–65
melatonin and, 289
risk of, and type of ERT, 253–66
breast-feeding. See nursing
Brenner, Paul, 235
bupropion (Wellbutrin), 82–83, 161, 239
Burke, Billie, 277

calcitonin, 72, 268–69
calcium supplements, 268
cardiovascular health, 238, 265
DHEA and, 292
castration
progesterone and, 175
removal of ovaries as, 223
cataracts, melatonin and, 289
Cetel, Nancy S., 235–36
chemotherapy, 226
childbirth, and oxytocin, 4
childhood. See also infants
sexuality during, 24–27
touch deprivation during, 107–8
chocolate, 5, 58
cholesterol, and testosterone, 123
circadian rhythms (body clock), 289
climacteric. See viropause
cognition
estrogen and, 253
vasopressin and, 102
collagen reduction, 257
commitment
chemical, involuntary, 92–93
lack of, 147–48, 219
oxytocin and, 94–95
sex differences in, 38
subliminal scents and, 106–7
testosterone and, 143–44
during thirties, 39, 41–42
during twenties, 38–39
vasopressin levels and, 93–94
communication, importance of, 303
compatibility
age combinations and, 20–24
emotional, 20, 21, 22, 23
during forties, 43
sexual, 20, 21, 23

Cooke, Cynthia W., 259
Coolidge Effect, 144, 219
cortisol, and music, 88
cosmetic benefits of estrogen, 257
courtship feeding, 166–67
Cutler, Richard, 291
Cutler, Winnifred Berg, 304
cycles, hormonal, 9–13, 189–94, 297–99. See also
menstrual cycle; testosterone, cycles
medications and, 298–99

Dabbs, James, Jr., 149–50
dehydroepiandrosterone. See DHEA
deprenyl, 161–62, 287, 306
depression
lack of touch and, 92
melatonin and, 289
progesterone and, 183
vasopressin and, 289
and viropause, 238–39
detachment, during sex, 36
DHEA (dihydroepiandrosterone), 73–88, 89
aging and, 211–12, 229, 292, 294–95
bonding and, 11
brain and, 77, 81–82, 103
bupropion and, 82–83
cycles, 11
effects on sexual drive, 77, 78–83
ERT and, 269, 285
estrogen and, 168
exercise and, 267
fat metabolism and, 79–80
influence on health, 85–87
influences on, 84
menopause and, 273–74
music and, 88
orgasm and, 14
pheromones derived from, 4, 65
production of, 77, 79
profile, 75–76
roles of, 75–76
sex drive and, 4, 77, 78–83, 110
during sixties, 49
supplements, 86–87, 89
vasopressin and, 103
DHEAS. See DHEA
diabetes, melatonin and, 289
diet
menopause and, 266–67
and serotonin, 7
diet drinks, 62
divorce
sexual stages and, 20
testosterone and, 148
dopamine, 5, 109, 111, 286–87
music and, 88
PEA and, 136
profile, 135–36
raising levels of, 161
role of, 7

sexual desire and, 83, 160, 161
testosterone therapy and, 160
drugs, 283
anti-aging, 286–95
delivery methods of, 299–301
dosage of, and estrogen, 299
interactions of, 283

ejaculation, refractory period, 142–43
endometrial cancer, and estrogen replacement
therapy, 270
endorphins, 279
positive thinking and, 306
touch and, 92
erections
aging and, 218
hormones involved in, 112–13
during sixties, 49
testosterone level and, 140
erogenous zones, and DHEA, 80–81
Esquivel, Laura, 58, 63
estradiol, 264
nonoral, 265
estrogen, 191. See also estrogen replacement
therapy
during adolescence, 31–32
during forties, 46
LHRH and, 128
in life cycle, 168, 169
lordosis and, 166
in males, 31–32
new drug delivery methods for, 300
oxytocin and, 102, 256
profile, 170–71
progesterone and, 178, 179
prolactin and, 8
psychological effects in twenties, 35
receptive sex drive and, 5, 21, 35, 41, 166–69,
172
role of, 5
sexual intercourse and, 304
vagina and, 215
withdrawal of, during menopause, 207–10, 213,
214, 215, 216, 253–57, 262–63
estrogen replacement therapy. See also hormone
replacement therapy
benefits of, 253–57, 265
breast cancer risk and, 264–66
calcium supplements in, 268
considerations and drawbacks of, 257–66
DHEA and, 269, 285
effects of, 252
endometrial cancer risk and, 270
melatonin and, 290
mood disorders and, 253–54, 255
progesterone replacement and, 255, 270
relative risk of, 259–60, 261–62
short-term trials of, 266
side effects, 257–58
testosterone and, 255, 269

estrone (Premarin), 262, 264–65
exaltolide, 71–72
exercise, 267
eye contact, 57–58, 103

fantasies
PEA and, 59–60, 61–62
of rape, 185–87
of seduction, 186
fat cells, and estrogen levels, 169
fatigue, chronic, 38
and gender differences, 47–48
Fisher, Helen E., 166
fluoxymesterone, 272
flutamide (Proscar), 246
forties, sex in, 43–47
FSH, 248

genetic disorders, 147
Greer, Germaine, 204, 284
growth hormone, 228–29, 273–74
benefits of, 290–92, 293–94
melatonin's effect on, 290
profile, 293–94
supplements, 274
G-spot, 226

healing, and touch, 117
health
DHEA and, 85–87
REM sleep and, 103
touch and, 95, 117
hormone replacement therapy, 13, 206–7, 233–38.
See also estrogen replacement therapy
considerations and drawbacks of, 285
lifespan and, 235, 260
new drug combinations for, 270–71
oxytocin and, 102
quality of life and, 235, 263
side effects, 247
value of, 208
when to begin, 248
hormones. See also specific hormones
overwhelming effects of, 2–3
sexual arousal and, 1–3
hot flushes, 207–8
hugging, 90, 117. See also touch
hysterectomies, 223–27, 258
effects of, on sexual pleasure, 226
osteoporosis and, 226
symptoms following, 224–26
testosterone replacement and, 270
types of procedures, 223–25

impotence, psychological, 217–19. See also sexual
dysfunction
incontinence, solutions for, 255–56
infants. See also nursing
scent-memory of, 100–101
sexuality of, 24

insomnia, melatonin as treatment for, 290
interdependency, 303, 304–5
  same-sex couples and, 305

Jensvold, Margaret, 298
jet lag, melatonin as cure for, 289

Kinsey, Alfred, 25, 280
Konopka, June, 198, 266–67

Landers, Ann, 171
Lawrence, D. H., 14
LH, 191–92, 229, 295
LHRH, 72, 295
  cycle of, 127–29
  estrogen and, 128
  PMS and, 199, 200
  proceptivity and, 174
  profile, 126–27
  progesterone and, 178
  replacement therapy, 159–60
  sex drive and, 159–60
  testosterone and, 139
LHRH-LH, 127, 156
Liebowitz, Michael, 58
liver, and estrogen, 169
longevity, 276–309, See also aging
  anti-aging drugs and, 286–95
  attitude and, 306–8
  DHEA and, 285, 292, 294–95
  and HRT, 235, 260
  masturbation and, 281–82
  melatonin and, 289, 290
  sex differences in, 296–97
  sexual activity and, 278–83
  sexual fitness and, 282–83
  testosterone replacement therapy and, 285
  touch and, 278, 279
  weight and, 305
lordosis response, 165–66, 173–74
love
  age differences and, 22–23
  as aphrodisiac, 89
  chemistry of, 53–54
  at first sight, 50–51
  obstacles overridden by, 32–33, 39
  during sixties, 50–51
love potions, 54–55

MAO inhibitors, 62, 254
marijuana, 63
masturbation, 37, 118, 119
  during adolescence, 30
  avoiding impotence using, 241
  during childhood, 25
  DHEA and, 78
  longevity and, 281–82
  PMS and, 200
  during premenstrual period, 194
  testosterone and, 122, 124

medications. See drugs
meditation, transcendental, 86, 161
medroxyprogesterone, 154–55, 179, 270–71
megestrol (Megace), 155, 246
melatonin, 289–90
menopause, 13, 169, 203–10. See also estrogen
    replacement therapy; hormone replacement
    therapy
  chemotherapy and, 226
  DHEA and, 79, 273–74
  diet and, 266–67
  as endocrine-deficiency disease, 235
  estrogen withdrawal during, 207–10, 213, 214,
      215, 216
  exercise and, 267
  hormonal changes during, 228–29
  hormonal replacement therapy and, 13
  male (see viropause)
  medical mismanagement of, 209–10, 216
  melatonin and, 290
  mood disorders and, 254
  "natural" approach to, 204–5, 258–59, 266–67,
      285
  physicians and, 204, 224, 234, 236–38
  premature, 227
  process of, 207–8
  progesterone and, 184
  relationship conflicts and, 205–7, 216–17
  sexual consequences of, 213–17
  surgical, 223–27
  symptoms of, 207–10
  testosterone supplements for, 271–73
  transition to, 46
  vaginal changes during, 214–15
menstrual cycle/menstruation, 12, 13
  animal reactions to, 67–69
  medications and, 298–99
  perspiration odor and, 66–67
  and receptive sex drive, 194
  sexual peaks during, 15–16, 191–94
  surgery outcome and, 298–99
  synchronization of, 66–67
menstruation, 12, 13
  animal reactions to, 67–69
methyltestosterone, 272
Mifepristone, 183
moods
  estrogen and, 253–54
  premenstrual, 196–97
  progesterone and, 182
  testosterone and, 9–10, 16, 144–46, 156
Morris, Desmond, 81
music, effects on romance, 87–88

nasal sprays, 72–73, 89, 115, 300
Newton, Niles, 97
nipple stimulation, 98, 111, 112, 256
Norplant implant, 181–82
nursing
  oxytocin and, 98, 115, 183

progesterone during, 183
prolactin and, 8

oophorectomy, 225
orgasm, 13–14
  DHEA and, 81–82
  hormones involved in, 113–14
  limbic activity and, 103
  oxytocin and, 4, 114
  PEA and, 5
  physiological changes during, 13–14
  during pregnancy, 181
  during premenstrual period, 193–94
osteoporosis
  DHEA and, 86–87
  estrogen derivation and, 226
  exercise and, 267
  prevention of, 268–70
ovaries, removal of, 223, 225–26
ovulation, 59, 74, 192
oxytocin, 72, 110, 111
  afterglow and, 114
  birth process and, 98
  bonding and, 4, 44, 93, 97–98
  childbirth and, 4
  commitment and, 94–95
  date rape and, 187–88
  defined, 97
  deprivation of, 107–8
  estrogen and, 102, 256
  hormone replacement therapy and, 102
  influencing, 95, 114–17
  longevity and, 288
  lordosis and, 166, 173–74
  during menopause, 214
  nasal spray, 115
  nipple stimulation and, 98
  nursing and, 183
  orgasm and, 114
  parenting and, 101–2
  profile, 96–97
  receptivity and, 111, 113
  "skin hunger" and, 108–9
  surges of, 10
  touch and, 33, 34, 37, 38, 40, 43, 44, 49–50, 88, 92, 93, 96–97, 115–17, 256

parenting, and oxytocin, 101–2
Parkinson's disease, 138, 161–62, 306
  melatonin and, 289
PEA (phenylethylamine), 191–92
  as aphrodisiac, 55, 56
  dopamine and, 136
  effects of, 56
  mania and schizophrenia and, 59
  medications influencing, 62–63
  methods of raising levels of, 59–63
  music and, 88
  orgasm and, 14
  profile, 56

roles of, 4–5, 56
  visual component of love and, 57–58
Pearson, Durk, 115, 288
penile implants, 246
performance anxiety, 217–18, 240
perfumes, and pheromones, 65, 69
perimenopause, 247–51
perspiration, 64, 66–67
  male, and menstrual cycles, 66–67, 74
  during menopause, 207
pets, and touch, 40, 49–50, 116
phenylethylamine. See PEA
phenylpropanolamine (PPA), 255–56
pheromones, 63–73, 129
  defined, 64
  DHEA as producer of, 74, 77, 78–83
  maternal role of, 4
  menopause and, 213
  sex differences in reacting to, 66
  territorial, 106–7
  types of, 64–65
  vomero nasal receptors and, 65–66
physicians, and women, 204, 215, 223, 224, 225, 226, 227, 234, 236–38
phytoestrogens, 266–67
pineal gland, 289
pitocin, 98
pleasure, sexual, and dopamine, 133–38
PMS. See premenstrual syndrome
pornography, 60–61, 219
pregnancy. See also nursing
  DHEA during, 74
  progesterone and, 179–80
  prolactin and, 180
  sex during, 180–81
premenstrual syndrome (PMS), 12, 37, 198–200
  alleviating symptoms of, 197–200
  charting, 197
  hormone replacement therapy and, 270–71
  masturbation and, 200
  melatonin and, 289
  sexual desire during, 193–94
  surgery outcome and, 298–99
professional help, seeking, 301
progesterone, 78, 175–84
  aversive sex drive and, 175, 177–79
  benefits of taking, 249
  birth control pills and, 181–84
  castration and, 175
  as depressant, 177–78, 183
  estrogen and, 178
  menopause and, 184
  menstruation and, 194
  new drug delivery methods for, 300
  profile, 176–77
  role of, 7–8
  sex drive and, 175–79, 182–84
  synthetic, 179
  testosterone and, 124–25

prolactin
  melatonin and, 290
  nursing and, 8
  profile, 99–100
prostaglandins, 197, 198, 200
prostate, enlargement of, 227–28, 245
prostate cancer, 155, 245–46
prostatectomy, 227–28, 245
Provera, in HRT, 270–71
Prozac, 6, 132, 133
puberty, 27–28

quality of life, and estrogen replacement therapy,
    235, 263
Quigley, M. E. Ted, 235

Raines, Howell, 210
rape
  date, 148–49, 151, 187–88
  forms of, 187–89
  sexual response during, 188–89
  violent, 149–50, 154, 188–89
  women's fantasies of, 185–87
Realm (Erox Corp.), 65, 69
receptivity, sexual. See sex drive, receptive
relationship conflicts
  careers and, 222–23
  estrogen deficit and, 254–57
  during forties, 44
  hormones and, 9–10, 12
  menopause and, 205–7, 216–17
  post-menopausal/-viropausal, 229
  sexual passages/stages and, 18–19, 20–22
  time-management and, 37–38, 40–41, 302
  during twenties, 36–38
  viropause and, 210–211, 217, 219–22, 234–35
relationship happiness
  interdependency and, 303
  maintaining, 301–3
responsiveness, sexual
  peak, during thirties, 41
  serotonin level and, 132
Robards, Karen, 1
Robbins, Tim, 73–74
romance novels, 59–61
RU-486, 183–84

scent(s), 63–73. See also pheromones; smell,
    sense of
  DHEA and, 80–81
  estrogen and, 5, 213
  infants and, 100
  pheromones and, 4
  subliminal effects of, 106
  territorial, 106
scent memory, in infants, 100
seduction, 164–66
  hormones involved in, 109–13
  proceptive sex drive and, 166, 169
self-esteem, and testosterone, 146–47, 150–51, 153

serotonin
  aggression and, 6
  aggressive sex drive and, 130–32
  alcohol and, 132
  dieting and, 7
  modulating function of, 131–32
  PCPA and, 133, 137
  profile, 130–31
  responsiveness and, 132
  sex differences and, 132
  sex drive and, 6
sex differences, 17
  hormones and, 11–12
  in longevity, 296–97
  melatonin and, 290
  regarding commitment, 38
  respecting, 302
sex drive. See also sex drive, aggressive; sex drive,
    aversive; sex drive, men's; sex drive,
    proceptive; sex drive, receptive; sex drive,
    women's
  DHEA and, 4, 73–89
  difficulties in studying, 14–15
  dopamine and, 7
  estrogen and, 5
  PEA and, 4–5
  peaks (see sexual peaks)
  pheromones and, 4
  progesterone and, 7, 182–84
  prolactin and, 8
  serotonin and, 6
  testosterone and, 5–6
  vasopressin and, 8–9
sex drive, aggressive, 118–63. See also testosterone
  aggressive behavior and, 140–42
  controlling hormones of, 125, 126–38
  dopamine and, 134–38
  LHRH and, 126–29, 159–60
  lowering, 153–55
  raising, 155–59
  serotonin as modulator of, 130–32
  testosterone and, 119–25
  vasopressin and, 125
  in women, 169, 172
sex drive, aversive, 174
  progesterone and, 175, 177–79
sex drive, men's, 121–24, 127–28, 129, 133, 138–40,
    142–46, 150–57. See also sex drive, aggressive
sex drive, proceptive, 165–66, 173–74
sex drive, receptive, 125, 164–202
  estrogen and, 5, 21, 35, 41, 166–69, 172
  menstruation and, 194
  oxytocin and, 111, 113
  peak, during twenties, 41
  progesterone and, 180
sex drive, women's, 164–201. See also sex drive,
    receptive
  bupropion and, 82–83
  complexity of, 169, 171–72
  during first two weeks of menstrual cycle, 191

during menstruation, 194
during ovulation, 191–92
peaks, 191–94
during PMS, 193–94
during second half of menstrual cycle, 192–93
types of, 169
sexual arousal, physiological effects of, 1
sexual cycles, 9–13
  DHEA and, 11
  length of, 9
  men's, 13
  testosterone and, 10, 16
  women's, 189–94
sexual development, adolescent, 27–34
sexual dysfunction
  bupropion treatment for, 82–83
  DHEA levels and, 212
  drug treatments for, 159
  estrogen replacement therapy and, 254–57
  menopause and, 213–17
  testosterone for, 156–57, 159
  various treatments for, 161–62
  viropause and, 217–23, 239–41
sexual fitness, maintaining, 282–84
sexual intercourse
  in adolescence, 33–34
  delay of menopause and, 304
  elderly and, 278–82
  estrogen levels and, 304
  hormones involved in, 113–14
  testosterone levels and, 304
sexual patterns, as changing throughout life, 19
sexual peaks, 13–16, 52
  versus cycles, 20
  hormonal, women's, 15–16, 189–94
sexual stages, 18–52
  childhood (birth to puberty), 24–27
  at fiftysomething, 47–48
  at fortysomething, 43–47
  hormonal forces and, 23
  in men versus women, 20
  psychological components of, 19
  relationship conflicts and, 18–19, 20–22
  as rites of passage, 23
  at sixtysomething and beyond, 48–51, 276–309
  in teenagers, 27–34
  testosterone in, 123–24
  at thirtysomething, 39–43
  transitions between, 20, 51–52
  traumatic events and, 23
  at twentysomething, 34–39
Sheehy, Gail, 209–10, 223, 234
single men, thirtysomething, 41, 42–43
Sintocinon, 115
sixties and beyond, sex in, 51, 276–309
skin, 80–81, 257
"skin hunger," 107–8
sleep
  melatonin's effect on, 289, 290
  REM, 103, 105

smell, sense of, 64, 65–68, 70–74
Sorel, Philip, 214
soybeans, 266, 267
Spitz, Rene, 92
Steinem, Gloria, 222–23
supplements
  calcium, 268
  DHEA, 86–87, 89
  growth hormone, 274
  melatonin, 290
  testosterone, 158–59
surgery, and menstrual cycle, 298–99

tamoxifen, 269
territorialism, 106–7, 143
testicles, 122–23
testopause, 151–52, 211
testosterone, 110, 111. *See also* testosterone in men;
    testosterone in women; testosterone
    replacement therapy *headings*
  aggression and, 5, 30–31, 140–51
  as antidepressant, 144–46, 156, 272
  blood sample measurements, 139–40
  cycles, 139
  forms of, 138–39
  levels of, and behavior, 9–10, 147–48, 154, 155
  masturbation and, 122
  moods and, 9–10, 16, 156, 158
  new drug delivery methods for, 300
  profile, 120–21
  role of, 5–6
  self-esteem and, 153
  sex drive and, 5–6, 119–25
  sexual intercourse and, 304
  vasopressin and, 8, 105–7
testosterone in men, 121–24
  during adolescence, 29–30
  cycles of, 9–10, 13, 16, 119, 122, 211
  deficiency in, and self-punishment, 221–23
  divorce and, 148
  effects of winning on, 150–51
  erections and, 140
  lack of commitment and, 143–44
  LHRH and, 127–28, 139
  during midlife, 46
  normal range in blood, 139–40, 157, 241
  peaks of, 122
  pecking orders and, 153
  prostate cancer and, 245–46
  relative deficiency of, and aging, 220–21
  during sexual development, 123–24
  for sexual dysfunction, 156–57, 159
  sexual stages and, 123–24
  during sixties, 49
  syndromes, 211
  types of, 244
testosterone in women, 20–21, 37, 124–25, 146–47,
    192
  during adolescence, 30–31
  aggression and, 30–31

and midcycle peak, 15
  during midlife, 46
  progesterone and, 124–25
  during sixties, 49
  during thirties, 41
testosterone replacement therapy for men, 155–57, 241–44
  considerations and drawbacks of, 156, 157, 158, 242, 243, 244, 245, 285
  DHEA and, 285
  methods of administering, 242–44
testosterone replacement therapy for women, 158–59, 271–73
  cautions for, 273
  estradiol and, 271–73
  estrogen replacement therapy and, 255, 269
thirties, sex during, 39–43
  and peak responsiveness of women, 41
  single men and, 42–43
  touching and, 39
time-management, and sexual difficulties, 37–38, 40–41, 302
Tomlin, Lily, 222
touch, 90–117, 301, 302
  during adolescence, 32
  bonding and, 90–91
  deprivation of, 91, 92, 101, 107–8
  endorphins and, 92
  health benefits of, 91–92, 95, 117
  inanimate surrogates and, 116
  longevity and, 278, 279
  lordosis response and, 173–74
  menopause and, 213–14
  mother-infant bonding and, 100–101
  oxytocin and, 33, 34, 37, 38, 40, 43, 44, 49–50, 88, 92, 93, 115–16, 256
  during thirties, 39
  during twenties, 36, 37, 38
transition periods between sexual stages, 20, 51–52
transurethral prostatectomy, 227–28
trazodone (Deseryl), 239
twenties, sex during, 34–39
  DHEA levels and, 74
  in men, 35–36
  relationship conflicts and, 36–38
  and shift from touch to penetration, 36
  touch and, 36, 37, 38
  in women, 35, 36

uterus, removal of. See hysterectomy

vagina

estrogen replacement and, 255
  menopausal changes in, 214–15, 272
  vitamin E as lubricant for, 267
vanity, value of, 305–6
vasopressin, 39, 72, 73, 102–7, 110, 111
  aggressive sex drive and, 125
  benefits of, 288–89
  as bonding agent, 93–94
  defined, 102
  DHEA and, 103
  depression and, 289
  fashion and, 109
  influencing, 115
  longevity and, 288–89
  nasal spray, 115
  orgasm and, 113–14
  oxytocin and, 102
  profile, 104–5
  receptive sex and, 174
  role of, 8–9
  sex flush and, 280
  testosterone and, 105–7
violence
  drug treatment of, 133
  PMS and, 197
  rape, 149–50, 154, 188–89
viropause, 13, 205, 210–12
  considerations before taking supplements for, 238–41
  as endocrine-deficiency disease, 235
  hormonal changes during, 228–29
  hormonal treatment of, 234
  melatonin and, 290
  physicians and, 234–35, 237
  reevaluating priorities during, 222–23
  relationship conflicts and, 210–11, 217, 219–22
  sexual consequences of, 217–23
  sexuality and, 239–41
  testopause and, 152
  testosterone and, 152
  time of, 46
visual stimulation, 57–58, 59–60
vomero nasal organ, 65–66, 71

weight
  gain, and menopause, 250–51, 262
  and longevity, 305
  loss, and DHEA, 79–80, 294, 295
widowhood, 49

Yen, Samuel, 292

ABOUT THE AUTHOR

Theresa L. Crenshaw, M.D., is a graduate of Stanford University; the University of California, Irvine Medical School; and the Masters and Johnson Institute, in St. Louis. The author of several books on sexuality, including *Bedside Manners* and *Sexual Pharmacology*, she lives in San Diego, California.